SCIENCE AND TECHNOLOGY POLICY

Editor: Rigas Arvanitis

Eolss Publishers Co. Ltd.,
Oxford, United Kingdom

Copyright © 2009 EOLSS Publishers/ UNESCO

Information on this title: www.eolss.net/eBooks

ISBN- 978-1-84826-059-7 (Adobe e-Book Reader)
ISBN- 978-1-84826-509-7 (Print (Full Color Edition))

British Library Cataloguing-in-Publication Data
A catalogue record of this publication is available from the British Library.

Library of Congress Cataloging-in-Publication Data
A catalog record of this publication is available from the library of Congress

Singapore

SCIENCE AND TECHNOLOGY POLICY

This volume is part of the set:

Science and Technology Policy Volume I
ISBN- 978-1-84826-058-0 (e-Book Adobe Reader)
ISBN- 978-1-84826-508-0 (Print (Full Color Edition))

Science and Technology Policy Volume II
ISBN- 978-1-84826-059-7 (Adobe e-Book Reader)
ISBN- 978-1-84826-509-7 (Print (Full Color Edition))

The above set is part of the Component Encyclopedia of *TECHNOLOGY, INFORMATION AND SYSTEM MANAGEMENT RESOURCES,* in the global *Encyclopedia of Life Support Systems (EOLSS),* which is an integrated compendium of the following Component Encyclopedias:

1. *EARTH AND ATMOSPHERIC SCIENCES*
2. *MATHEMATICAL SCIENCES*
3. *BIOLOGICAL, PHYSIOLOGICAL AND HEALTH SCIENCES*
4. *SOCIAL SCIENCES AND HUMANITIES*
5. *PHYSICAL SCIENCES, ENGINEERING AND TECHNOLOGY RESOURCES*
6. *CHEMICAL SCIENCES ENGINEERING AND TECHNOLOGY RESOURCES*
7. *WATER SCIENCES, ENGINEERING AND TECHNOLOGY RESOURCES*
8. *ENERGY SCIENCES, ENGINEERING AND TECHNOLOGY RESOURCES*
9. *ENVIRONMENTAL AND ECOLOGICAL SCIENCES, ENGINEERING AND TECHNOLOGY RESOURCES*
10. *FOOD AND AGRICULTURAL SCIENCES, ENGINEERING AND TECHNOLOGY RESOURCES*
11. *HUMAN RESOURCES POLICY, DEVELOPMENT AND MANAGEMENT*
12. *NATURAL RESOURCES POLICY AND MANAGEMENT*
13. *DEVELOPMENT AND ECONOMIC SCIENCES*
14. *INSTITUTIONAL AND INFRASTRUCTURAL RESOURCES*
15. *TECHNOLOGY, INFORMATION AND SYSTEM MANAGEMENT RESOURCES*
16. *AREA STUDIES (REGIONAL SUSTAINABLE DEVELOPMENT REVIEWS)*
17. *BIOTECHNOLOGY*
18. *CONTROL SYSTEMS, ROBOTICS AND AUTOMATION*
19. *LAND USE, LAND COVER AND SOIL SCIENCES*
20. *TROPICAL BIOLOGY AND CONSERVATION MANAGEMENT*

CONTENTS

VOLUME II

VOLUME I

POLICY-MAKING PROCESSES AND EVALUATION TOOLS: S&T INDICATORS

Rémi Barré
Director, Observatoire des Sciences et des Techniques (OST), France, and Visiting Professor, CNAM, Paris, France

Keywords: science and technology indicators, decision-making, data sources, criticism, democracy

Contents

1. Introduction
2. S&T Indicators: Definition, Terms of Reference and Categories
3. The Production and Use of S&T Indicators in Practice – the Question of the Data Sources
4. Indicators in the Decision-making Process: Limitations and Criticism
5. Perspectives and Conclusion
Glossary
Bibliography
Biographical Sektches

Summary

In the first part, we define S&T indicators are inputs of quantitative knowledge into the decision-making processes. They bring pieces of information into the process and help the actors to interact meaningfully with each other. S&T indicators can measure activities at different scales (micro, meso, macro), dealing with the different aspects (allocation of resources and definition of objectives) and the different contexts of decision (scientific, operational, strategic).

In a second part, it is stated that S&T indicators must have reliability and relevance to be useful in the decision-making processes. Then, we distinguish four categories of indicators: the human and financial resources aspects (inputs), the S&T production aspects (outputs), the interactions (co-operations, linkages, knowledge flows…) and the performance aspects.

In a third part, we state that S&T indicators production is an activity which deals with public policy debate and decision making; it is also, of course, a technical activity. The former has to do mostly with the relationship to the users of the indicators, the latter with the question of the source data which will be used to produce the indicators. The various source data for the indicators are presented, with their characteristics of level of aggregation, statistical, legal and technical status.

In a fourth part, we make a criticism of the indicators and show they cannot be used alone for decision-making, proposing a 'mixed' approach using both qualitative and quantitative elements.

Finally, in a fifth part, we suggest indicators are a useful device for public policy decision-making, provided they are considered not as results, but as entry points for debate. We conclude saying that indicators ultimately help build linkages and debates among an extended group of social actors, the result being that science and technology can enter in democracy

1. Introduction

1.1 The National Research and Innovation System

The national research and innovation system is the set of institutions, such as innovating firms, universities, public research organizations, knowledge based services, governmental institutions, which are involved in the production, diffusion and use of knowledge. The goals of such a system may be:

- the production of scientific knowledge,
- the contribution to higher education,
- the participation to industrial innovation and, more generally, to the scientific and technology base of industrial competitiveness,
- the production of the scientific expertise needed for the conception and implementation of public policies concerning health, environment, food safety, transportation, energy, etc.,
- the contribution to the strategic objectives of the state, expressed in terms of defense capabilities or technological self reliance in key areas.

In a 'knowledge society', the national research and innovation system plays a major role in the competitiveness of nations. It is a determinant of the quality of life enjoyed by the citizens of a nation. The efficiency of the national research and innovation system of a nation depends both on the quality of its institutions, and on the quality of their interactions.

In what follows, we will concentrate on that part of the system that consists in public institutions, so that we will deal mostly with public research and with public policy making activities.

1.2 Scale and Object of the Decisions on Research Activities

The efficiency of the national research and innovation system is rooted into the relevance of the decisions each one of its actors is making day after day. Such decisions affect each institution but also, in the longer term, the dynamics of the whole system and determine its capability of evolution and adaptation.

These decisions can take three different scales, corresponding to different kinds of institutions:

- micro scale: individual scientist, teams of researchers, research project;
- meso scale: research institutes, universities, research programs, scientific disciplines;

- macro scale: national policies.

At each one of these scales, the decisions may have two different objects:

- decision on the allocation of resources (as, for example, the funding of a research project);
- decisions on the definition of objectives and general organization (as, for example, the launching of a new program, or the reform of an organization).

Decisions concerning allocation of resources often suppose some sort of explicit or implicit ranking of the entities among which the choices are to be made, while decisions on the definition of objectives and general organization usually take the form of a process involving the management of the organization concerned.

We obtain a typology of six kinds of decisions concerning S&T activities by combining the scale and the object of decisions (Table 1). These decisions encompass what is called policy-making and evaluation, the former referring largely to the definition of objectives and general organization, the latter to the allocation of resources.

Scale	Object of decision	Example of decision
Micro (individual researcher, research team, laboratory, project)	Allocation of resources	Choice of project to fund
	Definition of objectives and general organization	Setting of the scientific objectives of a laboratory
Meso (research institution, university, program, scientific discipline)	Allocation of resources	Budgetary priorities within an institution
	Definition of objectives and general organization	Orientation of a program or of an institution
Macro (national level, S&T policy)	Allocation of resources	National budgetary priorities
	Definition of objectives and general organization	Orientation of national policy

Table 1. Scales and objects of decision concerning S&T activities

1.3 The Nature of the Decisions and the Actors Involved

Another way of looking at the decisions concerning S&T activities is to distinguish decisions dealing with issues internal to the scientific community, those dealing with the operationalization of research activities and those dealing with the relationship between scientific institutions and society.

Decisions of scientific nature will put forward the criteria of the scientific quality of a piece of research. Only scientists will be involved in the decision, according to the process of 'peer review': only scientists recognized as peers are legitimate to decide on the scientific issues.

Decision of operational nature will be concerned by criteria of efficient and proper use of resources. The actors involved are those in charge of the management of the resources, having operational objectives.

Decision of strategic nature will rely on criteria of socioeconomic relevance of a scientific activity. They will be concerned with determining the relevance and objectives of research and technological development. Here, a variety of actors representing social actors will also be involved, in particular those of the political decision-making system.

Nature of decision	Criteria of the decision and actors involved	Example of decision
Scientific nature	• Scientific quality • Scientists ('peers')	Choice among research projects for funding
Operational nature	• Proper use of resources procedural and regulatory validity of management • Managers, accountants,	Procedures of preparing, executing and monitoring the budget; contract, industrial property and personnel management
Strategic nature	• Relevance of objectives and overall organisation • Decision-makers, politicians	Budgetary priorities Institutional strategy

Table 2. Nature of the decisions concerning S&T activities

Decision-making regarding S&T activities will involve groups of persons, organized in committees or panels: decisions in S&T activities are characterized by their collective nature, rooted in the fact that either the peers or the political system are the source of legitimacy: in both cases the decision making process will be based on collective work.

Being so collective and complex, decision-making processes regarding S&T activities need knowledge inputs. S&T indicators are inputs of quantitative knowledge useful for the decision-making processes. Their role is both to bring pieces of information into the process and to help the actors of the process interact meaningfully with each other. All decisions regarding S&T activities, whatever their scale, object and nature, are concerned by this process. In the field of research activities, decisions, whatever their scale, object or nature, are often called 'evaluations'. In what follows, we will stick to the concept of decision-making, since evaluation concerns a particular aspect of the decision-making processes.

2. S&T Indicators: Definition, Terms of Reference and Categories

2.1 Definition

S&T indicators are quantitative measurement of parameters describing research activities and actors. Like all measurements, indicators explicitly relate to a conceptual model or a representation of how things work. The most frequent model is the input-output model: resources flow to research activities, which in turn produce outputs (scientific and technological results), which ultimately will produce impacts on society and the economy.

Indicators differ from statistical measurements in the sense that indicators are explicitly related to a policy question or a decision to be made. S&T indicators will therefore refer to the various possible decisions in the field of research, which can be synthesized as in Table 3 (below). S&T indicators can measure activities at different scales (micro, meso, macro), dealing with the different objects of a decision (allocation of resources and definition of objectives) and the different contexts of a decision (scientific, operational, strategic contexts).

In all cases, the *raison d'être* of indicators is to test an hypothesis: if a decision is to be done, the preparation of that decision consist in understanding the way the sub-system related to this decision will function and in imagining alternative actions and checking their relevance. All this points to an attitude of formulation and testing of hypothesis. It is very important to notice that such an attitude requires both a knowledge of the demands of stakeholders, and the construction of a conceptual framework, which will help identify the relevant parameters and useful hypothesis.

We can determine the form of S&T indicators and the circulation they will have by distinguishing them by their purpose:

- General interest S&T indicators: usually at national or regional scales, can be published widely in the form of an 'S&T indicators report', both for purposes of communication and support for general policy making;
- S&T indicators prepared for a particular question or decision to be made: they usually take the form of a dossier or a report of small diffusion, mostly for the use of those directly involved;
- S&T indicators related to an institution: they often take the form of a report both for internal use and for interaction with stakeholders.

Many countries publish every year, or every two years, a national S&T indicators report. Some have it published it entirely on a web-site. Research institutions publish partly their indicators on their web-site. This stresses the communication role of S&T indicators which must be recognized as tools that accompany the decision making process.

Scale	Object of decision	Nature - criterion of decision		
		Scientific evaluation **Scientific quality**	Operational evaluation **Operational efficiency**	Strategic evaluation **Societal relevance**
Micro (individual researcher, research team, laboratory, project)	Allocation of resources	Scientific quality of a researcher, research team, laboratory or project	Operational efficiency of researcher, research team, laboratory or project	Societal relevance of a researcher, research team, laboratory or project
	Definition of objectives and general organization	Scientific interest of the orientation of a researcher, team or laboratory	Operational efficiency of the organization of the project, team or laboratory	Societal relevance of the orientation of a researcher, team or laboratory
Meso (research institution, university, program, scientific discipline)	Allocation of resources	Scientific quality of research institution, university, program	Operational efficiency of research institution, university, program	Societal relevance of research institution, university, program
	Definition of objectives and general organization	Scientific interest of the orientation of a research institution, university, program	Operational efficiency of the organization of the research institution, university, program	Scientific interest of the orientation of a research institution, university, program
Macro (national level, S&T policy)	Allocation of resources	Scientific interest of the orientation of the research and innovation system	Operational efficiency of the research and innovation system	Societal relevance of the S&T policy orientations
	Definition of objectives and general organization			

Table 3. S&T decisions and indicators.

2.2 Characteristics of S&T Indicators Useful for Decision-making

S&T indicators, if they are to be of any use in the decision-making processes should tend toward two basic characteristics: their reliability and relevance.

2.2.1 The Reliability Criterion

The reliability of an indicator is the confidence one can have that it measures exactly what it pretends to measure. There are two components to reliability:

- the accuracy of the computation of the indicator,
- the coherence between what is measured in reality and it is measuring.

The accuracy of computation is obtained through the transparency of the data collection and treatment processes. Source data must be referenced, treatments must be made explicit and reproductible. The production of the indicators has to be challengeable and disputable. Criteria of scientificity apply to the production of indicators.

In practice, limitations to accuracy come from difficulties to have full transparency on the production of source data and the complexity of the treatments which make it almost impossible to explicit and share the various steps. Some sort of validation of computation occurs through comparison of results obtained by different indicators producing units, hence the importance of the diversity of indicators production capabilities.

The coherence between what is measured in reality and what is said to be measured depend on the conceptual aspects of indicators production. Two questions can be mentioned:

- How precise are the definition of the measured parameters. For example, there are many ways to define what is a researcher (one can include, or not, the post-doctoral students), so that risks of non comparability are high between two measurements, unless it is very precisely indicated what is the definition used.
- How is made the correspondance between the parameter actually measured and it is measuring. This is the whole problem of the 'proxies': for example, a classical 'proxy' for measuring the technological orientation of a public research institution or a university, is to build an indicator of its patenting activity. But in the case the patenting is contrained by the resources devoted to the patenting bureau of the institution or university, the proxy will measure the evolution of the budget of the bureau, and not the technological capability of the institution!

2.2.2   The Relevance Criterion

In the definition of an indicator, its relevance for decision-making processes is of particular importance. S&T indicators have to address the questions which are at stake in all the possible contexts as shown in Table 3. They need to measure parameters of entities of various scales, from laboratory to national level (micro, meso, macro scales) and parameters describing the human and financial resources aspects (inputs), the S&T production aspects (outputs) or the institutional, financial or cognitive interactions (co-operations, linkages, knowledge flows…).

They have to address allocation of resources issues or definition of objectives and general organization. In the first case, the indicators will have to help ranking entities or at least make comparative statements; in the second case, they will rather help positioning an entity in its context, comparing the positions and evolutions with other entities.

In general, measurement of the evolutions along a sufficient time period, comparisons among entities (regions, countries, laboratories...) and geographical or institutional dimensions of the indicators permit to help their relevance.

In brief, the relevance of an indicator will depend on:

- the proper understanding of what is at stake and of what is the demand of the stakeholders and decision makers,
- the quality of the underlying conceptual model, which helps define both the parameters to be measured and the hypothesis to be tested and discussed.

2.3 The Categories of S&T Indicators: Input, Output, Interaction, Performance Indicators

It is convenient to distinguish four categories of indicators: the human and financial resources aspects (inputs), the S&T production aspects (outputs), the interactions (co-operations, linkages, knowledge flows…) and the performance aspects.

2.3.1   Inputs Indicators: The Human and Financial Resources

Input indicators refer basically to the Research and Development (R&D) expenditures and to the human resources. Definitions of R&D expenditures measurement have been codified by the OECD *Frascati Manual* and human resources by the *Canberra Manual*. Research equipment and infrastructures (libraries, large band electronic network, data bases…) also enter input indicators.

The assessment of financial resources is often more difficult than it can seem at first sight, since it may involve public budget analysis, checking what is included in the given figures, sorting out questions such as the value added tax, the way of counting investments, the way of separating education from research…

Human resources indicators often raise equally difficult problems such as the way to account: part time and multi-institutionally based personnel, doctoral students, technicians and engineers working in research projects, visiting scientists, and so on.

2.3.2   Outputs Indicators: The Production of S&T Activities

The production of research activities are of many kinds:

- industrial innovation (definitions and classifications have proposed by the OECD *Oslo Manual)*
- higher education, measured through the delivery of diplomas, particularly PhDs
- scientific and technological production measured through bibliometric techniques.

Bibliometrics deserves some precision (see also *Bibliometrics and institutional evaluation*). The basic idea behind bibliometrics is that is that the production of a 'certified' publication is the marker of the production of scientific or technological knowledge, recognized as being at the frontier of knowledge. A 'certified' production is typically, a publication in a peer reviewed journal in the case of scientific production, and a patent, in the case technological production. The reviewers of the journal and the patent examiners provide a validation of the knowledge produced.

Bibliometrics is the set of statistical techniques and methods by which indicators of S&T production are computed from scientific publications and patents data files. For bibliometric use, the scientific publications data file must contain the following items: journal name, year, volume, pages, name and address of laboratories, name of authors, title, eventually keywords and citations. The patents data files must include: year of priority and year of publication, reference number of the patent, technological classification codes, name and addresses of grantees, names and addresses of inventors, eventually legal status, and citations.

Bibliometric data are extremely rich, detailed, but also up to date, complete and accurate since they are produced primarily for the immense market of documentation needs. S&T indicators production through bibliometry uses those existing files for its own specific purposes. Also, bibliometric techniques permit access to micro-level information (on a researcher, a laboratory, a university, a firm) without legal restriction. Finally, bibliometrics allows not only descriptive measurements, but also cognitive indicators (see below): it allows to address the substance of the science or the technology involved.

These advantages have to be balanced against some difficulties. The main difficulty is that bibliometric data need very large files, of a complex nature: they are costly to handle and, for publications data, costly to access. This requires both resources and great care about the data files used. Another difficulty is that there exist a possibility of significant biases, so that a methodological capability is required for the proper production and interpretation of bibliometric indicators.

Bibliometrics is an extremely lively field of methodological research and experimentation, developing methods for a more precise characterization of the cognitive content of the publications. The *Journal of Information Science* or *Scientometrics*, for example, publish those methodological research results.

2.3.3 Interactions Indicators

In S&T production, interactions are a major element of the production process and therefore of strategy. Such interactions may be financial, institutional or cognitive. Financial and institutional interactions can be measured through financial flows or the identification of 'who owns whom', that is, the ownership of the capital of firms (identification of subsidiaries and parent companies of firms). Cognitive interactions are depicted through the working links between scientists. Here again, bibliometrics provide very powerful measurement opportunities, for example, by looking at the co-publications patterns. These indicators permit the identification of networks of S&T production. Bibliometrics address many the key issue of cognitive interactions: identification of citation flows, common key-words, proximity of research themes ('mapping of research'), relations between science and technology.

2.3.4 Performance Indicators

This set of indicators refers to performance in various senses, such as the productivity or the efficiency of the research activities, but also the scientific and socioeconomic impacts.

The productivity or efficiency indicators can be computed in the form of a ratio of a volume of output divided by the volume of input which has been needed to produce that output. Such ratio have to be handled with very great care since there is always a multiplicity of both outputs and inputs in a research activity.

Socio-economic impacts are in principle very important indicators, but there is hardly any practical and sound methodology to compute them, so that they derive mostly from opinion surveys (see also *Technology assessment*).

2.4 Types of Indicators: Descriptive, Cognitive and Opinion Indicators

Indicators of S&T activities may be of three types:

- Descriptive indicators: they measure essentially the volume of the parameter involved (funding, human resources, scientific output, linkage…), through a direct 'objective' observation. They are akin to the size and weight of an object.
- Cognitive indicators: they identify the substantive scientific or technological content of the activity. They are akin to the 'chemical' composition of an object.
- Opinion indicators: they measure what stakeholders (researchers, users, decision-makers…) think of a research activity. They are akin to the opinion of a person about an object.

2.5 Overview: Categories and Types of Indicators

In brief, indicators may measure what are the various inputs, outputs, interactions and performances of research activities, in a descriptive, cognitive or opinion framework. Furthermore, they specify who performs that activity, when and where (see Table 4).

Type Episte-mological status Category	Descriptive type Volume and broad category of activity	Cognitive type Substantive – thematic nature of the activity	Opinion type Opinion of stakeholders on the activity
Human and financial resources (inputs)	Volume of resources used as an input for the research activity	Thematic nature of the skills and knowledge input for the research activity	Opinion on the resources
Production (outputs)	Volume of output produced by the research activity	Thematic nature of the knowledge and skills produced	Opinion on the production
Interactions	Volume of resources flows and number of linkages	Thematic nature of the interactions, thematic distances, thematic knowledge flows	Opinion on the interactions
Performance	Efficiency ratio (output/input), volume of impact and effect	Cognitive impacts and effects	Opinion on the performances

Table 4. Categories and types of indicators

3. The Production and Use of S&T Indicators in Practice – the Question of the Data Sources

S&T indicators production is very much an activity which socially embedded, since it deals with public policy debate and decision making; it is also, of course, a technical activity. Such sociopolitical and technical aspects are closely linked. The former has do mostly with the relationship to the users of the indicators, the latter with the question of the source data which will be used to produce the indicators.

3.1 S&T Indicators Production Activity

3.1.1 The Relationship with the Client: Understanding the Demand and Conceiving the Product

As we have seen in the definition of the indicators, there must necessarily be users of the indicators, having a certain explicit or implicit demand. In general, the demand for S&T indicators is borne by the ministry in charge of research, which is thus the client. The demands can also come from Directors of research institutions or agencies or from regions. In practice the demand can result in a variety of forms of products, such as:

- a fairly broad demand for general presentation of the situation and perspectives, resulting in the need for a national S&T indicators report;
- a demand for facilitating the communication on some particular topic with a large number of external actors or constituencies, resulting in a well presented, simple and short leaflet
- a demand for accompanying a complex process involving the actors of research, which can result in a full analysis and indicators report of restricted circulation.

It is important to note that the demand will be very seldom explicit and detailed. It is clearly the role of the indicators builder to translate statements of a general nature into specific indicators which are relevant to the needs, and realistic in terms of technical production. This task requires both an excellent technical command of S&T indicators, as well as a good attention to the client and an understanding of the sociopolitical context of the research policy.

3.1.2 The Production of the S&T Indicators

In most cases, the S&T indicators production will consist in the statistical treatment of existing data from various sources. Hence the sequence of tasks of S&T indicators production:

- identification of the relevant data sources negotiation of access to those data (see below),
- preparation, reformatting, validation of the data files so that they can be treated,
- data treatments programming, execution and validation, to get the indicators
- presentation of the indicators in a document, with technical annexes and interpretations.

The first and last of those tasks requires a good overview of S&T indicators data sources, methods and possibilities. The second and third require computer specialists and statisticians.

3.1.3 Bringing the Indicators to the Client and the Users

Since indicators are involved in decision making and communication, the process of delivering the indicators to the client and users is an essential task. This can take many forms, such as: contacts with journalists, organization to workshops and seminars, participation to conferences, contribution to working groups, writing articles in the general press, and so on. Indicators do not 'live' by themselves, by require to be pushed towards their users in order to contribute to better-informed decision making.

3.2 Source Data for S&T Indicators Production

3.2.1 From Source Data to Indicators

Indicators are in most cases built from existing data sources. Beyond the particular case where the data collection has to be done ('raw source data'), it is useful to characterize such data sources according to the level of agregation of the information they contain (Table 5).

Level of data	Content	Examples	Comments
Raw source data	Data known by the individual item, prior to any collection of data	Contract files of universities, informations of the firms on their RD activities	One has to build one's own source data through a survey or data collection: very heavy work to be avoided except in particular cases
Primary source data	File resulting from the survey or data collection, prior to any aggregation (file related to the individual items : scientific article, patent, laboratory, researcher, firm…)	Public researchers file, scientific publications and patent files, innovation survey file	Preferred source, although it may have accessibility problems and raise technical difficulties The interest is the existence of source data and the large margin of maneuver for ad-hoc indicators building
Secondary source data	File of tabulated (aggregated) numerical - statistical results computed from a survey or data collection file	Tables from surveys published by the national statistical office; aggregated figures of financial flows; data published by OECD	If the tables are detailed enough (not too much aggregated), there is some margin of maneuver for indicators building

Tertiary source data	Indicators report accessible in paper or electronic form	US Science and Engineering indicators report, EC indicators report, French Observatoire des Sciences et des Techniques report.	Indicators found already prepared must correspond to what is needed; Little value added

Table 5. The four levels of source data useful for indicators building

The interest of the raw source data as well as primary source data is that the indicators builder is free to define the classifications considered relevant (scientific, technological, geographical, by type of institution, by size…). The drawback is that the technical work is much more difficult and heavy.

3.2.2 Statistical Status of the Data

The reliability of the data, hence of the indicators, will depend very much on the statistical status of the data source used. From exhaustive to statistically significant, there will be a degradation of the reliability from excellent to almost nil (Table 6).

Statistical status	Examples of data bases	Comments
Exhaustive	Patents, RD firms, PhDs delivered	Best situation, not always possible
Statistically significant sample	innovation surveys	Must be professionally done, usually by the national statistical office for its surveys having of very large number of individuals
Selection of the most important / significant items	Scientific publications	The criteria for selection of such most important, significant items must be explicit and performed in a transparent, credible way
Non statistically significant sample (significance unknown)	Ad-hoc surveys with partial response rate, internet searches	In principle, use of such data should be avoided, since no interpretation is possible; in practice: very careful uses, in terms of hypothesis setting

Table 6. Statistical status of data

3.2.3   Legal Status of Data

The legal status of the data will define important aspects of indicators activity, namely the rights, constraints and conditions of publication of the results. It will also influence the cost of indicators production, since privately produced and owned data will entail specific expenses for the right of use (Table 7).

Legal status	Examples of data bases	Comments
Open public data Produced by public services	Patents, national budgetary data, international organizations data, open reports data	Best situation but the information is usually very aggregated except for patent data or some national budgetary documents
Legally restricted data produced by public services	Detailed researchers data, detailed contract data, tax credit data, defense institutions detailed rd data	When the legal restriction is for individuals privacy protection, one can get anonymous files; otherwise, one can only get aggregated statistical files where institutions or specific informations do not appear anymore
Data produced by public services and covered by statistical secret	Surveys performed by the national statistical office Industrial RD or innovation surveys	Extremely constraining situation One must get a formal approval after an administrative process, and the primary source data once obtained, can only be used for the specified work, to be done by one specified person
Data produced by public services and restricted in other ways	Administrative surveys or files normally used only by the administration itself Public research survey	Access to be negotiated on a case by case basis
Privately produced and owned data	Scientific publications, economic data on firms	Data accessible only through a license, at a cost, which specifies what one can do and not do with the data, in particular in terms of publication and contract work
Data from one's own survey	Ad-hoc surveys of institutions, laboratories…	Often, confidentiality guarantees are given to the respondent, which have to be respected

Table 7. Legal status of data

3.2.4 Technical Status of Data

The sources will also influence the technical conditions for the data treatment and indicators computation. One should note that to produce one's own survey or data collection is a methodologically difficult and operationally heavy and costly endeavor (Table 8).

Technical status		Examples	Comments
Raw source data		Survey to be performed (opinion of research project managers in a program)	Technical status of data to be treated depends on the way data collection will be made and formatted
Primary source data base	Files to be formatted and cleaned before extraction and/or to be completed	Management or ad-hoc surveys files (files of laboratories involved in a program)	To work may be quite heavy and complicated; changes from year to year to be mastered
	Files to be directly extracted and reformatted	Professionally made survey file or commercial CD-ROM (patent data files, scientific publication files)	In fact, there will be usually the need to unify the names of the institutions
Secondary Source data base	Tables of statistical figures in paper or electronic form	Statistical tables published in a survey report (industrial RD or innovation survey reports or international commerce reports)	Need to have a file suitable for data treatment: it either exist or has to be built from existing survey report
Tertiary source data base	small size tables of indicators	a few figures copied directly from an existing indicators report (OST or OECD or European commission indicators report)	not much value added

Table 8. Technical status of data

3.2.5 Data Sources Providers

The data have a variety of institutional origins, and it is an important part of the indicators production activity to negotiate with the data source providers the conditions of access to their data. This has legal, technical and methodological aspects (Table 9).

Data sources providers	Examples
Commercial data base producers and servers	ISI for scientific publications files, Dunn & Bradstreet for data on firms,
International organizations	OECD for macro indicators on RD, UNESCO for education data, European Commission for macro and meso indicators
National and regional institutions (ministries, agencies...)	Ministry of Research / Higher Education for budgetary, national resources and students data, Research Agencies for research programs and project funding
Research institutions, universities, large laboratories	Research institutions for external contract, research use and equipment data
National statistical office	National surveys on industrial RD, innovation, qualifications

Table 9. Data sources providers

3.2.6 Overview of Data Sources for S&T Indicators

Since indicators are linked to specific demands, there is no standard set of source data to be defined. The indicators production activity is by definition an exercise of creativity by which the understanding of an implicit demand is matched with existing data sources. The 'art' of indicators production is in this creative matching, for which we can give some rules and provide some advice – but which must be done and invented on a case by case basis by those involved.

4. Indicators in the Decision-making Process: Limitations and Criticism

In the definition, we have stated that S&T indicators are aimed at bringing quantitative knowledge into the decision making processes regarding research activities. It is important to be more specific regarding the way indicators can enter the decision making process, which will lead us to make a critical assessment of the indicators.

4.1 The Possible Extreme Options for the Use of Indicators in Decision-making

In case of decisions regarding allocation of resources, we present the two extreme options, the first ignoring indicators, the second making exclusive use of them (Table 10).

In option 1, the decision is purely based on qualitative grounds (pure opinion). This is the standard, or classical way of deciding, in S&T activities, like in any other context. In practice, if a decision on resources allocation needs to be taken (like funding research projects or institutes), this mode of decision-making will be based on qualitative descriptions; a non-explicit process of getting at an opinion will take place.

In option 2, the decision purely based on indicators (pure computation). In this other, extreme case, the indicators provide directly and mechanistically, the decision, which results from the measurements of descriptive parameters of the entity. The idea is avoiding any subjectivity by using only quantitative and 'objective' elements for getting at the decision. In this case, there is no need of a committee, nor of a decision-maker as such.

	Option 1 **Decision purely based on qualitative grounds** **(pure opinion)**	**Option 2** **Decision purely based on indicators** **(pure computation)**
Characterisation of the parameters describing the entity	Qualitative characterisation	Quantitative description indicators (mostly output and input)
Scoring of the entity on a criterion	Scoring given by a committee after a non explicit process	Measure of performance as a ratio Output / input
Overall scoring of the entity	Scoring given by a committee after a non explicit process of agregation of the criteria	Explicit weighting of criteria and computation of a weighted average
Decision based on…	The opinion of a committee or a decision maker No need of indicators At the limit : no need of a dossier	Quantitatively defined descriptive parameters No need of a decision-maker

Table 10. Options for the use of indicators in decision making

Immediately there is valid question that can be raised: if indicators are reliable and relevant measurements of parameters regarding research activities, why should this second option be considered 'extreme'? If that second option appears possible, this would open very interesting opportunities for 'objective' decisions, for quicker and cheaper processes. For both technical and epistemological reasons, we will see that such an option is not only unrealistic, but also very dangerous, if it was to be taken seriously.

4.2 Criticism of the S&T Indicators: What do they Really Measure?

The first implicit hypothesis behind the postulate that indicators can measure the parameters of research activities, is that a subsystem called 'scientific production' can be identified and its resources separated from all other flows, without any feedbacks nor interactions from the exterior of this subsystem. The second implicit hypothesis is that we have a model of S&T production, defining all the relevant parameters involved, including those involving knowledge flows. The third implicit hypothesis is that we can measure all these parameters.

We do not dispose of such a full model of S&T production. But even we had one, it would not be relevant since the first implicit hypothesis – which is the possibility to separate research activities from all the others social and economic activities – is contrary to what is the 'new mode' of research: a context involving a variety of actors with differing aims and strategies, interacting in networks of suppliers and users.

In the real world, descriptive indicators cannot fully characterize research activities; productivity indicators cannot be defined solely by the ratio of an output variable to an input variable. In other words, the impossibility to have indicators fully describing research activities is not linked to a technical difficulty in finding the 'good number', because there is no 'good number'. The reality is never fully grasped by measurements, however accurate and numerous. If the indicators do not fully measure research activities, then what can they be used for, and how?

4.3 Towards a Realistic Role of S&T Indicators in Decision-making

Coming back to the resource allocation case presented above, we suggest indicators should be used in the context of a third option, which we call the 'mixed' option. It is mixed in the sense the decision-making process is based both on a quantitative dossier and on judgement. Judgment results largely from interactions among those concerned, and discussion and criticism of the indicators is an aspect of such interactions.

	Mixed option Quantification where needed in order to improve analysis, understanding and transparency
Characterization of the parameters describing the entity	quantitative indicators, and qualitative characterization
scoring of the entity on a criterion	scoring given by the committee after discussion and criticism of the indicators
overall scoring of the entity	explicit weighting of the criteria
decision based on…	quantitative et qualitative dossier explicit procedures, substantive debates

Table 11. Use of Indicators in decision-making – the mixed option

5. Perspectives and Conclusion

The fact that the scientific productivity of a research system cannot be captured by a single number derives from what we have learnt these last years about research activities: they are characterized by a systemic and interactive dynamics among a variety of actors, linked to institutions, themselves embedded in society.

5.1 Towards an Ethic of Indicators Production

S&T indicators should be viewed as elements of expertise injected into the knowledge producing and decision-making processes. As any form of expertise, S&T indicators do not stand alone any more; they cannot tell any 'truth' per se; they do not give any answer as such. The only significance S&T indicators can have, is through the discussions they can foster, especially discussions starting from the criticism of the indicators and debates stemming from tentative interpretations.

This sets the stage for the analytical knowledge and S&T indicators production and use. All analytical work must explicit its underlying conceptual model of research and innovation, its source data, as well as the variety of possible interpretations. Monopoly on source data and restricted access can only be considered as threats to the whole process. Transparency of the production of S&T indicators, explicitation of underlying concepts, identification of approximations, argumentation of the validity of the proxies become part of the indicators building mandate and point towards an ethics of indicators production.

5.2 Towards 'Socially Robust' S&T Indicators

The most valuable aspect of analytical work is the criticisms and debates regarding both its methodology and the interpretation of its results. In such context, analytical knowledge does not close the debates, but contribute to it, and, in turn, the debates open new avenues for analytical work itself. This calls for the production of 'socially robust' indicators, that is indicators which have been co-produced by the social actors in an open and public way.

We suggest there is room for indicators as a useful device for public policy decision-making, provided indicators are considered not as results, but as entry points for debate; provided also that such quantitative inputs is understood as a multi-stage exercise involving both analysts, research actors and policy makers. To make sense, it should be seen as a joint learning process.

5.3 Indicators as Tools for Democratic Decision Process in the S&T Area

Indicators become one of the means which help actors shape their debate through the expression of their diverging views about the relevance, significance, underlying assumptions and interpretation of the indicators. It is the very process of such criticism which enables actors to address the deeper questions related to science policy. It is the exchange of arguments going along with criticism which leads to revealing substantive points which are important to the actors.

Therefore, indicators are not supposed to provide answers, nor 'the thrush', which we consider would be an illegitimate claim, but they provide the cornerstone of the debates. There will be a sequence of quantitative and qualitative analysis phases, each one feeding from the previous and into the next, in virtuous circle of understanding. Indicators appear as central features of processes like benchmarking, foresight and strategic evaluation, which precisely aim at opening and widening the decision making arena.

In this sense, indicators ultimately help build linkages and debates among an extended group of social actors, the result being that science and technology can enter in democracy.

Glossary

R&D: Research and development
S&T: Science and technology

Bibliography

Barré R. (1994). Do not look for scapegoats! Link bibliometrics to social sciences and address societal needs, *Scientometrics,* **30**(2-3), pp. 419–424. [This articles stresses the role of societal demand in S&T indicators.]

Barré R. (1997). Les outils bibliométriques, instruments essentiels pour les recherches concernant les économies fondées sur la connaissance, *Revue d'économie industrielle,* **79**, 1er semestre, pp. 119–128. [This article reviews the advantages and difficulties of bliometric indicators.]

Callon M. and Courtial J. P. (1996). La scientométrie au service de l'évaluation, in Callon, M., Laredo P. and Mustar P., eds. *Le management stratégique de la recherche et de la technologie*, Paris: Economica. [This chapter shows the role of indicators in evaluation and decision making, in the case of programs management.]

European Commission (1997). *Second Report on S&T indicators*, Publications Office of the European Community, Luxembourg, 729 pp. [S&T indicators report stressing both the european Union as a whole and as a set of nations; particular attention is given to central and eastern european countries as well as to mediteranean countries.]

Foray D. (2000). *L'économie de la connaissance*. Paris: La Découverte. [A short but complete overview of the concepts of the knowledge economy.]

Gibbons M., Limoges C., Nowotny H., Schwartzman S., Scott P., Trow M. (1995).*The new production of knowledge*, London: Sage Publications. [This book describes the emerging features of research and innovation activities, stressing their networking and interactive characteristics.]

Narin F., Hamilton K., and Olivastro D. (1997). The linkages between US technology and Public Science, *Research Policy,* **26**, pp. 317–330. [This article shows that the citations to scientific publications in patents provide a way to link science and technology in a cognitive way.]

Nationa Science Board (2000). *Science and Engineering Indicators*, Washington DC: NSF. [This report presents a wealth of indicators, detailing about all the facets of S&T in the USA. Much less on other countries.]

OECD (199x). *Frascati Manual*, Paris. [This report states the official definition of what are the research and development activities and how they should be measured.]

OECD (199x). *Canberra Manual*, Paris. [This report states the definition of what are the human resources in the area of science and technology and how they should be measured.]

OECD (199x). *Oslo Manual*, Paris. [This report defines what is a technological innovation and how innovations should be measured.]

OST (1999). Collection *Science et Technologie, indicateurs*, Rapport de l'OST (editor) Paris: Editions Economica, 563 p. [This is the S&T indicators report for France, with an emphasis of the regional scale, both at French and European levels.]

Pavitt K. (1988). Uses and abuses of patents statistics, in Van Raan, A. F. J. (Ed) *Handbook of Quantitative Studies of Science and Technology*, North Holland: Elsevier. [This article stresses the possible bias of patents when used as technology production indicators.]

Roqueplo P. (1997). *Entre savoir et décision, l'expertise scientifique,* collection 'sciences en questions', Paris: INRA Editions. [This book shows that the concentration of various disciplinary scientific knowledge on a particular question is unable to provide the answers society needs: the reality is always beyond the addition of partial knowledge.]

UNESCO (1998) *World Science Report – 1998*, Paris: Editions UNESCO, pp. 22–30. [This is UNESCO science report, which contains some indicators at world level, but also assays of qualitative nature to describe the scientific activities of different regions of the world, including the less developed parts of the world.]

Van Raan, A. F. J. (1997) Scientometrics: state of the art, *Scientometrics,* **38**(1), pp. 205–218. [This is an overview of bibliometrics in its cognitive perspective, in particular through mapping of research themes.]

Biographical Sketch

Rémi Barré holds a Civil Engineering degree (Ecole des Mines) and a Doctorate in economics, Ecole des Hautes Etudes en Sciences Sociales (Paris). He is currently Director of Observatoire des Sciences et des Techniques (OST) and Visiting Professor at CNAM (Paris), in charge of the program 'prospective and evaluation of research and technology'. He is member of the Executive Committee of ESTO (European Science and Technology Observatory), linked to the IPTS (CCR-Sévilla) and a member of the CREST (Advisory Board to the European Commission S&T policy) Evaluation sub-group. He is Editor of the *Science and Technology indicators Report*, published every 2 years by OST, co-Editor (with B. Martin, J. Maddox, M. Gibbons, P. Papon) of *Science in tomorrow's Europe*, Economica international, 1997, co-author (with B. Amable and R. Boyer) of *Les systèmes d'innovation à l'ère de la globalisation*, Economica, 1997, and editor of the '*Science et Technologie Indicateurs*', Economica, 2000.

EVALUATION PRACTICES IN A MODERN CONTEXT FOR RESEARCH: A (RE) VIEW

Emilio Muñoz, Juan Espinosa de los Monteros and **Víctor M. Díaz**
Unidad de Políticas Comparadas, CSIC, Madrid, Spain

Keywords: science policy, accountability, evaluation methodologies, R&D programmes, knowledge production, knowledge impact, knowledge "modes", evaluation models, biology metaphors, functional evaluation, societal evaluation, comparative perspectives, Europe-USA.

Contents

1. Introduction
2. Basic Definitions of Research and Evaluation Methodologies
3. Relationship between Science Policies, Promotion and Management of R&D Activities
4. A General Frame of Reference for Evaluation from the European Perspective
5. Emerging Issues on Evaluation from the United States
6. Conclusions
Acknowledgements
Glossary
Bibliography
Biographical Sketches

Summary

The increasing relevance of science (and technology) for development has raised the interest in the process of evaluation of research and its impact. The concept of research as well as evaluation practices has been experiencing an evolutive process leading to the merge with other concepts – technology, industrial development, innovation.

The influence of European institutions has been decisive in fostering the application and development of evaluation methods and practices to public policies, and among them to research and innovation policies. Some of these forms of evaluation are presented in relation to the different types of research programs and activities. Some innovative approaches based on new models have been developed to take into account either the role of research laboratories (the 'compass-card' model), the societal quality of research or the functional performance of research programs (the 'transducing model'). Some case studies are discussed to illustrate these approaches.

The comparative perspective with respect to the situation of research evaluation in the United States reveals the dynamics of that process with respect to the increasing need for accountability of science towards society.

1. Introduction

Science and technology are instruments of increasing strategic value in society for the attainment of wealth and for enabling competitiveness in a globalized world. Changes in the economic context influence the performance of science and technology. Important social changes are occurring in the developed world in the transition from a production to a services society, which open new challenges in several areas of science and technology. An important consequence of this situation is the changes taking place in the design and assessment of science policy. The principle articulated by Vannevar Bush that basic research and its practical benefits 'accrue to society through an apparently unrelated process' is questioned every day. The performers of basic research have to demonstrate user relevance, relationships with industry or the utility of their work.

Evaluation has become a critical factor since government sponsors are calling for greater accountability on the part of researchers and they are looking for funding research areas and projects of strategic relevance. In spite of this trend, evaluation is still showing its limits as the indicators and methods applied to evaluate science and technology outputs and their effects on innovation, are still based mainly on the concept of linearity implied by the principle mentioned by V. Bush. The bibliometric methods used to measure scientific and technological production, i.e. scientific articles and patents, concentrate on productivity. The economic and human resources devoted to science and technology are viewed as inputs into the system. This output/input model of evaluation is leaving aside an assessment of the interactions between science, technology and innovation and their actors. Such a model, inspired by econometrics, does not underscore the role of human players in the process of generation and use of knowledge, as well as the influence of cultural values and the environment on it. However, citations and referencing data have been used in some fields like biotechnology and biosciences to detect the links between science and technology.

In the present review, we describe the trends orienting the (new) science and technology policies and report the efforts undertaken by different countries and institutions to adapt the evaluation practices to this new context.

2. Basic Definitions of Research and Evaluation Methodologies

The concept of research is as old as science, but has experienced an important evolutionary process as it merged with other concepts, such as technology, industrial development and innovation. This combination took place when it was recognized, after the two World Wars, that research and development are essential in an industrial production system.

Therefore, **research and development** (R&D), a concept unheard of until the fifties, became a universal watchword in developed countries. Consequently, a series of indicators providing some grounds for assessing research performance and development needed to be established. Moreover, the efforts to promote R&D activities were shared by different institutions: government, business, and higher education, in such a way, that these three sectors were incorporated into the statistics which accounted for expenditure efforts (economic and human resources) and so were their results (outputs such as scientific production and patents).

The way in which R&D performance was understood and measured, came from an interpretation of the intimate relationship between research and subsequent development and how they intervene in a pipeline leading to industrial production through innovation, starting from a laboratory discovery, through prototype production or manufacturing start-up, to full-scale production and market introduction. However, the difficulties to correlate the R&D efforts in a direct and simple manner with the innovation practices and market successes has led to a shift from the linear interpretation of the R&D influence on industrial production to a more systemic one.

Evaluation practices experienced a similar evolutionary process, starting with the traditional 'peer-review' system applied by the scientific community to assess research projects, to the need to evaluate the strategies of integrated goals through programs usually designed and funded by governments.

In order to be able to approach a review of the different evaluation practices and their influence on the different instruments used to promote research and development it seems logical to point out some definitions and to check them against the evaluation methods currently used for each one.

Innovation might be defined as the application of an invention at any stage of the production process, either in the technical or in the organizational patterns, addressed to a significant market need. Innovation outcomes are difficult to measure and the current trend is to assess the innovation capacity of a firm through a survey which employs quantitative and qualitative methods.

Research is the process of careful, focused inquiry, frequently carried out by trial and error. As a result of the evolutionary process of understanding and measuring the research procedure, several distinctions have emerged throughout the 20th century.

Basic research is defined as the research process carried out by scientists and collaborators who lack a conscious goal, other than the aim to unravel nature's properties and components. Basic research has sometimes been referred to as 'pure' research, but this term misses the fact that a research program's objective might be to address issues on technology, or problems that may be of interest to the government (health, environment) or a given industry (information technologies, pharmaceutical laboratories).

Usually, basic research is funded through calls from the public sector and the funds are allocated following the evaluation of the projects presented by the researchers. This process of '*ex-ante*' evaluation is carried out by experts, the 'peer-review' system, performed anonymously and free of external influences. The acceptance of the idea that research is influencing economic and social development of the nations has led to some reshaping of the 'peer-review' system. In the European Union research, the peer-review system is applied with some modifications: a) the individual evaluations are carried out independently, but in the Commission headquarters; b) these evaluations are contrasted and discussed jointly with a panel made up of three or four independent evaluators who have previously assessed the project; this leads to a 'consensus' evaluation. In the United States, some agencies, such as the National Health Institutes, are considering including some non-experts into the review process. The debate remains open on the pertinence and qualification of these modifications of the peer-review system.

Basic research has been unable to escape from the increasing demand of accountability. The outcomes of this type of research are scientific publications. Bibliometrics has evolved as the methodology chosen to carry out 'ex-post' evaluation of these research activities (see *Bibliometrics and Institutional Evaluation*).

Applied research addresses the results of basic research to a point where they can be used to meet a specific need or, alternatively, it attempts to solve a specific problem through research. Applied research can be funded in different ways: a) as in the case of basic research, by calls from public agencies which base their funding decisions on the expert evaluation report. These calls may be part of independent research projects or of a research program; b) by calls from public agencies related to the attainment of specific objectives of a research program; c) from industrial requirements through contracts.

'Ex-ante' evaluation of applied research projects from public calls follows the same pattern as basic research evaluation, although the need for reshaping the 'peer-review' system appears more obvious for this type of research. As a matter of fact, even critics of this reshaping in the case of basic research, may accept it for this type of research activities. The research carried out by industries may be evaluated by experts, internal or external to the companies, depending on the firms' strategies. The current trend is to apply evaluation procedures in most cases as companies are becoming more familiar with the practices followed by the scientific community.

The *'ex-post'* assessment of the outcomes of applied research may be supported by combining the measures of bibliometric outputs with the information from surveys addressed to the stakeholders whose interests should be met by the research in question. A pressing problem stems from how to include, among the merits of researchers from the public system, the results of the research they have carried out while under contract with industries.

The development stage of R&D refers to the steps required to bring a new or modified product into the market. There are no specific methods to evaluate this part of the R&D process, except in what is related to aspects of dissemination and utilization of the results of the research process. In general, this aspect of the R&D fits the philosophy of a research program better than that of isolated research projects.

Research Program has emerged as a way of gathering the tools necessary to attain specific goals through research. A research program usually reflects an agency's strategy, either public or private, to fulfil the aims of a given policy. There are several ways in which a research program can be launched, and each of these can be evaluated 'ex-ante' by different mechanisms, essentially resorting to experts resources. The strategic value of a *'research program'* and its complexity, as well as its relevance for the implementation of scientific and technological polices, has raised the interest for the 'ex-post' evaluation of research programs and their outcomes. The idea of 'strategic research' has co-evolved, along with the concept of 'research program'. The notion of 'strategic research' was coined in order to qualify research activities dedicated to providing knowledge and/or techniques for solving problems of (social or industrial) relevance.

3. Relationship between Science Policies, Promotion and Management of R&D Activities

At different times throughout the history of science and technology, several types of science policy models, which emphasize different criteria and methods for the evaluation ('ex-ante' and 'ex-post') of research activities.

The different types of science policy present throughout the second half of the twentieth century can be summarized as follows.

3.1 Policy for Science

Policy for science began after the Second World War with support from political and expert leaders of the United States (President Roosevelt and engineer Vannevar Bush). This policy was essentially aimed at fostering research through government funding. This aim was achieved through research projects and public calls. Evaluation was strongly rooted on the 'peer-review' system on the basis of the principle of 'self autonomy' of the scientific community.

3.2 Policy by (means of) Science

Policy by (means of) science was inspired by a linear model of science push and technology pull. It emerged as a consequence of the identification of science as a promoter of economic growth and development in the industrialized world. The management of this type of policy was carried out through research projects of basic and applied science as well as by research contracts. The notion of research programs begins to emerge as an instrument to develop strategic options in science and technology. Evaluation based on experts still predominates, though modifications to the 'peer-review' system and introduction of new evaluation tools, such as panel reviews, are beginning to be considered.

3.3 Science Policy in the 80s. The Systemic Concepts

The changes characterizing the new orientation of science policy are the following:

(a) The model on which science policy actions and instruments are based has evolved from the idea of a linear link between generation of knowledge and its conversion into technology, to a systemic concept which has paid attention to the socio-historical context, the specificities of the national (or regional) areas, and the nature of the relevant interactions among the actors of the so-called 'systems of innovation'.

(b) These systemic models are changing the framework in which scientific and technical development is understood. A new social contract emerges, in which science loses some of its autonomy and restricts its own limits in response to stronger requirements imposed by the need for greater responsibility towards society. Researchers are no longer shielded from political, economic and social pressures. Research in the fields of electronics, life sciences, biotechnologies, biomedicine, and many other fields is not carried out in isolation from society, neither is it free from the influence from political, historical and economic milieus. The tight link between the generation of knowledge in some areas of research and its applications, is changing the drive of knowledge pursuit, and is modifying the rules for obtaining funds and gaining public recognition.

(c) The idea that a new mode of production has emerged and that its nature should be fully recognised has gained much acceptance in academic circles thanks to the work of Gibbons and others. The generation of knowledge according to this new 'mode' of production, or 'mode 2' in Gibbons' terminology, is characterised by socio-economic and institutional change and, from a cognitive perspective, insists on the use of concepts of multidisciplinarity and interdisciplinarity. An *interdisciplinary* activity deals with knowledge deriving from more than one discipline. Interdisciplinary topics comprise objectives from two or more disciplines, in such a manner that they give rise to an 'entity which is new, simple, and intellectually consistent', so that the result is more than the sum of its parts. A *multidisciplinary* activity produces knowledge which accumulates in an additive fashion. This knowledge arises from more than one discipline, although each of the disciplinary components maintain their identity and, as a result, their inter-relationships and their non-linearity are not evident.

(d) The description of science as an endless frontier has arisen in order to gain increasing support from the public sector for science and technology. However, the idea that a stronger research system would yield even greater benefits for society can be challenged by confronting the promises implicit in that idea and the results obtained. The promise held by Vannevar Bush that more research 'would mean more jobs, higher wages, shorter hours, more abundant crops, more leisure for recreation, for study, for learning how to live' remained unchallenged during periods of economic prosperity and wealth; the United States were the paradigm during the first twenty years after the Second World War as they were the major economic power to finish the war period with its industrial base intact. However, these arguments entered into contradiction with the real situation under periods of economic crisis and recession; after the crisis of the seventies which led to negative changes in the social well-being indicators regarding employment, wages, leisure and gap between wealthy and poor.

4. A General Frame of Reference for Evaluation from the European Perspective

The methods and practices of evaluation, applied to public policies, are experiencing many developments and experimentation. The European Union and the European Commission are contributing in a decisive manner to this type of practice. There are, nevertheless, relevant differences between European countries with respect to the tradition and intensity of the evaluation practices. The Northern European countries possess greater experience in the use of evaluation practices. Other countries, such as Belgium, the North of Italy and several regions of France, are beginning to consider evaluation as a 'help system' for planning and managing public participation. This development is due to the implementation of a series of European programs of socio-economic nature and significance, such as the Structural Funds Programme and several programs of Control and Regulation. In most of the Southern European countries, the evaluation practices are still considered a regulatory obligation constraining the process of decision-making. In many cases, the relevance of the evaluation practices in these 'less-developed' countries, in terms of evaluation capacity and skill, is even more limited by the absence of quantitative (or measurable) objectives, as well as by the lack of good quality information, essential in order to obtain good results in the evaluation process. The work performed thanks to the stimulus and support of the European Commission has allowed the identification of a series of necessary components in order to carry out efficient evaluation activities of programs under evaluation:

1. analysis of the capacity to acquire and use the economic resources dedicated to the program;
2. identification of the tangible products obtained as opposed to the products anticipated;
3. analysis of the program's efficiency (i.e. assessment of the costs of unit of product);
4. evaluation of its impact;
5. evaluation of the impact in socioeconomic terms; and
6. identification of the difficulties confronted in the course of the program.

Based on these analyses, the evaluators may provide adequate proposals enabling them to give recommendations regarding the issues related to the program and to propose changes in the next programs.

The lack of adequate information may hamper the quality, depth and relevance of the processes of analysis and evaluation. Therefore, public administrations share the responsibility to establish effective follow-up systems to provide this information with sufficient levels of quality and a certain degree of continuance.

4.1 Procedures for the Evaluation of Socio-economic Programs

There are different approaches to the evaluation of programs, mainly of socio-economic character. The 'bottom-up' approach is applied with the aim of collecting and valuing information at the project level, in order to process and include it into the analysis of the program, i.e. all the projects addressed to a specific goal. When the number of projects is high, this approach rests upon the selection of a sample. An advantage of this approach is that it allows the gathering of information, by means of surveys and interviews, on the results and impacts of every project. In this 'top-down' approach, the analysis of the impact of the program requires statistical data and information. The existence of databases may permit the performance of longitudinal studies based on the history of each case, by correlating the results of the program with the basic objectives. When the database is lacking, several other methods can be used; for example, the use of focal groups.

The 'top-down' analysis does not require samples and allows the processing of information with a high degree of aggregation. This approach is based on the data given by the respective agencies, though additional tools, such as case studies or in-depth studies, are needed for specific and more detailed information analysis.

It is also possible to employ a mixed approach, combining the 'bottom-up' and 'top-down' approaches. This hybrid system of analysis seems best suited for more complete evaluation practices and thus, is adequate for the application to the 'ex-post' assessment of programs or for their follow-up according to a subject perspective.

In any of these approaches, the existing shortcomings in information, which can be detected in official reports and sources, can be overcome by additional sources of information from the clients, participants or managers of the programs. To this end, the evaluation process rests on in-depth analyses, surveys, interviews, protocols and questionnaires aimed at obtaining the opinions and reactions of these participants.

4.2 The Research and Development (R&D) Programs

The briefly outlined methodology has been applied to the evaluation of socioeconomic programs. The challenge is to adapt these procedures and instruments to the public R&D programs which have specific properties and characteristics as opposed to socio-economic programs which are focused on short-term objectives. The R&D programs fostered by public agencies are also different from the research and development strategies of companies which are more likely to be highly-focused on the economic success of the company. However, the publicly funded R&D programs are characterized by the following:

- The activities of these R&D programs are, in essence, addressed to influence economy and society in the long run: they act in an interactive and indirect way with society and the economy.
- The economic indicators used as inputs, are essentially macro-economic ones, collected from official statistical databases, such as the percentage of Gross Domestic Product devoted to R&D activities as well as the amount of human resources employed in those activities. They are distributed in large subsections such as: Government, Higher Education and Business. It is extremely difficult to correlate these macro aggregates with data from specific projects related to R&D programs.

On the other hand, the indicators of performance ('output') relate mainly to the dynamics of the scientific community either from indicators -research publications and other documents- or from its links with industry, such as patents in different international contexts. These indicators are obtained and analyzed according to classic bibliometric methodology.

However, there are structural problems inherent to the production of a public good for the establishment and introduction of economic output indicators to measure the economic performance of R&D programs. As a matter of fact, economic theory says is that markets provide poor incentives for the production of a public good, because providers cannot take advantage of the benefits derived from its use. Nevertheless, the application of indicators, such as technology trade and number of international patents has allowed the evaluation of economic performances of some R&D programs from the European Framework Program. The recognition of the difficulties in correlating the influence of basic and applied research (public good 'knowledge') as source of growth has led to the establishment of an innovation survey on a European scale to try to further examine the economics of R&D activities.

The functioning of R&D programs is shaped by the sociological and behavioural patterns of the European science and technology administrations and their cultural relationships with the scientific community. There are differences with respect to independence, expertise, methodologies employed to assess the market failure of S&T government support for industry between European countries depending on their size, demographic and economic relevance, as well as their scientific and technological cultures. Consequently, there are divergences in the way national agencies managing R&D programs are run and controlled in relation to the part scientists and technologists play in their dual role as clients and managers.

In the R&D programs, unlike the socioeconomic programs, there is a frequent mix in the planning and management levels. There is little separate information on the role played by agencies involved in these two levels of intervention, planning and management.

4.3 Innovative Approaches to Evaluation of Research Activities

The European Commission, apart from its role as administrative agency responsible for the self-control of its research programs, has been acting as a funding agency for projects aimed at exploring new methodologies on evaluation and other science and technology policy tools. A great deal of these projects have been involved in performing comparative studies between different European countries comparing and assessing national experiences in order to build a common evaluation culture in Europe.

4.3.1   The Different Roles of Research Institutes and Laboratories

The evaluation of research institutes has been approached by the methodologies introduced by Laredo and collaborators, such as the 'compass card model' (Figure 1) which considers research a 'professional activity which is inserted in many different contexts, each having its own regulatory mechanisms and ways of producing reputation and/or wealth'. This model is based on the fact that the relationship between the funds allocated through a research policy and the actual outputs and outcomes is not linear. Furthermore, the model, drawing from the analysis of the dynamics of science proposes to evaluate the mission of research institutes on different grounds:

- certified knowledge --which is the knowledge transmitted by a code or conventional system of structured signs and symbols acting as a vehicle for communication;
- embodied knowledge -inscribed in the minds or bodies of researchers and technicians or in instruments and machines-;
- participation in competitive advantages and links with the business sector;
- development of new public goods or services (public policies);
- participation in public debates of scientific issues.

Each research unit or laboratory defines its 'strategic profile' as a mixture of the positions held in the five different dimensions mentioned above. This model has been validated firstly with two tests carried out on a national scale (in a mission-oriented research institution and in a diversified institutional setting at a regional level), and secondly, as part of a European project trying to complement, by a 'bottom-up' approach, the analysis and evaluation of the dynamics of the public sector research. The field of human genetics was chosen on the basis of a series of criteria that facilitated convergence for a comparative analysis.

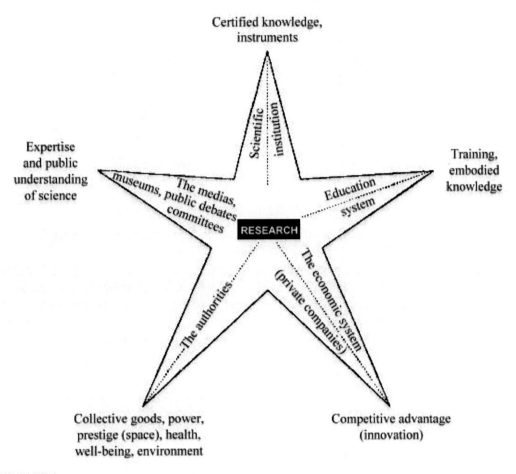

Source : CSI

Figure 1. The Compass-Card model. Taken from Callon, et al. (1997)

Laboratories studied with this methodology were able to identify different and contrasted involvements in each dimension and these activities did not combine at random, thereby creating a limited set of configurations. The use of the same method on a research field, namely that of human genetics, has enabled the identification of the role of research units related to new trends and areas for research. Human genetics has been repeatedly shown to be a high priority research field by science and technology policies. The research units studied in six countries (Germany, Ireland, Norway, Spain, Sweden and the United Kingdom) created or redirected their research agenda towards human genetics during the late 1980s or early 90s. This finding confirms that scientists prefer to work and are capable of moving into fields which emerge as priorities.

The field of human genetics also offers a clear example of the links between R&D policy and the shaping of a scientific organization which adapts to a given policy. Resource allocation largely takes place on a competitive basis. Most researchers have temporary contracts or are employed to work on a specific project, and the bulk of the research activities are funded through competitive mechanisms. The 'myth of infinite benefit', as described in the argument of science as an endless frontier, leading to the assumption that more science and more technology will lead to more public good is questioned by the strategy of resource allocation revealed by the studies in the field of human genetics.

The units involved in research on human genetics are transdisciplinary and/or interdisciplinary and are located in a variety of institutional settings. This fragmentation seems to respond to some institutional drawbacks and to some specific needs of a field that represents a typical case of 'new mode of generation of knowledge'. The research is problem-oriented and researchers look for the best suited setting in which to carry it out, since human genetics is not recognised as an independent medical discipline neither in teaching nor in health care environments. The recruitment of new skills and technologies appears to be a critical factor for the future success of the unit. Research collaborations fill gaps in skills, expertise and techniques, and is also vital for access to biological samples and patient histories. Funds for research are provided by a variety of agencies: government and research councils, sectoral agencies with mission-oriented objectives, foundations and charities, European Union funds, and the industry. Flexibility and ability to adapt are critical assets in this emerging, highly dynamic research area, where the flow of knowledge, techniques and applications are advancing at a much faster pace than the traditional slow pace of academic and research institutions.

This provides grounds for the argument that the classical self-autonomy of a research enterprise is gradually being reduced. Scientists follow 'a niche or diversification strategy to ensure organizational survival, through the selection of the study of a particular disease or the application of a given technique to reach a leading position in both scientific and social aspects'.

A last, but not less important, point concerns the level of awareness of the scientists on the existence of a social debate on the uses of human genetics (and other new biotechnologies), the repercussion on their activities, as well as their willingness to participate in the debate. There were no common patterns of action from the different countries in the study. The institutions have attempted to approach the issue by establishing ethic committees. This is a strong indication that the issue is still pending in the research agenda of research institutions and organizations. Their lack of efficacy and sense of opportunity (there is no conscience that we are in a 'risk society') to cope with this critical issue may hamper the future development of science and technology in the industrialized world and may obstruct the participation of experts in the necessary social debate.

4.3.2 Evaluation of Societal Quality of Research

The concept of 'societal quality' of research was introduced into the Dutch practices of evaluation in the early 1990s to suggest an analogy with 'scientific quality' of research in order to broaden the scope of evaluation. For science policy, the concept of "societal quality' is defined in broad terms by incorporating diverse phenomena; from the direct use of results of policy-oriented research to long-time collaboration with industry partners and the funding of answers to urgent problems. For evaluation purposes, 'societal quality' has to be traced through indicators. Along this tracing process, two categories, *relevance* and *impact*, have been taken into account for evaluation in the Dutch process of evaluation. *Relevance* is concerned with the actual and envisaged linkages and promises of proposed, ongoing or concluded research projects or programs. *Impact* stands for the uptake of research and the effects of such uptake.

The evaluation of societal quality, as carried out in the Dutch research system, is based on established indicators and criteria. Indications and indicators have been drawn from an evaluation repertoire concerning the missions and strategies of the unit or program to be evaluated.

The conclusion at this stage is that the concept of societal quality (and related concepts) needs to be further articulated and methods to evaluate societal quality should be developed and applied.

4.3.3 The 'Transducing Model' as an Alternative to R&D Programs Evaluation

This approach uses quantitative techniques via questionnaires, addressed to the main actors of the research and technology activities (main researchers and/or the managers of the projects), and compares the results with the track record of the funding agencies, mainly by means of their annual reports and their tracing with respect to the predicted goals of the programs. The use of surveys permits to explore several dimensions of the outcomes of the program in similar ways to those described previously for the 'compass-card model': generation of knowledge and its relevance; links with other elements of the R&D system management of the program; degree of satisfaction of the main actors with the program.

The model is based on an analogy with the biochemical pathways (Figure 2), which combine the notions of structural and functional biology. The term *operational biology* was coined for this purpose and provides the basis for the model for evaluation of R&D programs. R&D programs are like biochemical pathways in which the structure and function of the active elements, proteins (agencies and researchers) transform and use nutrients, energy, (resources in the R&D metaphor) through the process of cellular transduction in which effectors (experts, actors) operate.

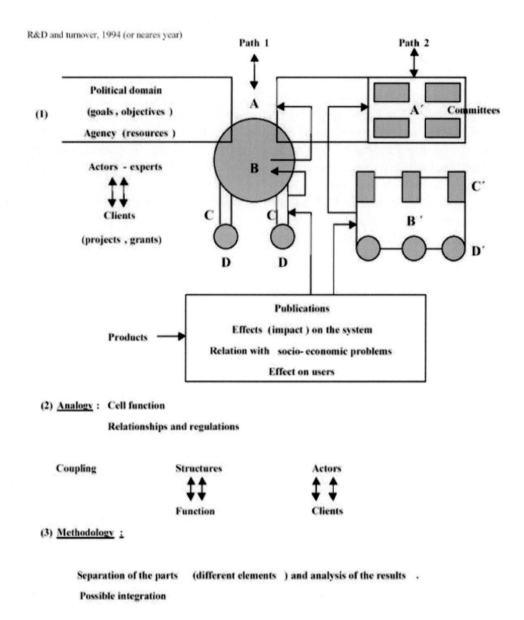

Figure 2. The 'Transducing' model of R&D Programs

The 'transducing model', as depicted in fig.2, offers some advantages and analytical possibilities. The model allows the distinction and assessment of the different levels of action in R&D programs. The first one is the planning stage in which the objectives are defined; the second one corresponds to the management and funding stages which allocate economic resources through the intervention of the agencies. Two paths are possible for the fulfillment of this second level: a passive approach where the agency (A) acts only as a transmitter and controller; an active approach in which the agency (A'), in the absence of targeted, ear-marked objectives, behaves as a transducer.

The third stage is the one in which the program is carried out; experts and clients are interacting. In path 1, there are two types of actors: the experts ('peers') and bureaucrats (B) who are allocating resources, in relation to the pre-established goals, to the clients (receptors) as groups (C) or individuals (D). In path 2, the experts are acting within the function of agency A', whereas the clients – beneficiaries – are represented in rectangle B' and they will receive grants-in aid, fellowships, as groups (C') or individuals (D') to perform research activities in a complex institutional and operational environment. The fourth stage refers to the production, with the products poured either into the scientific-technical 'milieu'-statistics and bibliometrics – or into the external 'milieu' – societal quality or evaluation. The assessment of the products deriving from path 1 should be carried out with the mediation of experts and clients, while in the case of path 2, the analysis of the products requires a more complex mediation with the participation of the agency and the different types of actors. The fifth and last level, corresponding to 'feedback', applies to either of the two paths and may influence the future decisions at every stage of the program.

The model attempts to create an analogy between the research programs and the functioning of a cell, where there is a coupling between structure and function. The parts or elements of the program can be separated, leading either to a reductionist methodology for the analysis or to an integrated view from the analysis of the parts, in an analogous manner to the methods applied in biochemical analysis.

This model has been used by carrying out a series of evaluations on various R&D programs and innovative technologies or sectors in Spain (New Materials, Health, Pharmaceutical Sciences, National Health Fund).

5. Emerging Issues on Evaluation from the United States

After the Second World War, when science became a significant item on the political agenda, the United States took the lead in issues related to evaluation. The establishment of the National Science Foundation and the subsequent incorporation of other agencies, such as the National Institutes of Health, led to the establishment of the peer review system as the basic instrument to make (good) decisions in the promotion of basic research. The system remains highly praised and essential for protecting the research enterprise, but is not free of problems. Most of the problems arise as a consequence of the changes in science and technology policies that have been referred to previously: loss of autonomy of science, a generation of knowledge that it is more tightly linked to economic or social objectives, the awareness of limits to the benefits of science, all these arguments combined with the never-ending competition for publications, supremacy and prominence.

Two conflicting issues are emerging regarding the evaluation process by peer-review in the United States: the need for social accountability of science and the cases of misconduct.

The first issue has attempted to be solved by the incorporation of non-specialist scientists and/or social representatives in the panels in charge of evaluating research projects for their funding. The experience is under way, and information regarding the outcomes is still pending.

The second issue is complex. The establishment of an Office for Research Integrity in relation to the funding agencies raised much interest, but its performance has not been at all satisfactory. The action of the 'whistle-blowers', responsible for arousing suspicions of misconduct, has raised interesting problems but, at the same time, caused problems by misinterpretations and over-judgments. In mid-October 1999, the U.S. Office of Science and Technology Policy published in the Federal Register a new government-wide policy defining scientific misconduct in order to protect the integrity of the research record. The political action shows the significance of the issue. Although the answer is not simple, it is worth reflecting on this issue in order to alert the newcomers to the scientific enterprise that they are sharing a sociopolitical responsibility. As Bloom stated in an editorial of the journal *Science*, 'No nation's scientific community is immune to the various obvious and not so obvious forms of misconduct; thus, there needs to be ongoing international discussion of these issues. Strengthening the integrity of the research enterprise requires recognition that ethical hooliganism, be it intended or not, is intolerable'.

Another new aspect regarding the evaluation of science refers to the increasing involvement of society and its individuals as agents in the production and application of science, and the risks derived from this (new) process. Particularly appealing to illustrate the challenge of the (new) research stake is the case of clinical research (see also *Ethics and Science Policy*). All countries are undoubtedly experiencing the difficulties that clinical research is facing as it is outpaced by the increasing complexity of research and the need for complex and costly oversight mechanisms. An article published in *Science* by two leading authorities of the Duke University Health System reveals that the evaluation by the members of the Office for Protection from Research Risks -established in 1972 with the aim of developing uniform standards for patient protection and assessing its application – through site visit has led to the identification of administrative deficiencies. While accepting and encouraging the need for greater oversight in clinical trials and research, the authors criticize the complexity of the system in institutional terms; they ask for a comprehensive review of the legislation that protects the subjects of research experimentation in order to adapt it to a new situation. The main message is the need to develop an 'effective, simplified system that is understandable, that works, and that is adaptable to change'.

This very brief review should mention that the institutional aspects of policy-making are rarely the object of social sciences research. Because of the strong ties between the scientific base and the productive sector, social sciences in the USA have mainly been interested in the economics of science. The analysis of laboratories and their participation in American public policy has been limited to a few, but relevant, studies from the point of view of administrative sciences.

In summary, the main trends for the evaluation of research activities and programs in the United States continue to focus on the most traditional paradigm of the 'republic of scientists', delving into the improvement of the peer-review system and in the measurement of the economic benefits of basic science. In this context, addressing the increasing need for social accountability and taking into account the great complexity of the scientific domains is gaining strength.

6. Conclusions

- Evaluation holds great importance and a strategic position in the domain of science policy studies. As science participates more in the wealth of nations, the critics against scientific progress also rise. Therefore, evaluation procedures need to accommodate the new course of the science and technology policies in accordance with the new demands from society.

- Evaluation of research activities depends on the significance of scientific and technological activities in the economic and political agenda. But, measures in order to demonstrate the usefulness of science seem less important to policy-makers. In this respect, Europe appears to be far more worried than the United States by such a demonstration.

- There is an important shift in the focus of evaluation from the individual scientist and the 'republic of science' paradigms, to a new paradigm addressed by the new sociology of science, which is paying more attention to the collective organisations, and to the societal needs in the analysis of the dynamics of science.

- In view of the complexity of the science and technology production processes, it is necessary to go beyond the concept of linearity in evaluation models. More complex indicators must be developed and it is necessary to give way to more complex models and metaphors. Finally it is necessary to take into account the international dimension in these approaches.

Acknowledgments

The authors wish to thank the Spanish R&D National Plan managed by the Interministerial Committee for Science and Technology (CICYT) and the IV European Framework Programme (TSER Programme), as well as the Spanish National Health Fund (FIS), for financial support at different stages of the original work performed by them and referred in this paper. Discussions and informations gathered from many colleagues: María Jesús Santesmases and Luis Sanz-Menéndez from our Research Unit, Jordi Camí of the IMIM hospital in Barcelona, José Conde and L.E. Claveria at the time they were involved in the management of the Spanish Health Fund, P. Laredo (CSI, Ecole des Mines, Paris) U. Schimank and Markus Winnes (Köln) and M. Mira Godinho (CISEP, Lisbon) have been extremely useful. However the authors are the sole responsible for the text. The continuous and valuable secretarial help by Mrs. M. Carmen Montalvillo is gratefully acknowledged.

Glossary

Applied research:	Research carried out with a solving-problems perspective.
Basic research:	Curiosity driven research with the aim of producing knowledge.
Development:	The process that uses research results to bring a new or modified product into production.
'Ex-ante' evaluation:	Evaluation of research projects (and programs) before they have been executed (proposals).
'Ex-post' evaluation:	Evaluation of the outcomes of research projects and programs.
Peer-review	The activity of evaluation carried out by independent experts

evaluation:	acting on their individual qualification and for general interest.
Republic of scientists:	The realm of science as controlled by scientists themselves.
Research program:	Research strategies developed by public or private agencies to attain targeted goals.
Societal quality of research:	Influence of research on social aspects.

Bibliography

General references

Bührer S. and Kuhlmann S. (eds) (1999). *Evaluation of Science and Technology in the New Europe*, Proceedings of an International Conference on 7 and 8 June, 1999, Berlin, 209 pp. German Federal Ministry of Education and Research, European Commission. [Reviews the role and function of evaluation at various levels of decision making in relation with the changes of the S&T systems in the wider European context.]

Callon M. Laredo P. and Mustar P. (1997). *The Strategic Management of Research and Technology: Evaluation of Programs*, 445 pp. Paris: Economica International. [Contains relevant information provided by top scientists on the links between science, technology and the market and on how to evaluate those links and the resulting hybrid activities.]

Gheorghiu L. Dale A. and Cameron H. (guest-editors) (1995). National systems for evaluation of R&D in the European Union. Special issue of *Research Evaluation* **5**(1), 1–108. [Contains relevant information on the different evaluation systems in European member countries as an outcome of the activities of the Network on European Science and Technology Policy Evaluation.]

Report of the Advanced Science and Technology Policy Planning Network (ASTPPN), authored by Kuhlmann S. Boekholt P. Georghiu L. Guy K. Heraud J. A. Laredo P. Lemola T. Loveridge D. Lukkonen T. Polt W. Rip A. Sanz-Menéndez L. and Smits R. (1999) under the title '*Improving Distributed Intelligence in Complex Innovation Systems*', 92 pp. Karslruhe: Fraunhofer Institute Systems and Innovation Research, ISI. [Discusses the need for a synergy of the combination of the different tools used for strategic decisions in innovation policy: evaluation, technology assessment and technology foresight.]

Specific References
On framework and conceptual aspects

Gibbons M. Limoges C. Nowotny H. Schwartzmann S. Scott P. and Trow M. (1994). *The New Production of Knowledge: The Dynamics of Science and Research in Contemporary Societies*, 177 pp. London/Thousand Oaks/New Delhi: Sage Publications. [Analyses the modes of knowledge production and identification of the distinctive characteristics of 'mode' 2.]

Gilbert L. (1998). Disciplinary breadth and interdisciplinary knowledge production, *Knowledge, Technology and Policy*, Spring/Summer 1998, **11**(1&2), 5–15. [Introduces definition of disciplinarity and interdisciplinarity in relation to the production of knowledge.]

Laredo P. and Mustar P. (1996). Laboratory configurations: an exploratory approach, 23 pp. Presented in *EASST/4S Conference: Signatures of Knowledge Societies*, Bielefeld 10–13 October. [Aims to delve into the characterisation of laboratory activities in the frame of the Compass Card model.]

Muñoz E. (1994). *Una Visión de la Biotecnología: Principios, Políticas y Problemas*, 145 pp. Madrid: Fondo de Investigación Sanitaria. [Discusses some concepts of the philosophy of biology that may help in the evaluation of the applications of life sciences.]

Sarewitz D. (1996). *Frontiers of Illusion: Science, Technology and the Politics of Progress*, 235 pp. Philadelphia, PA, USA: Temple University Press. [Offers critical insights on the myths that have guided the science policy and presents new challenges that are being confronted by the science and technology enterprise.]

On empirical and case studies

Bell J. I. (1999). Clinical research is dead; long live clinical research, *Nature Medicine,* **5**(5), 477–478. [An analytical exploration of the relations between basic biomedical and clinical research.]

Bloom F. E. (2000). Unseemly Competition. *Science,* **287**(5453), 28 January, p. 589. [Presents US policy to deal with the problems of scientific misconduct and the limits of the evaluation by a peer-review system.]

Espinosa de los Monteros J. Martínez F. Toribio M. A. and Muñoz E. (1994). *El Programa Nacional de Nuevos Materiales en el periodo 1988-1992. Su evaluación mediante una metodología dual.* Documento de Trabajo 94-10, 86 pp. Madrid: IESA, CSIC. [The first of a series of evaluation of Spanish National R&D programs by a blend of quantitative and qualitative methods.]

Espinosa de los Monteros J. Martínez F. Toribio M. A. Muñoz E. and Larraga V. (1995). *El Programa Nacional de Investigación y Desarrollo Farmacéutico durante el periodo 1988–1993. Una evaluación mediante una metodología dual.* Documento de Trabajo 95-08, 74 pp. Madrid: IESA, CSIC. [The application of the hybrid methodology to the evaluation of the Spanish National Programme on Pharmaceutical R&D.]

Espinosa de los Monteros J. Martínez F. Toribio M. A. Muñoz E. and Larraga V. (1995). *El Programa Nacional de Salud durante el periodo 1989–1993. Una evaluación mediante metodología dual.* Documento de Trabajo 95-09, 71 pp. Madrid: IESA; CSIC. [Extension of the methodology to the Spanish National R&D Programme on Health.]

Espinosa de los Monteros J. Larraga V. and Muñoz E. (1996). Lessons from an evaluation of Spanish public – sector biomedical research, *Research Evaluation,* **6**(1), 43–51. [Assesses the inputs and outcomes of the funds devoted to promote research in biomedicine in relation to the objectives of the patient-care system.]

Espinosa de los Monteros J. Díaz V. Toribio M. A. Rodríguez Farré E. Larraga V. Conde J. Claveria L. E. and Muñoz E. (1999). La investigación biomédica en España. (I) Evaluación del Fondo de Investigación Sanitaria (FIS) a través de los proyectos de investigación financiados en el periodo 1989–1995 a centros de investigación a instituciones sanitarias asistenciales (hospitales). (II) Evaluación del Fondo de Investigación Sanitaria (FIS) a través de los proyectos de investigación financiados en el periodo 1989-1995 a centros de investigación, facultades y escuelas', *Medicina Clínica,* **112**(5), 182–197 and (6), 225–235. [The two articles aim to assess the funding activities of the Spanish National Health Research Fund through an analysis of the research projects by surveys addressed to the principal investigators and the managers of the research centres.]

Laredo P. and Schimank U. and Winnes M. (1999). *An Approach to Public Sector Research Through its Research Collectives. Overview, Interim Report B,* 25 pp. plus two Appendices of 123 and 74 pp. Paris/Köln: Armines/CSI and Max-Planck-Institut für Gesellschaft. [Presents a synthesis of the work undertaken for the international comparison of public sector research in Europe and some OECD countries using an exploratory approach based on the quantitative and qualitative analysis of laboratories or research units acting in the field of human genetics selected as a test.]

Muñoz E. (dir) (1999). *Memoria Final Proyecto FIS 96/1803* (mimeo), 517 pp. Madrid: IESA, CSIC. [Contains a series of contributions by the members of the multidisciplinary team that evaluated the different activities-research projects, training schemes, and scientific production-derived from the funding activities of the Spanish National Health Research Fund along the period 1988–1995.]

Muñoz E. in collaboration with Díaz V. M., Espinosa de los Monteros J., and Santesmases, María J. (2000). Targeted research and technological innovation and their relationships in a new socio-political context. Approaches to their evaluation. Presented in the workshop on *Innovation and Diffusion in the Economy: The Strategy and Evaluation Perspectives,* Lisbon 24-25 January, Lisbon: CISEP (ISEG,

Technical University of Lisbon). [Introduces the comparative view within a European frame of the different evaluation exercises carried out in Spain under the direction of the main author.]

Narin F. and Noma F. (1985). Is technology becoming science? *Scientometrics,* **7**(3–6), 369–381. [In this article the authors apply a quantitative methodology based on publication records to establish the links between science and technology.]

Sanz Menéndez L. (1995). Research actors and the state: research evaluation and evaluation of science and technology policies in Spain. *Research Evaluation,* **5**(1), 79–88. [A review of the development of research evaluation practices in Spain in the last twenty years.]

Snyderman R. and Holmes R. W. (2000). Oversight Mechanisms for Clinical Research. *Science* **287**(5453), 28 January, 595–597. [Debates on the consequences for human subjects of the increasing complexity of clinical trials and on how to assess them and adapt the situation to a changing environment.]

Van der Meulen B. and Rip A. (2000). Evaluation of societal quality of public sector research in the Netherlands. *Research Evaluation,* **8**(1), 11–25. [An important effort to evaluate the relevance and impact of the research activities in the Dutch public sector.]

Biographical Sketches

Emilio Muñoz. Born on January 13, 1937, is Research Professor in Biology and Biomedicine at the Consejo Superior de Investigaciones Científicas (CSIC), now in the Unidad de Políticas Comparadas (Grupo de Ciencia, Tecnología y Sociedad), former Instituto de Estudios Sociales Avanzados, C/ Alfonso XII, 18, 28014 MADRID, Spain. From 1982 to 1991, he served different managerial positions in the Spanish R&D Administration (General Director for Science Policy, General Secretary of the National Plan on R&D, Chairman of the National Research Council, CSIC). He has been member of the CREST European Committee and Chairman of the COST Program. He has published more than 350 articles on biochemistry, molecular biology, science and technology policy as well as public understanding of science with special emphasis in the field of life sciences. He has authored or co-authored several books, among them: *Establecimiento de la bioquímica y biología molecular en España*, Centro de Estudios R. Areces (with María J. Santesmases); *Biotecnología, Industria y Sociedad: El caso español*, Fundación CEFI; *Changing structure, organisation and nature of public research systems. Their dynamics in the cases of Spain and Portugal*, IESA-CSIC (with María J. Santesmases and J. Espinosa de los Monteros). He is now involved in the analysis and evaluation of scientific and technological policies with special emphasis on the problems of R&D Regional Policy and coordination between administrations as well as in the biotechnology field, analyzing the socioeconomical and political implications of the uses of biotechnology. He has been member or associated partner in about a dozen of European projects funded by DG XII on different programs (MONITOR, BRIDGE, STRIDE, FAIR, TSER).

Juan Espinosa De Los Monteros. Born in Madrid (Spain) on June 1, 1938. He holds a Ph.D. in Chemistry granted by the Universidad Complutense of Madrid (Spain) and a postdoctoral research career at the University of Sheffield United Kingdom). He is a Research Associate Professor at the Consejo Superior de Investigaciones Científicas – CSIC (National Research Council of Spain), now in the Unidad de Políticas Comparadas, Grupo Ciencia, Tecnología y Sociedad, Alfonso XII, 18, 28014 Madrid (Spain). From 1982 to 1992, he served different managerial positions in the Spanish R&D Administration as Responsible for Management of several R&D Programs: Acuaculture and New Materials at the Interministerial Committee of Science and Technology (CICYT); and manager in charge of R&D Programs at the Instituto de Salud Carlos III (Ministry of Health and Consumption Affairs of Spain). He is now involved in the analysis and evaluation of scientific and technological policies with special emphasis on the problems of R&D programs and their relationships with regional and international policies. He has published a series of articles on evaluation of public R&D policies and programs in national and international journals (*Science and Public Policy, Research Evaluation, Medicina Clínica....*). He is the author or co-author of a series of books, among them: *Evaluación de las actividades de investigación y desarrollo tecnológico en acuicultura en el periodo 1982-1997*. Tomo I, *Directorio de publicaciones de investigadores españoles en acuicultura*. Tomo II (Ministerio de Agricultura, Pesca y Alimentación, Centro de Publicaciones).

Víctor Díaz. Born on October 22, 1970, holds a degree in Political Sciences and Sociology, speciality of Industrial Sociology, granted by the Salamanca Pontifical University at Madrid. He is performing studies for a doctoral degree in the Department of Sociology, section of Human Ecology and Population, at the Madrid Complutense University. He has carried out research in the Institute of Demography of the Spanish Research Council (CSIC) on the prospective of the evolution of Spanish population until 2025. He possesses diplomas of specialist in several areas related to the evaluation of training programs, human resources and labor law and social security. He is now a member of the Grupo Ciencia, Tecnología y Sociedad in the Unidad de Políticas Comparadas (CSIC), where he is directing and collaborating studies on the evolution of public policies, in particular in the fields of health and biomedicine and biotechnology, being in charge of the monitoring during the last five years of the evaluation of biotechnology sector in Spain. He has published several articles in peer reviewed national and international journals. He has contributed to the first analytical and prospective study on industrial biotechnology in Spain sponsored by the Ministry of Industry and Energy (1999-2000). He has authored or co-uthored several books, among them: *Salud y ancianidad en Segovia* and *Evaluación de las actividades de investigación y desarrollo tecnológico en acuicultura en el periodo 1982/1997.*

BIBLIOMETRICS AND INSTITUTIONAL EVALUATION

Jane M. Russell
Centro Universitario de Investigaciones Bibliotecológicas, Universidad Nacional Autónoma de México, Mexico City, Mexico

Ronald Rousseau
Department of Industrial Sciences and Technology, KHBO, Oostende, Belgium

Keywords: Bibliometrics, Citation Measurements, Journal Impact Factors, Relative Impact Indicators

Contents

1. Introduction
2. Bibliometrics as an Evaluation Tool
3. Output Evaluation
4. Citation Measurements
5. Journal Impact Factors
6. Relative Impact Indicators
7. Future Trends and Perspectives
Glossary
Bibliography
Biographical Sketches

Summary

All evaluations are dependent on the availability of adequate and reliable data relating to the outcome of the activities under scrutiny. Literature-based or bibliometric indicators which quantify the production and use of bibliographic material, have been used extensively in the assessment of research performance. Their use is based on the assumption that the immediate purpose of research is to produce new knowledge and that publication is the primary form of output. Publication counts serve as an indicator of the amount of new scientific knowledge produced by researchers. The impact of this new knowledge can be measured by the number of times publications have been cited by other scientists in subsequent work. Impact, however, cannot be automatically equated with quality. A particular form of estimating the potential quality of scientific papers is to relate this to the prestige and impact levels of the journals in which these are published. These journal impact factors can also be used to compare the citation performance of research groups within specialist fields.

The validity of bibliometric indicators is much greater at the aggregate levels of research groups, university departments and research institutes and should be applied with extreme caution when measuring or comparing the performance of individual scientists. Bibliometric indicators are not intended to replace peer review, but rather to make research visible and debatable, ensuring that experts are sufficiently informed to make sound judgements. Publication-based evaluation, however, considers purely the research aspect of institutional scientific activity and should, therefore, be seen as only a partial indicator of overall scientific performance.

The main constraint to the general validation of bibliometric techniques is the limited availability of databases and other information sources providing reliable and comprehensive raw data for analysis, particularly with regard to research carried out in developing countries. The potential of web-based electronic sources for providing comprehensive and accurate production and citation data for bibliometric analysis coupled with the capacity of the Internet to integrate information from a large number of different sources, promises to revolutionize the way indicators are constructed by eliminating many of the methodological constraints experienced today.

1. Introduction

The worldwide preoccupation with 'value for money' in science requiring the rationalization of dwindling support for scientific research, has led to increased use of quantitative data by policy makers. Indicators based on the statistical analysis of quantitative data provided by the scientific and technological literature have been used to measure scientific activity since the beginning of the 20th century. The term 'bibliometrics' was first introduced in 1969 as a substitute for statistical bibliography used up until that time to describe the field of study concerned with the application of mathematical models and statistics to research, and quantify the process of written communication. Evaluative bibliometrics is a term coined in the seventies to denote the use of bibliometric techniques, especially publication and citation analysis, in the assessment of scientific activity.

Research evaluation is not the only area of science studies where bibliometrics has a traditional role to play. These techniques are also used extensively for studying the interaction between science and technology, in the mapping of scientific fields, and for tracing the emergence of new disciplines, as well as in the development of foresight indicators for competitive advantage and strategic planning. Bibliometrics is also relevant to other fields. Economists and historians of science, for example, use bibliometric indicators to measure productivity and eminence.

Bibliometric analysis of scientific activity is based on the assumption that carrying out research and communicating the results go hand in hand. Scientific progress is attained by researchers getting together to study specific research topics, steered by the previous work of colleagues. The classic input-output model used to describe the scientific research process suggests that publications can be taken to represent the output of science. Publications, most commonly in the form of the refereed article and the scholarly monograph, are regarded as the definitive statements of the results of research projects. This production can be quantified and analyzed to determine the size and nature of the research carried out. Studies can be performed at macro level to measure global, regional, or national trends or at the micro level of institutions or groups.

Indicators of scientific research can be divided into two main groups: the input indicators such as money spent, equipment used or personnel employed while output indicators such as the literature-based indicators already mentioned, represent the results and outcomes of the research process. Indicators are either absolute or relative. Absolute indicators refer to one particular characteristic of research activity such as number of articles published, number of citations or the amount of money spent while relative indictors show the relationship between two or more aspects such as number of articles per research group or the number of citations per paper. The latter set of indicators is generally more useful in research evaluations due to their ability to establish compound relationships between inputs and outputs such as the amount of money spent per group per article or the productivity of research groups in terms of the number of articles published per group.

Bibliometric indicators are more powerful at higher levels of aggregation and are more suitable for analyzing patterns in a large set (a faculty or large research team) and less suitable for the evaluation of individuals or small research teams. Consequently, the validity of bibliometric indicators when applied to small data sets is questionable making peer review judgements imperative at this level. Whatever their level of aggregation literature based indicators should not be used by non peer policy makers who do not have the necessary background knowledge of the research area or research groups concerned. Interpretation of quantitative data must go hand in hand with qualitative assessment procedures.

At all levels of evaluation no indicator should be taken in isolation. A series of indicators representing the different facets of scientific activity should be employed. When these partial indicators converge to give a unified picture, their validity is strengthened. Some examples of these partial indicators refer both to input into the research process, such as the level of research funding, and also to impacts resulting from the research process. Examples of the latter are non-bibliometric impact indicators such as recognition in the form of prizes or invitations as keynote speakers in major international meetings.

Conceptual and methodological problems associated with finding appropriate output measures arise from the intangible nature of much of the output of basic research activities. Nonetheless, publication and citation data have proved meaningful for measuring scientific output and its impact on the course of scientific research. The number of publications that a research group produces is taken to represent their scientific production and their primary contribution to the generation of new knowledge. Contributions to scientific knowledge take the form of new facts, new hypotheses, new theories or theorems, new explanations or new synthesis of existing facts. The number of times this new information is cited by the authors of later publications measures the impact of their work on the advancement of research in their specialized field, and sometimes, even in other areas of knowledge. It is also indicative of the amount of recognition they enjoy from other members of the scientific community. The reward system theory of science implies that scientists must share their results in order to gain recognition from their peers. Furthermore, the number of publications and citations a research group receives is associated with their visibility as scientists. However, not all published scientific work is equally visible. The level of visibility depends greatly on the place and language of publication as well as the field in question. Work that is not internationally visible will have little chance of being picked up by scientists other than those in close communication with the authors in question. The inclusion of the group's publications in international databases is also a factor affecting their visibility, particularly as these sources are used extensively for the generation of bibliometric indicators.

Over the last decade impact factors (IF) of scientific journals have gained importance in scientific work and information management, as well as in research management and policy. IF is used as an indicator of journal performance and as such has a role to play in the evaluation of research groups, institutes and even countries. Quality journals in science generally contain coherent sets of articles with respect to content as well as professional standards. This coherence stems from the fact that most journals are nowadays specialized in relatively narrow sub-disciplines and their 'gatekeepers' (editors and referees) share views on questions like relevance, validity and quality with the invisible college to which they belong.

An important consideration, therefore, in bibliometric studies are the channels used for the dissemination of research work and their coverage in widely accessible bibliographic databases. This latter point is even more important when considering impact indicators due to the fact that only one series of databases, the Citation Indexes produced by the Institute for Scientific Information (ISI) (Philadelphia, USA), is available for citation analysis and for the production of journal impact factors. This service includes only a small proportion of journals published worldwide, restricting its coverage to a few thousand highly cited, mainstream journals.

In the present study we look at one important application of bibliometric indicators, institutional research evaluation based on the analysis of the publication and citation outputs of groups of researchers. The role of journal characteristics, such as the journal impact factor, in literature based evaluations is also described. We concentrate our discussion on the natural and life sciences where bibliometric indicators have reached a higher level of development than in other areas of human knowledge. Special attention is paid to the theoretical foundations of indicator production and the different methodologies available for their construction.

2. Bibliometrics as an Evaluation Tool

With the advent of 'Big Science' bibliometric techniques found a new application in the realms of science administration as a research management and policy tool. Previously, bibliometrics had been the little known domain of librarians, sociologists and historians of science. The need for a relatively quick, easy and inexpensive alternative to peer review for evaluating research performance led to the 'discovery' of bibliometrics by science policy specialists and the emergence of a new field of study dedicated to the quantitative study of all aspects of science activity. This new field of scientometrics attracted specialists from different backgrounds, such as mathematicians, information professionals, computer scientists, psychologists, as well as researchers from the natural and medical sciences with a special interest in the study of their own disciplines. The widespread interest in this new field led to the creation in 1977 of its own journal, aptly named *Scientometrics,* and in 1995 to the formation of its own international professional society, the International Society for Scientometrics and Informetrics. Early in the 1960s the introduction of the *Science Citation Index* (SCI) had given bibliometrics a great methodological push. Science indicators research has also been instrumental in the development of the field of scientometrics from the seventies onwards.

Apart from the theoretical and applied research aspects of the field, bibliometrics and scientometrics also give support to countless evaluation exercises performed by tenure, promotion, and awards committees all over the world, as well as by government science policy-makers. While never intended to replace peer review the adjunct of bibliometric indicators make for better-informed expert decisions with respect to budget allocations and in the definition of research agendas and strategic goals. Most bibliometric evaluations of papers, journals and institutions correlate well with peer review appraisals suggesting that bibliometric indicators are generally accordant with the intuitive notions of knowledgeable scientists, as well as with the cognitive state of the art of particular research fields. Nonetheless, rather than bibliometrics being championed as a cheap alternative to peer review, the two methods offering different viewpoints on a common problem, should be considered complimentary and, wherever possible, used concurrently, especially in small scale evaluations.

The expansion of automated bibliographic information services linked to the exponential increase in the volume of scientific literature has presented greater opportunity for the application of bibliometric indicators in research evaluation. This, in turn, has required the design and implementation of better systems design and software development for the handling of large quantities of data and the application of algorithms for the calculation of a wide range of indicators. As these indicators have become more accessible, their weaknesses and strengths have become better understood.

An important and relatively recent application of bibliometrics is in program evaluation. Mapping a field, for instance, before a program is launched, immediately after the end of the program and, perhaps, a few years later furnishes relevant information on many aspects of the field under study, such as the occurrence of cognitive and structural changes. In funding programs too, analysis of scientific publications before and after the funding period can give important insights into its effect on the generation of publishable results.

Although bibliometrics is now a routine tool in evaluations, its use still has its critics. The fact that hard techniques are applied to one important field of human activity, namely the search for new knowledge that are subject to certain social control and coercion, is frequently the basis for censure. Quantitative studies of science then are often reproved for a reputed lack of theoretical foundation. In particular, the absence of a theory of citing is frequently debated, suggesting the need for a more secure epistemological footing to support this practice. Nonetheless, the extensive body of experience gained in the application of bibliometrics in different disciplinary contexts has proved effective for the provision of reliable and useful data for science policy decisions. Interestingly enough, applied techniques, such as the mapping of science, when based on clearly formulated assumptions, have given rise to new theoretical perceptions of the structure and development of science. Useful insights have come also from an increasingly critical user group. Given that the applied side is an important driving force in scientometrics, user feedback has undoubtedly helped to advance the field. For this reason, current research is focussed on the development of new and more powerful literature based indicators required by the user population, as well as on the advancement of fundamental aspects of the field validating it as a *bona fide* research area respected by the broader scientific community.

2.1 Role of Bibliometrics in Institutional Evaluation

In many countries stagnating expenditure on higher education coupled with a growing intake of students in many universities, limit the possibilities for research funding. Furthermore, a growing culture of accountability in research environments is forcing scientists and teachers to become more and more productive. Funds are assigned according to performance. Research evaluation and research excellence are bywords in today's academic climate. Traditionally, assessment of scientific research has been limited to peer review during the grant awarding process or during evaluations for promotion or tenure. Today bibliometric techniques are increasingly used as an intrinsic component of a wide range of evaluation exercises. The present tendency is for institutions to be graded more on the visibility of their products then on their long-term reputation or resources.

The ability of publication and citation analysis to encompass different levels of aggregation makes it a technique ideally suited to national and institutional studies. Nonetheless, literature based indicators are appropriate only for institutional settings that reward publication and only for those activities that produce written knowledge. The fact that the role of written knowledge is influenced by cultural and socioeconomic aspects, as well as cognitive determinants that vary between fields of science and between different institutional settings, is considered their main theoretical constraint. Some institutions, for example, recompense behavior that reinforces the reward system of the international scientific community with their own internal reward structures. Others may set their own standards and goals.

Some indicators established globally for the evaluation of scientific performance might not be adequate for a fair or realistic assessment of certain research scenarios. Scientific output indicators based on mainstream publication in international journals should not be taken as the only bibliometric indicator for the evaluation of applied research in developing countries where publication in national journals in the local language is the norm. Disciplinary considerations are paramount. For researchers in the social sciences and humanities, monographs and books are important dissemination channels for research results. Technological research results are published mainly in congress proceedings, reports and patents, and are better represented in this type of gray literature than in mainstream journals. The output of technological and innovation research, in many cases, is not written up as such but appears as designs, applications, models or know-how. In these instances, literature based indicators, clearly, have little meaning.

An important consideration in any exercise of institutional evaluation is that results and recommendations to policy makers should have the general acceptance of the researchers concerned. Consequently, scientists and research managers should be included in the team responsible for the planning, execution and analysis of the research activity. Without the involvement of these key players, the evaluation exercise is unlikely to receive validation by the other members of the research community.

Institutional evaluation should be a continuous process. Ideally, procedures should be in place for the systematic monitoring of research performance and other fundamental scholarly activities. To accomplish this, institutions should develop their own data-system and make it available through the local intranet. In this way information is continually available for consultation by academic staff and other internal users, as well as for providing the raw data for the periodic generation of bibliometric and other scientometric indicators required for evaluation exercises. In practice, most evaluations are focussed on the short-term, often covering only three or four years. This is understandable otherwise results span too long a period for them to be useful for science managers. Nonetheless, their ultimate value can be measured only over the medium and long term.

In institutional evaluation exercises, scientific output and impact are related to input measures, such as research expenditure and the number and categories of academic staff. When carrying out comparative studies, other factors are considered, such as differences in the institutional academic and administrative structures, educational models, etc. Consequently, before deciding upon the procedure for collecting bibliometric data it is necessary to consider the internal institutional research structure. While research administration of many universities follows the traditional departmental structure, the increase in multi and interdisciplinary research, often organized in a program structure, has given rise to research groups formed by members of different departments. Research groups, rather than individual scientists, are today targeted for the allocation of research funds. For this reason, the research group is the most common unit for bibliometric analysis in institutional evaluations. This in turn has produced a wave of interest in scientometric research focussed on the identification of research groups by co-author analysis and its corroboration by expert opinion. Notwithstanding, the research performance of any aggregate of scientists can be assessed using bibliometrics. This aggregate is often termed a 'unit' which can be taken to represent any given set or sets of scientists depending upon the objective of the evaluation.

While no absolute quantification of research performance is possible, valid and useful comparisons can be made between research groups working in the same fields. When making comparisons between groups it is essential to apply indicators to matched groups, comparing like with like, as far as possible, and to give careful thought to what the various indicators are actually measuring. It is also important to study not only the similarities between groups but also the differences, especially those that could be directly influencing the research performance.

2.2 Methodological Considerations

Different data sources are available for bibliometric analysis. Before the advent of bibliographic databases quantitative analysis of the scientific literature was done manually directly from the published material. This is still possible in small-scale studies but it is labor-intensive and a lot of time is spent entering and checking data. In institutional evaluations it is possible to use internal documents such as research reports or annual reports which are often available through the institutional libraries. Alternatively, where researchers are expected to produce periodic progress reports on their research work as part of an institutional or program requirement, these can be used for analysis. When information is required in addition to that contained in the records of bibliographic databases, such as funding sources or acknowledgements, the original publications have to be located and consulted.

Large-scale analysis, on the other hand, requires the use of commercial, public or institutional bibliographic databases either online through the Internet or using a database vendor service. The most viable offline option involves consulting the CD-ROM versions. Online access requires a fast option for downloading records unless offline downloading and subsequent delivery by email is an option. The mainframe databases of the producers contain the most up-to-date information that will not typically be the case with the CD-ROMs. Special software is now available for the downloading and handling of bibliographic records, in some cases *ex professo* for bibliometric purposes.

Databases providing numerical data for indicator production, the so-called science indicators databases, are also available for purchase usually corresponding to records with specific characteristics retrieved from the mainframe database. Many of these can be customized with respect to the characteristics of the records required (all records from a particular institution or for a particular group of scientists). Also special provision can be made by the database producer or vendor for certain analytical flexibility that is not possible using records recovered from the online version. Likewise retrieval software in the CD-ROM version generally places certain limitations on the types of data that can be recovered and there are fewer options for data manipulation. Many databases offer abstracts in addition to bibliographic data that can then be searched for more specific information regarding research content. Nowadays more and more full text databases are becoming available.

A third option is to contract the services of a handful of agencies throughout the world that specialize in gathering and organizing literature-based indicators for evaluation purposes such as the Institute for Scientific Information (ISI) in Philadelphia; CHI Research, Inc. in New Jersey; the Information and Scientometrics Research Unit (ISSRU) in Budapest; and the Center for Science and Technology Studies (CWTS) at Leiden University in the Netherlands. These agencies also carry out innovative research on the development of new indicators and indicator data sources. Little standardization exists with regard to precise methodologies and sources used for indicator generation between the different producers making comparisons generally invalid. Although most indicators are generated using records provided by ISI, the precise data sources employed are often based on different journal sets. Variations also occur in specific datafields, particularly in the address field and types of documents.

Although created initially as a new information retrieval tool based on citations, the particular attributes of the SCI have made this the most frequently utilized bibliographic file for bibliometric analysis. SCI is a multidisciplinary database indexing cover to cover approximately 3,500 international journals, in most subfields of the natural and life sciences. This is in contrast to most other international databases, such as Chemical

Abstracts, BIOSIS, or Medline that cover specific disciplines. SCI includes all author addresses while the majority of the best known international databases include the addresses of only the first authors of papers. This is an important consideration in collaboration studies and when comparing the scientific production of different countries and regions. However, it is the inclusion of all papers cited in each of the articles indexed in the database that makes the SCI unique.

ISI also produces citation indexes in the social sciences, *Social Sciences Citation Index*, and in the arts and humanities, *Arts and Humanities Citation Index*. In these fields, however, the applicability of citations to assess research impact is questionable for which reason these indexes are not used as extensively as in the natural sciences (see *Citation Measurements*). In addition, the fact that SCI processes few books, proceedings and other types of documents limits its usefulness in disciplines where the journal article is not the main vehicle for communication of results. The disciplinary orientated international databases tend to cover a wider range of topics within the discipline, of document types and process a far greater number of national journals. This gives them a broader information base for the bibliometric analysis of research performance in all fields and in countries whose journals are poorly represented in the SCI, such as those from the developing world.

In general, bibliographic databases designed not for bibliometric purposes, but rather to provide users with a way of accessing and controlling the abundant literature available in their specialized fields, are by no means ideal sources of data for analysis of research performance. Lack of standardization in searchable fields and the inability to recover information on important aspects of the published documents for bibliometric analysis are just two of the limitations. Such is the case of author affiliations, so important in institutional analysis. Assigning authors to their institutions is somewhat problematic as addresses are often incomplete and sometimes incorrect. There is a special problem with the names of institutions from non-English speaking countries as most of the big international databases are developed in English. Furthermore, articles do not necessarily have a specific field assignation. This difficulty has been resolved to some extent by assigning articles to the disciplinary areas of the journals in which they are published. Special classifications of journals have been developed for this purpose by organizations such as CHI Research. ISI and others have developed over the years novel indicators to enrich the battery available for use in evaluation studies. A well known example is ISI's journal impact factor (IF) which determines the impact of journals based on the number of citations they receive in relation to the number of articles published over a specific time period (see *Journal Impact Factors*).

To achieve a higher degree of standardization in bibliometrics is by no means an easy task. Responsibility rests not only with bibliometricians and database producers, but also with journal publishers and editors who need to formalize and standardize their procedures and methods more rigorously. While some journals link individual authors with their institutions, others list all institutional addresses without making explicit the relationship between the authors and the addresses. In many journals it is difficult to identify different document types and some have different definitions of a particular document type. Letters are a case in point.

The bibliometrician, then, must know fully the data sources available and carefully select those most likely to answer the questions under study. For instance, it is well known among specialists that different results will be forthcoming when consulting the SCI in CD-ROM version compared to doing a search online via the SCISEARCH database. Adequate strategies then have to be developed for the downloading of the required records and their subsequent manipulation. Some aspects of each record that may not be included in an independent field in the original database, such as the countries that normally form part of the author address, need to be separated out and individually coded. Data need to be validated as far as possible and adjustments made where necessary. This often implies checking with the original journals, when practically feasible, or with reference works. The advantage of using the few 'bibliometric' databases available for consultation is that this alleviates the problem of extensive cleaning up of data, classifying and coding fields necessary before initiating the process of bibliometric analysis.

The best strategy for collecting a complete and accurate publication list for the unit to be evaluated is, perhaps, to use a combination of data sources. The use of publication lists provided by the members of the unit, together with references to documents retrieved from a suitable bibliographic database by searching on the names of the group members, should cover most of their scientific output. Annual reports are often incomplete and should therefore be used only as complementary sources of publication data. Citation data should be collected for all published material and not restricted to publications in journals covered by SCI. Work published in journals and series not covered by SCI can well be cited by SCI covered sources.

The next decision to be made is what type of output and impact indicators should be applied to the unit under study. This will clearly depend upon the precise objectives of the evaluation as well as the research fields concerned. Large differences in applicability of bibliometric data have been found even within fields in the natural and life sciences. The particular characteristics of the unit to be evaluated such as the relative prominence given to teaching and research activities, also have to be taken into consideration especially in comparative studies. New indicators are continually being devised, some of which may only be relevant in particular institutional settings. Below we describe only those that are well established and which are generally applicable to the majority of research scenarios. In most instances a combination of production and citation indicators is employed.

3. Output Evaluation

The number (quantity), type and distribution of publications are the most commonly applied bibliometric indicators of scientific output. The production of a research unit is its number of publications. Productivity is expressed as the number of publications per person-year equivalents for research.

The production of research publications is highly skewed. In any given field of research, a small number of high producers will be responsible for a significant percentage of all publications in the field. The distribution of individual productivity was first examined by Lotka whose power law [equation (1)] is one of the major regularities studied in bibliometrics:

$$x^n y = c \qquad\qquad\qquad (1)$$

where y is the proportion of authors making x contributions each; n and c are parameters depending upon the time span covered by the bibliography, and probably the field being analyzed.

Important decisions have to be taken before embarking on publication analysis, the most fundamental of which refers to the types of publication to count and their relative worth. Other important considerations are what time spans to use and how best to assign credit to individual authors or institutions in multi-authored papers or books.

3.1 Weighted Values of Publications

Only appropriate kinds of publication should be counted depending upon the prevalent communication channels of the fields under study. While many researchers in the humanities prefer to publish results in book form rather than as articles, in the applied sciences, such as engineering research reports and patents are more common forms of output. Where, for example, counts of articles and books are mixed, the values assigned to books need to be weighted. Review articles also need special consideration. Output comparisons between scientific and medical fields, on the one hand, and the social sciences and humanities on the other, depend to a large extent on the relative weightings assigned to the different types of publications used by researchers in these two distinct areas of knowledge. One book has been deemed equivalent to anywhere between two to six articles.

Even in the experimental sciences where refereed articles are the obvious output medium, the relative importance for the evaluation exercise of other document types need to be assessed. The SCI database classifies journal papers into several types of which usually only three or four are considered as significant contributions to research. Articles, notes, reviews and letters are commonly included in both scientific output and citation counts. Letters are sometimes excluded depending upon their function in the particular field under study. Other publication types such as meeting abstracts, discussions and editorials are useful for bibliographic retrieval but are generally excluded from bibliometric analysis.

Distinctions are often made in evaluation exercises between papers published in SCI journals and those published in non-SCI journals. The first category refers to highly cited, highly visible journals for which citation data are available. Publication in the second category of journals is unlikely to produce the same level of citation as in the more prestigious SCI journals due to reduced visibility within the scientific community. The fact that authors compete for publication in SCI journals bestows on these a certain intrinsic element of quality regardless of their citation levels.

Even when considering the output of a single author or group of authors in a given field each publication may not represent an equal contribution to science. Not all articles and more especially not all books, contain original research. Some are aimed at discussing current developments or matters of professional concern. Books in the humanities are more likely to represent original research contributions than those in the sciences where textbook writing is more prevalent. For this reason, some solid conceptual basis is needed to decide what should be counted as a publication. Furthermore, the amount of work contained in a single article or book varies not only between fields but also by individual choice. Consequently publications are sometimes weighted according to the number of published pages, as well as in relation to the number of authors.

The increasing trend for research performance to be measured principally on production output and citation impact has produced a phenomenon known as salami slicing, whereby some scientists spread their results over several less notable papers instead of concentrating these into one more weighty and significant article. The importance of peer review in these cases is paramount. When the unit to be evaluated consists of a large number of scientists with a significant number of papers, the 'salami slicing' effect will cancel itself out unless of course this practice is widespread.

3.2 Time Span

A fixed time span of one year or more is appropriate for the analysis of the publication output of research units such as groups or institutions. Four-year periods are often used (see *Role of bibliometrics in institutional evaluation)*. An important consideration when comparing publication histories of groups of scientists, is that their professional ages (normally taken as the number of years since being awarded their PhD degrees) must be similar. In comparing the publication outputs of research groups, significant differences in the years lapsed since the groups were formed should be taken into consideration.

Time series analysis can be applied to supplement average values calculated for the entire time span. Indicators are studied on a year-to-year basis in order to trace the evolution of a group's production (and impact). Examples of these are total number of publications published annually; number of SCI publications published annually; percentage of SCI publications published annually.

Trend analysis is also useful for determining the effect on scientific output of variations in the research environment, such as changes in research budgets and in institutional goals, and the implementation of reward and incentive schemes. Publication output can vary considerably with time also for individual reasons. This appears to be true for both high and low producers.

3.3 Assigning Publications to different Research Units

One of the frequent aims of institutional research evaluation is to analyze the contributions made by particular scientists or groups of scientists. In today's climate of increasing collaboration in science the evaluator has to decide the most equitable way of assigning credit to each and everyone of the authors signing a paper. This situation arises when a research unit publishes articles in collaboration with scientists from other groups both from within their institution and from outside.

Several procedures have been put forward, two of which are the more frequently applied. Co-authored publications are counted either as one publication for each of its co-authors (total author counting) or as fractional counts (fractional counting) assigned to each author based on the number of co-authors. When a paper has A authors, for instance, each author is attributed with one point. Alternatively, each of the A authors receives a score equal to $1/A$. Another method involves giving credit to only the first author of the paper (first author counting). The popularity of this method is due to the fact that in most formats ISI citation data (not publication data, however) are based on the first author only.

	Authors		
Article 1	a_1	a_2	a_3
Article 2	a_4	a_2	a_3
Article 3	a_3	a_4	
Article 4	a_5	a_4	

Table 1. An example of a small set of four articles, and their respective authors.

First author counting (FA), total counting (TO), and fractional counting (FR) lead to the following, different, rankings between authors:

FA: $a_1 = a_3 = a_4 = a_5 = 1$; while $a_2 = 0$
TO: $a_3 = a_4 = 3$; $a_2 = 2$: $a_1 = a_5 = 1$ (a_2 moves from last to third)
FR: $a_4 = 8/6$; $a_3 = 7/6$; $a_2 = 4/6$; $a_5 = 3/6$; $a_1 = 2/6$

In this example fractional counting refines the ranking obtained by total counting, while first-author counting does not give an accurate picture of the relative contribution of the authors.

Which of these methods to use is an important decision as different accrediting procedures may produce significantly different results. Help may be available in the form of publication policies of a particular group, department or institution. In some academic environments, for example, it is customary for the names of senior faculty to appear at the end of the author list, more in virtue of their position than any contribution they have made to the study. The order in which authors' names appear in a paper is often field dependent. In some fields alphabetical listings is preferred while in others the first author is the scientist making the greatest contribution to the research. When deciding upon a particular procedure these practices should be taken into consideration. In the absence of these, different counting methods should be applied and compared. Although the simplest method to apply, first-author counting is not recommended. It is particularly unfair when authors are systematically ordered alphabetically, or practice 'noblesse oblige' (putting the senior author last).

4. Citation Measurements

The aim of citation analysis is generally held to be that of injecting a 'quality factor' into the evaluation of scientific performance. The level of interest that colleagues show in the published research of a group of scientists is seen as one way of estimating quality. However distinction should be made between quality, importance and impact. While the first is an inherent property of the work, the last two are based on external appraisals. Importance refers to the potential influence on surrounding research activities while impact reflects the actual influence. Judgment of the basic quality of scientific research can be made only by peers. To operationalize the general concept of quality in scientific research is virtually impossible as this may refer to a variety of values - cognitive, methodological and esthetic. But even if citations do not provide an indicator of the quality of scientific papers, they do reflect, at least partially, the impact they have on the scientific community. 'Imperfections' in the scientific communication system result in the importance of a paper not necessarily being identical with its impact.

The most important measure of impact is the number of citations, or, on an aggregate level, the average number of citations per paper. Short-term impact is indicated by counting the number of citations received by the publication during the first few years following its appearance, for example during the first three years or during the third year after publication. Three-year or four-year citation counting periods are often used. The period over which citations are counted is termed the citation window. Long-term impact considers a longer time span.

There is a time lag between work being published and its coming to the attention of researchers in the field who may wish to cite it. Obviously the time that elapses between work being completed and its being cited will be partially determined by the speed and place of publication. Although scientists are naturally interested in their results being published expeditiously rapid publication is usually regarded as less important than the prestige or readership of the journal selected for publication. Interest in speed of publication varies with discipline. In areas where research is developing rapidly how quickly results can be communicated to the specialist community is of the utmost importance for laying claim to results before other groups do. Rapid publication is also a requisite for the prompt citation of research work.

Age distribution of citations within a specific research area will determine the speed with which publications are normally cited. There are substantial differences in the year when maximum citations are achieved, both between fields of research and within fields. Some papers make a quick but fleeting impact on their fields, others have a slower but more durable impact. Important paradigm defying research will continue to be cited until it is superseded or becomes part of the established body of knowledge in the field. Short-term impact is related to the visibility of research groups at the research front and can be ranked with other visibility indicators, such as international contacts, awards, invitations to participate in prestigious international meetings, among others. Short-term impact does not necessarily guarantee long-term impact as many attractive theories fail to withstand the test of time.

Long-term impact is not a very useful indicator in the evaluation of research groups, partly because certain of these may have ceased to exist or have changed their research interests. Furthermore there is no guarantee that groups will perform equally well (or badly) over prolonged periods. Research groups cannot be required to carry out work with long-term impact but they can be expected to take part in scientific discussions in the field. For this reason, short-term impact is of primary importance in institutional evaluation.

Following the argument that it is not the great mass of small impact papers that contribute most to scientific progress, but rather those few key papers having a great impact, the presence of highly cited papers is an important performance indicator. This can be expressed as the annual percentage of highly cited papers (those perhaps receiving more than a certain number of citations per year) and is useful when comparing the citation performance of matched research groups or institutions. Nonetheless, this indicator is seldom calculated in purely bibliometric analysis as the presence of highly cited papers is an attribute that peer review will highlight.

Most authors include citations to their own work in their papers. This is to be expected as scientists build on their previous work when designing experiments making internal citations essential, unavoidable and justifiable. Self-citations account for some 10% of all citations. Although citations by one of the co-authors of a paper does not constitute impact on peers, these are rarely eliminated from citation counts. The final estimation of the impact performance of a research group, for instance, will be the result of a comparison with that of the field in general. This is normally done using ISI data on field and journal impact factors (see *Journal Impact Factors*) and citation scores which include self-citations. For most evaluation exercises the elimination of self-citations from ISI data is not a practical option.

The variability to be expected in citation counts is enormous. Review publications and work on innovative techniques and methods are usually highly cited. The absolute number of citations depends greatly on the particular characteristics of the research topic or field in question. Scientists publishing in fields with a large volume of research output, such as molecular biology, are in a better position to be cited than those working in disciplines such as mathematics, where a research paper is longer in the making and is likely to contain only a few references to previous work. These differences will be reflected not only in the number of citations per paper but also in the impact factors of journals in these different fields (see *Journal Impact Factors*).

Objections raised regarding the use of citation data for evaluation purposes have focused on assumptions regarding the citation process. Many random factors unrelated to the scientific importance of research contribute to the final outcome of papers being cited, such as certain references being close at hand or the choice made between equally relevant papers. Thus when small numbers of papers are involved, chance factors may obscure a real difference in impact. As the number of papers analyzed increases, the relative contribution of chance factors decreases and that of real differences increases. Consequently, studies involving larger sets of papers give more reliable results.

As with production data citation analysis is considered a fair evaluation tool for those scientific sub-fields where publication in the journal literature is the main vehicle of communication but only when the limitations of the databases used for the evaluations are fully understood and taken into account for the final conclusions. Furthermore there is powerful evidence to show that citation measures correlate well with faculty performance ratings.

Citation analysis is subject to a certain level of inaccuracy. Errors and variations occur in the details of the cited articles in the SCI many of which can be related to inaccuracies in the original publications. The presence of a specific mistake in a citation to a particular article occurring in more than one citing publication suggests that authors may copy references directly from lists in other papers thus propagating errors. Other discrepancies appear to be due to irregularities in the internal standardization process at ISI or are simply keyboarding errors.

5. Journal Impact Factors

The study of the use and relative impact of scientific journals is one of the important applications of citation analysis. Research on citations was given a significant impetus by the publication of ISI's Journal Citation Reports starting in 1976. These reports, currently available on the web, in CD-ROM and on microfiche, provide a wealth of indicators and statistical data on the citation patterns of journals covered by ISI's series of citation indexes. Frequently applied in research evaluation exercises are the international journal lists grouped by discipline and ranked according to impact factor. The ISI impact factor (IF) is basically a ratio between citations and citable items published in journals. It measures the frequency with which the average cited article in a journal has been cited in a particular year.

The ISI impact factor [equation (2)] of a journal in the year Y is calculated as:

$$\frac{CIT(Y, Y-1) + CIT(Y, Y-2)}{PUB(Y-1) + PUB(Y-2)} \tag{2}$$

where the number of citations in year Y to papers published in the year Z is denoted as $CIT(Y,Z)$, the number of publications in the year Y by $PUB(Y)$.

In order to calculate an analogue of the ISI impact factor for a non-ISI journal, this journal is simply added to the pool of ISI source journals. How often this particular journal is cited by ISI journals (during the period under investigation) is calculated and added to the number of times the journal cites itself. This is then divided by the number of articles published by the non-ISI journal. Although this is a simple procedure there are two caveats. Firstly, ISI always includes journal self-citations, so this must be done here too. Secondly, if this new impact factor is used to compare this non-SCI journal with ISI-journals, the ISI-journals' impact factor must be recomputed, because the pool of journals has changed.

Selection of appropriate citation and publication windows is crucial for calculating impact factors. Mean citedness, i.e. impact factors, can be calculated using either a synchronous or a diachronous approach, and with different time windows for publication and citation data. The ISI impact factor is synchronous involving a single citation year and two publication years. The term 'synchronous' refers to the fact that the citations are all 'given' in the same year. These impact factors 'mix' different publication years rendering them more robust when the object under study is the journal itself. Synchronous indicators better represent the permanent impact of journals whereas diachronous indicators characterize their actual impact. Consequently when the article (or the scientist who wrote it) is being evaluated diachronous impact factors are more commonly employed. Furthermore they can also be calculated for one-off publications such as books containing contributions by different authors and conferences proceedings.

The following formula is used to calculate the *n*-year synchronous impact factor of a journal J in the year *Y*:

$$IF_n(Y) = \frac{\sum_{i=1}^{n} CIT(Y, Y-i)}{\sum_{i=1}^{n} PUB(Y-i)}$$

The *n*-year diachronous impact factor of this journal for the year *Y* is:

$$IMP_n(Y) = \frac{\sum_{i=k}^{k+n-1} CIT(Y+i, Y)}{PUB(Y)}$$

where $k = 0$ or 1, depending on whether the year of publication is included or not.

Many specialists warn against the use of journal impact factors as surrogates for actual citation performance. One of the main reasons put forward is that large differences exist in quality and citation scores between papers published in the same research journal. The presence of highly cited papers tends to artificially bump up impact factors. Furthermore, all citations are treated as equal regardless of the citing journal. Some fields that are useful for science do not receive many citations. Consequently, if the impact factors and other citation-based measures were to become the main criteria for journal quality then whole sub-fields would be eliminated and others would have their worth undermined.

As is true for the number of citations received by an individual paper the IF of journals is similarly affected by factors unrelated to the scientific quality of the articles. Such is the case of foreign journals handicapped for their inclusion in the SCI by a lack of English content. A high quality paper in a less visible foreign journal is likely to receive fewer citations than a poorer quality paper published in a highly visible SCI journal. Consequently, impact factors measure only the (international) use of journals at the research front of their respective fields. Evaluators should bear this fact very much in mind when applying IFs in assessment exercises especially at low levels of aggregation.

6. Relative Impact Indicators

A time series analysis alone is not sufficient to obtain a complete picture of the impact of research groups. In the absence of any absolute standards their performance needs to be compared to that of other research groups in the field. One way of doing this is by calculating the ratio of the average of the group's citations (per article) with the average of the journals in which they have published. Alternatively the ratio of the average of the group's citations with the average of the field (or fields) in which they are active can be determined.

A group publishes articles in a number of journals: the so-called journal package of the group. It is possible to obtain an IF for each journal of this package, i.e. an average number of times an article published in this journal is cited. The impact factor is denoted as **JCS** (Journal Citation Score). The **JCS** or expected impact of a group of articles is calculated using publication and citation data from the JCR. Weighted average IF for the journal package in which the group publishes can be constructed using impact factors of journals (symbol: **JCSm**).

However this does not answer the question of whether or not a group is more cited than similar groups nor whether the group's journal citation score is high or low when viewed from an international perspective. To do this we need to calculate, per sub-field, the average number of citations an article from that field receives, called the world average per field, and denoted as **FCS** (Field Citation Score). Usually a group is active in several sub-fields, and consequently a weighted average is used (symbol: **FCSm**), similar to the calculation of **JCSm**.

The indicator **JCSm/FCSm** compares the weighted-average-impact factor of the journal package with the world citation average of the fields in which the group is active. It corresponds to the ratio of the average impact factor of the journal package to the average world impact of the fields in which the group is active. When the **JCSm/FCSm** is larger than one, the group publishes in journals having an IF that is larger than the world average in the field. In this way we can compare the actual number of citations per publication of a research group with the expected number based on the average JCS values to determine whether this actual number of citations is high or low. The validity of the indicator is based on the assumption that the set of journals used in the evaluation yields an adequate representation of the discipline.

Figure 1 is an example of the impact of the research publications of a Belgian university using this indicator which compares citation scores with the field. Since groups publish in different fields, the figure contains weighted scores. A three-year average line is used to smooth out irregularities and make the general trend more obvious.

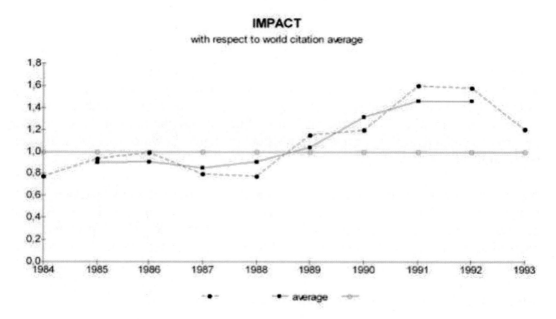

Figure 1. An example of the impact of the research publications of a Belgian university.

When calculating the impact of a field, two approaches are possible: either the average of the impact factors of all journals in the field is calculated (this is called the average impact of this set of journals), or, alternatively, a global average. The latter is the better approach. The difference between these two approaches is shown as follows. If C_i denotes the number of citations (over a certain period) of journal i, and if P_i denotes the number of publications in journal i, then I_i will denote the impact of journal I (citations per publication). The average impact factor is then defined as:

$$AIF = \frac{1}{n}\sum_{i=1}^{n}\frac{C_i}{P_i} = \frac{1}{n}\sum_{i=1}^{n}I_i$$

The global impact, on the other hand, is calculated as:

$$GIF = \frac{\sum_{i=1}^{n}C_i}{\sum_{i=1}^{n}P_i} = \frac{\mu_C}{\mu_P}$$

where μ_C and μ_P denote the mean number of citations, and the mean number of publications.

7. Future Trends and Perspectives

The development of bibliometric and scientometric techniques has been strongly influenced by the availability of adequate data, as well as by the facilities available for computing large quantities of information. Consequently the transition from print format to electronic media in scientific communication, will also bring about significant changes to the way production and citation indicators are constructed. The Web as a place of integrated scientific and technical communication promises to become not only a major new and improved source of production and citation data but also a legitimate one. For this reason, any advanced bibliometric data-system for institutional monitoring must be sufficiently flexible to take into account rapid developments in electronic publishing and in relevant Internet facilities

Major changes to bibliographic and bibliometric methodologies will emerge with the increased presence of academic journals and other scholarly works on Web. The seemingly limitless capability of the Web for storing and integrating information will make available unified (and hopefully, uniform) data sources for bibliometric analysis, accessible from any PC anywhere in the world. A case in point would be the emergence of local and regional citations indexes and their integration into a virtual world citation atlas. The presence of every scholarly work ever written linked to every work it cites or is cited by in a universal Web based bibliographic and citation database would solve many of the problems plaguing the construction of output and citation measures in today's non-electronic environment.

New indicators will emerge such as those already proposed to legitimize the importance of intellectual input of colleagues recognized in the acknowledgements section of research papers and reports. The 'seamless' all-embracing Web environment could lead to a better assessment of multidisciplinary work as well as the identification of new and significant relations between different disciplines.

The problem of the under-representation in international scientific databases of studies from developing and other countries written in non-English languages could eventually become a thing of the (non-electronic) past. The prospect of on the spot translation of non-English articles on the Web, for example, would lead to an increase in the visibility and citation levels of non-English speaking scientists. Online publishing ventures such as the establishment of global repositories for research would also increase the holistic presence of studies from developing and other peripheral countries.

The fact that the Internet accentuates the value of individual pieces of information puts increasing emphasis on the individual article and rather less on the journal, a trend that could diminish the value given to journal impact factors in the short-term and likely to cause their demise in the long-term. Clustering articles on the Web by subject would allow these to be ranked according to their importance for different fields, a process which could identify citations made to inappropriate references as a result of scientists playing the 'citation game' of preferentially citing their own or their colleague's work. Furthermore, with escalating networking in science, scientists will become increasingly aware of what their peers are doing. This could bring about a possible increase in the speed with which results are incorporated into the work of others thus reducing citation lags.

We can also look forward to the opportunity for better accuracy in bibliometric data with the automatic softbot checking of bibliographic elements such as titles, authors, directly from their original Web sources. We can also hope for greater access to the results of research from the social sciences and the humanities due to increased international presence through web publishing.

In anticipation of these and other major paradigmatic shifts in scientific communication, a new field of study is already emerging. Cybermetrics or webometrics is the name given to the innovative metric study of electronic communication. At this transitional stage there are numerous discussions on the validity of adapting traditional methods for the construction of production and citation data using electronic formats. One particular concern is the definition of what constitutes a valid publication in Internet. Another issue is how to calculate Web impact factors and their validity in terms of measuring the impact of a particular web space.

But one aspect that will change little in the face of the electronic revolution is the human element. Our poor understanding of the psychological and social processes underlying peer review as well as those involved in the publication and the citation processes, will not be automatically improved by wider access to output and impact data. Notwithstanding we will be able to provide peers with more comprehensive and more reliable bibliometric data to guide and support their decisions and to better defend these before different non-scientific sectors of the community, such as science managers, politicians and the general public.

It should be kept in mind that research evaluation is not an end in itself. It is only an aid to the real goal that is providing those people and institutions with the talent and motivations to carry out scientific research, with the best conditions possible under which to do so. Budgetary and other kinds of constraints make evaluations necessary for the equitable distribution of resources. The evaluation of short-term strategic research as well as the long-term curiosity-driven search for new knowledge demands the same accountability and rigorous standards as scientific research requires of itself. For this reason the challenge is not only for the application of bibliometric and scientometric techniques in research evaluation to keep up with the rapid changes occurring in scientific communications patterns and practices but also to constantly improve the theoretical foundation for the construction of output and impact indicators as an adjunct for peer review.

Glossary

Bibliometrics:	The application of quantitative methods to books and other communication media.
IF:	**Journal Impact Factor :** Basically a ratio between citations and citable items published in journals.
ISI:	**Institute for Scientific Information :** Philadelphia-based company providing scholarly information services including the citation indexes and the Journal Citation Reports.
JCR:	**Journal Citation Reports :** A publication of ISI providing bibliometric data for journals covered by their citation indexes including the journal impact factors.
Mainstream	Generally used to refer to a group of highly cited and highly

journals:	visible journals covered by the ISI citation indexes.
Peer review:	Evaluation of a scientist's work by his/her colleagues with similar scientific interests and rank.
SCI:	**Science Citation Index** : A publication of ISI indicating who cites whom, used as a source of data for citation analysis and journal impact factors in science, technology and medicine.
Scientometrics:	the study of the measurement of scientific and technological progress.

Bibliography

Cronin B. (1984). *The citation process: The role and significance of citations in scientific communication*, London: Taylor Graham. [An essay, still valid today, debating the meaning of citing practice focussing on science as a social process.]

De Bruin R. E., Kint A., Luwel M., and Moed H. F. (1993). A study of research evaluation and planning: the University of Ghent. *Research Evaluation* 3(1), 25–41. [Provides a detailed description of how to perform a systematic evaluation of research performance.]

Egghe L. and Rousseau R. (1990). *Introduction to informetrics, Quantitative methods in library, documentation and information science.* Amsterdam: Elsevier. [Part III of this book describes the basic ISI citation tools and provides definitions of citation and citation-related concepts.]

Garfield E. (1979). *Citation indexing-its theory and application in science, technology and humanities.* New York: Wiley. [Seminal work by the 'Father' of citation indexing.]

Martin B. R., and Irvine J. (1983). Assessing basic research: Some partial indicators of scientific progress in radio astronomy. *Research Policy* 12, 61–90. [Convergent partial indicators provide a reasonable estimate of a group's relative contribution to scientific progress.]

Moed H. F., Burger W. J. M., Frankfort,J. G., and Van Raan A. F. J. (1985). The use of bibliometric data for the measurement of university research performance. *Research Policy* 14, 131–149. [The first complete description of the evaluation of university research groups indicating the potential of bibliometric indicators as tools for university research policy.]

Proceedings of the Biennial Conferences of the International Society for Scientometrics and Informetrics. 1987 through 1999. [Proceedings series reporting original studies in the field of Bibliometrics, Scientometrics and Informetrics, many of which are related to indicator production.]

Van Raan A. F. J., ed. (1988). *Handbook of Quantitative Studies of Science and Technology.* Amsterdam: Elsevier. [The main purpose of this compilation is to present a wide range of topics in the domain of quantitative studies, incorporating theory, methods and applications.]

Van Raan A. F. J. (1997). Scientometrics: state-of-the-art. *Scientometrics* 38(1), 205–218. [This review emphasizes the duality of scientometrics as both a basic and applied field, in which the applied side is considered the driving force.]

Biographical Sketches

Jane Russell is a senior researcher at the University Centre for Library Research (CUIB) at the National Autonomous University of Mexico (UNAM) and professor of the postgraduate programme in Library and Information Science in the Faculty of Philosophy and Letters at the same university. Her specialist field is the production, communication and impact of research in Latin America. She is a member of the International Society for Scientometrics and Informetrics (ISSI) and of the Latin American Network for Science and Technology Indicators (RICYT), and also of the editorial committees of the journals "*Scientometrics*" and "*Cybermetrics*".

Ronald Rousseau is professor at the KHBO, Department of Industrial Sciences and Technology, and guest professor at the Antwerp University (UA), School for Library and Information science. He has

SCIENCE AND TECHNOLOGY POLICY - Vol. II – *Bibliometrics and Institutional Evaluation* - Jane M. Russell, Ronald Rousseau

64

written numerous articles dealing with citation analysis, research evaluation, informetric models, information retrieval, library management, Internet-related research, applications in ecology, and undergraduate mathematics. He is a member of the American Society for Information Science and Technology, and an active member of ISSI. He has received the Prize of the Belgian Academy of Sciences, as well as the 2001 Derek de Solla Price award.

SCIENCE AND TECHNOLOGY POLICIES IN AFRICA

Roland Waast
Research Unit on Knowledge and Development, Institut de Recherche pour le Développement (IRD), France

Keywords : Science policy, National science, Brain drain, International cooperation, World scientific agenda, World bank, Regional institutions, NGO's, Africa, North Africa, Southern Africa, East Africa, Morocco, Nigeria, Senegal, Ivory Coast, Republic of South Africa.

Contents

1. Introduction
2. Historical Background
3. Current Policies: A Typology
4. Conclusion
Acknowledgments
Bibliography
Biographical Sketch

Summary

The forces which govern today the construction of S&T policies in Africa cannot be understood without taking a look at the past, which explains the transformation of colonial sciences to a national science model, which still moulds present attitudes and institutions. The article examines the research capabilities that have been built up, draw attention to the sharp contrasts between different regions (North Africa, South Africa and "intermediate" regions). This unequal landscape has been subject to a profound shock due to the deep changes in the world scene: economic liberalisation, globalization and the instauration of a worldwide scientific agenda.

The trend that Africa experiences today is toward a more radical shift towards a "mode 2" science; science becomes highly fragmented, internationally-oriented and demand-driven. The demand in question is often that of powerful NGOs, international organizations and of bilateral cooperation schemes, which act under the promise that local capacities will be enhanced by their field programmes. Different countries react with a whole new range of practices and policies, depending on their political regimes and the alternatives left by previous strategies. A quick typology of policy regimes are discussed in the article, from *laissez-faire* to policies recommended by international or foreign bodies (the "donors", especially the World Bank). Some room seems left for emerging policies, as demonstrated by South Africa, or some North-African countries, either with a national range or a regional range.

1. Introduction

Is there still room for national science policies in Africa? This question is raised at a time where several sub-Saharan countries have seen their scientific organizations bereft of budget and deprived of any hold over those charged with implementation. Public establishments are left with neither programmes nor reliable and loyal research staff. The research profession is exercised more on a temporary, short-term basis than in the context of a career.

The forces which govern the construction of policies today cannot be understood without taking a brief look at the past, at the colonial model and its transformation to a national science model which has moulded attitudes and institutions. A sharp contrast can be seen between different regions (North Africa, South Africa and "intermediate" regions). In this already highly unequal landscape Africa has had to face the shock created by the deep changes experienced worldwide: economic liberalism, globalization and the set-up of a worldwide scientific agenda. The trend is an anarchic and radical move towards a "mode 2" science, which is highly fragmented, internationally-oriented and demand-driven. This is a paradox in countries where farms and enterprises are hardly geared to research and development. However, the demand in question might well be that of the powerful NGOs, international organizations and of bilateral cooperation schemes, which all function under the promise of the enhancement of local research capacities under their field programme objectives.

The article investigates how different countries react to these changes with a whole new range of practices and policies, depending on their political regimes and the alternatives left by previous strategies.

2. Historical Background

2.1 From Colonial Science to National Science

Colonization generated research centres, but not universities. Formed by circumstance rather than explicit policy, these research centers have left a specific type of knowledge types and, above all, models that still today serve as reference for public action. This model is a choice of research fields (mainly in agriculture and health), a choice of institutions (mainly full-time researchers, employed by specialist institutes), a choice of organizational set-up (subordination to technical departments, favouring technical innovation diffusion). The research centers were maintained by the former colonial countries, which assured financial support and cooperation which substituted to local authorities. However, a dual network then took shape. At the time of independence, the elite in Africa was limited. The strong expansion of the universities was to increase the local elite. In the space of one decade, nearly all the countries had set up their higher education system. Standards were high and research was a normal part of the teaching staff's duties. Their research work, monitored by the academic authority, supported by cooperation schemes, concerned basic sciences, whereas the technically-based ministries valued the applied nature of technical work conducted by specialized agencies (particularly in agriculture). The 1970s saw the start of Africanization of all these establishments. The new researchers moved in with a different approach. Their institutional culture did not now amount just to being part of a learned community; rather it embodied the concepts of a "national" endeavour and promoted "autocentred" communities. In this way a new model of "national science" appeared, whose main characteristics were as follows:

- Science is a public good.
- The state bears most of its funding.
- Science is geared toward the needs of the country.
- Researchers are civil servants and have a right to forge their careers.
- They defend national as well as scientific values.
- Apart from scientists, scientific products are destined to governmental bodies. The direct users of the research are hardly involved, and certainly not by way of any commercial links or marketable values, considered "impure" in the eyes of these researchers.

The governments expected a great deal from research and education in terms of usable results. Budgets allocated increased tenfold in ten years. So did the numbers of researchers. From 1975 national steering bodies began to spring up. In English-speaking Africa, scientific "Councils", specialized according to field (Health, Agriculture, Industry, Energy) were given the task of defining priorities, managing incentive budgets and assessing results. A national "Council" for research was heading the specialized councils in order to make their objectives consistent with the national Plan and forge links with universities. In French-speaking Africa, a direction of research was set up in each ministry which was in charge of the research establishments; an inter-ministerial department (sometimes a ministry or specialized governmental office) coordinated the system as a whole. It is true that the interministerial agencies were often unheeded with since they had no control over budgets (which were assigned to the operating ministries). However, with or without centralized planning, good results ensued: in 1985, scientific publications were quite visible on the international scene, eminent figures emerged; certain leading-edge establishments gained a good reputation and notable innovations stemmed out from research done.

2.2 North Africa, South Africa

South Africa and North Africa each have their specific characteristics. The former is an old colony of settlers. Its "modern" scientific history goes back more than two centuries. Its first universities date from 1870 (but in 1950, there were only 2,000 "coloured" students out of the 30,000 in the country as a whole). When in 1945, South Africa gained autonomy from the British Empire, research became a priority. The brand new Council for Industrial Sciences drew up a national policy, founded laboratories in strategic subject areas, formed links with companies and administered the incentive fund which was to bring with it academic research. The apartheid regime went on to reinforce the system by favouring a concentration on military and security-related fields, the basic sciences and advanced technology. The post-apartheid regime is taking care not to weaken this complex, whose capacities extend from aeronautics to nuclear systems, from chemistry to metallurgy, from food processing to highly advanced medical specialties. It is trying to redirect it, to serve basic needs and the competitiveness of civilian industry, and to encourage appropriation by black Africans, long kept out of any such enterprise.

On the southern shores of the Mediterranean, the beginnings of science can be traced back even further. The first universities date back 1000 years. The "Reform" in the Ottoman countries gave rise to the creation of high-level Schools of Engineering and

Medicine: their graduates performed outstandingly throughout the 19th Century. European domination closed the doors of higher education on local people and at the same time precluded any jobs in the public sector or qualified employment. However, members of the elite continued to set their hearts on superior studies for their children and subsequent independence unlocked an enormous desire for education. In Egypt free education was extended to higher education in 1962. In 1990, there were 1 million working graduates and the student population made up 15% of the 20-24 year age group. In the Maghreb, comprising Morocco, Algeria and Tunisia, the explosion of education came about after 1975.

With some delay, independence brought the opportunity to create a national system of research. This was to be divided between the national centres (in Egypt powerful, in nuclear research and experimental sciences; and in Tunisia, active in agriculture) and the university system (which was especially successful in experimental and engineering sciences, without a great deal of government support).

2.3 Capacities Established

The major bibliographic databases reveal the difference in capacities between countries. They also indicate the unexpected downturns that have occurred between 1985 (the peak time for national sciences) and the present. Scientific output is only a small part of world production (1.5% considering all disciplines), but is substantial in the disciplines that are crucial for that region (agriculture and tropical health: 10 -15%). There is a strong hierarchy between countries. Only 7 or 8 countries have a significant capacity in experimental and engineering sciences (and South Africa and Egypt share three-quarters of the potential). In 1985, the leaders were South Africa (which alone represented 33% of the production of the African continent), Egypt (22%) and Nigeria (13%). Quite far behind came Kenya and the Maghreb countries (around 4% each). If we add three French-speaking countries (Senegal, Côte d'Ivoire, Cameroon) and four English-speaking ones (Tanzania, Zimbabwe, Ethiopia, Ghana, all around 1.5%), that accounts for 90% of Africa's production. The other countries are tiny in terms of scientific production (which does not exclude the existence of small circles of specialists, around an eminent figure or in a breeding ground that a prominent establishment shelters, which create some unexpected centres of strength: embryology in Ghana, pharmacology in Madagascar, for instance).

For a decade, the scene has been undergoing profound changes. The most spectacular breaks in the curve concern the abrupt decline of Nigeria (but also, in agriculture, that of Kenya or the Côte d'Ivoire). Conversely, the Maghreb countries show a spectacular rise (a growth of 9% per year). Although South Africa remains far ahead, North Africa as a whole from now on carries more weight (35% against 30%). Morocco has become the third scientific producer in Africa, ahead of Kenya, Tunisia and Nigeria, all at about the same level. Other variations reflect the reduction of fruitful international cooperation actions (frequent, like in Niger, Gabon, Mozambique), or their resumption (Uganda or Ghana); but also the erosion, disintegration or resurgence of scientific communities confronted by dramatic changes.

2.4 Recent Change, and the Factors Behind

The upheavals of a profound change, destabilizing the national sciences, began to make their effect from about 1985. These were in no way confined to Africa. Increasing emphasis on the market economy brought states to reduce the degree of their intervention. Expectations for progress hinged no longer on the discoveries of science but on the innovations of industry. And the well-being of all was perceived not as the anticipated result of planning but of the free rein of the market. In the scientific world the preoccupation became, both in the industrialized countries and in the South, to associate clients to funding, guidance and, if possible, performance of research.

Gibbons and his colleagues describe this shift in the professional practice of research and the reorganization of institutions and policy changes as the spread of a new mode of knowledge production ("Mode 2" as opposed to the older, "Mode 1"). It can be summed up in five points:
- it is firms and not the academic establishment that become the centre of research activities
- research is performed in worldwide networks
- the search for profit (rather than knowledge) becomes the guiding principle
- the market rather than evaluation by scientific peers is the regulator of research
- the profession is exercised according to short-term contracts (and not in the context of a life-long career).

While we are examining the signs of the emergence of this new mode of knowledge production in the industrialized countries of the North, the paradox is that the cultural and professional revolution on which the wave of change is supposed to be based is more intense in the South. The tendency is at its strongest in the poorest countries, notably in Africa, affected now by 15 years of far-reaching economic crisis. The devaluation of scientific careers has caused researchers either to emigrate or change profession. Those who stay in their countries look for all sorts of contracts on an individual basis in order to survive professionally. There is not really much demand from producers; but there are many calls from international organizations, NGOs and multi-lateral co-operation schemes seeking to further their own programmes.

3. Current Policies: A Typology

3.1 Laissez-faire

In this context, many governments in Africa have abandoned any attempt to control their research scientists' survival strategies. Nigeria stands as a typical example. Between 1965 and 1985 it was one of the countries which most fully backed research and higher education. The economic crisis, generated by the fall in oil prices and the growing debt burden, added to the arrival to power of a succession of dictatorships, completely upset the situation. Education was no longer a priority. Higher education, deemed too budget-hungry, was left on the sidelines. The intellectual professions were regarded as parasites and many of their members were accused of subversion and were persecuted. Strikes in the universities were suppressed. From 1987 the budgets for higher education and research were restricted to payment of salaries. However, inflation reduced these 20-fold. Buildings were no longer maintained and equipment was not renewed. Overall upkeep was not ensured. Even today the University of Ibadan, the prime and most celebrated university of the country, has neither telephone nor electricity because it cannot pay the bills. It cannot anticipate relief from public aid of the industrialized countries in the North, which is everywhere in decline, and is withdrawing from the so-called social sectors, especially in the "intermediate" countries.

The primary result of this degradation (1986-1990) was a large-scale exodus of scientists. First to go were the highly-reputed scientists who had contact with the best networks and who were internationally competitive in their discipline. Subsequently the flow ebbed and changed direction; the countries of the North and the international organizations became saturated. Africa still offered, however, some attractive destinations: today, Ghana and even the University of Saint Louis in Senegal offer Nigerian scientists posts, which are eagerly accepted. Besides these moves, many scientists have converted to other professions (banking, farming, civil services). In the universities, numerous positions remain vacant. In the prestigious Agricultural Research Institutes, the average turnover is 1 and a half years, which makes any planning of experimental projects (monitoring of cultivation trials) impossible.

Nevertheless, researchers do stay on and some of them prosper as much as they ever did. This is the case in areas where field research is an essential complement to laboratory work in the North. In this category are political sciences, ethno-botany and plant chemistry, investigation of certain diseases (such as sickle-cell anaemia), trials of medicines, vaccines or of health-care protocols. Some well-trained and informed researchers have managed to capture or create the demand from science and industrial companies in the North. Others have adapted themselves as consultants for local industry, for oil extraction or for NGOs seeking sources of information.

Research is therefore undertaken as much as before. It has even become a good way to earn a living. But the nature of it has changed. Much closer to development than to investigation, it is oriented less towards education and does not much lend itself to publication. A split has formed between researchers sticking to the "national science" ethos and practices but doomed to impoverishment; and those open to the "market", paid for their service and connected to worldwide circles of research working on the latest leading subjects.

Some benefit from research centres which are almost tailor-made for them, well equipped and built off-campus with funds from abroad. There is a shift in the research topics: there is no demand for inorganic chemistry and scientists are no longer interested in it; conversely, physical chemistry flourishes, when it applies itself to biology. The hierarchy of disciplines is altered and the rules of promotion have been turned upside down. The mark of success is no longer academic achievement (careers have been blocked and withered), but material affluence. Young contract researchers can sometimes outdo eminent professors in this respect.

The technical ministries observe bitterly that, as the work of the research centres under their supervision is not financed by them, they no longer have any control over their programmes or over the personnel, which has become volatile. In an attempt to preserve a kind of equity among the teaching staff, the university authorities sometimes try to forbid the use on campus of electricity generators, telephone lines or larger equipment paid for with funds outside the faculty's control. The main effect is to push the high-performing research centres outside their territory. As for the government it just contents itself with the fact that such research becomes "profitable". No general policy document has been issued for ten years; and no public debate is on the agenda concerning this subject. The National Research Councils are lying dormant. Not one measure of the structural adjustment plans has dealt with the promotion of science. Disengagement by the state, erosion of salaries and persecution of intellectuals have singularly devalued the scientific professions.

On this basis the laissez-faire policy has led to a changed landscape. The initiative for research work no longer resides with the universities or even with the research centres. It is the private companies and sponsors who place their "orders" to match their own agenda. Belonging to a renowned establishment provides a good label, but a living is earned through contract work. The profession is flexible and internationally mobile. It is exercised in the context of world-wide networks which bring in new studies. Scientific production is no longer monitored by the body of peer referees or by organizations responsible for scientific planning, but by a market for research.

Although they do not represent such a pure, typical example, several other African countries are showing similar trends. The situation is not much different in Egypt and indeed that has long been the case. Since 1990 public funding for research has virtually disappeared. The United States are now withdrawing their powerful cooperation, which had been assured since the Camp David agreement. The Academy of sciences and other central organizations are thus losing their sources of financial backing along with their associated power and privileges. To avoid dealing with the bureaucracy of these bodies, financial sponsors and clients for services from industry, which are abundant, contact directly the most well reputed researchers and the more efficient laboratories, as they would do with any private enterprise.

The East African countries provide another example. There is little demand from industry. However, a kind of privatization of science is sustained by orders from numerous NGOs, which extrapolate the concerns of social movements prevalent in the industrialized countries (such as the environment, feminism, governance). Beyond the official stances, a spirit that denigrated research, portraying it as too slow and too reflexive, became fixed in the minds of political authorities. Economic crisis, which has been raging since 1985, has justified the reduction of funding of research to nothing, put a freeze on recruitment and devalued the profession. In Tanzania, careers have retracted so much that a professor, after several years' work and achievements with publications, earns no more than an assistant lecturer - $40 net per month (in 1999). Members of the teaching staff each depend on grants obtained through external aid; each establishment on a fragile combination of different funds; and each research scientist on consultancy contracts to survive. In Uganda, half of the teaching staff of the University of Makerere were laid-off in 1994. Since then, even though the economic situation is improving, posts have remained frozen. In the past decade, scientific production and the number of participating authors have declined in the Agricultural institutes in Kenya by 40%, which was a model in scientific terms in the eighties. In the absence of a national policy, certain leading establishments engage in devising institutional strategies. The University of Zimbabwe has embarked on a reorganization plan geared to "service to users and clients": this appears to be inspiring the confidence of cooperation schemes from abroad. In the same country, the Council for Medical Sciences has set up a satisfactory dialogue between private and public-sector scientists, NGOs and financial sponsors. Many other African countries are showing similar trends (such as Ethiopia, Madagascar, Ghana). With very mixed success science is being reinstitutionalized, founded on the values of a privatized, worldwide science.

3.2 Policies Recommended by Outside Bodies (especially the World Bank)

Some donors are anxious to preserve a less precarious research activity in Africa: European countries, the European Union and some American Foundations are in the forefront. Their wish is from now on not to invest in isolated projects, but in a sustainable process of rehabilitation. In anticipation of greater consistency of their doctrines and programmes they enter into agreements with the World Bank to support institutional reforms that could take the place of new national policies. It is a question of bringing research back up to international standards; and to accustom scientists and institutions to doing their own fundraising. Two countries can be regarded as test beds, one for the reorganization of national centres of research (Côte d'Ivoire), the other for that of the universities (Senegal).

Research in agriculture in the Côte d'Ivoire was highly productive for a long time. It was founded on a dozen or so institutes inherited from colonial days, attached to fundamental work (sustained by the French research institute, Orstom) or more applied and specialized according to categories of product (like cotton, citrus fruits, or oleaginous crops). These institutes were in the hands of the French institutes that had created them, and since the seventies "Africanization" transferred the management of the Centres to nationals of the country. The new Ministry of Research assigned to them a growing number of national scientists and appointed directors and managers. In a new role as partners, the French institutes contributed financially and their personnel cooperated in the implementation of programmes henceforth decided on and evaluated in bi-lateral committees. Scientific publication and technical achievements blossomed. This harmony began to disintegrate from the 1980s. Economic crisis hit the country and funds for research (even in agriculture) became meagre (less than 0.3% of the agricultural GDP, which served mainly to pay the devalued salaries of the country's research scientists). The French institutes had to accept ever-increasing obligations in terms of imposed research subjects and larger financial contributions demanded. A situation developed of recurrent dipping into budgets, political interference and dissension about programmes. Orstom and Cirad (two French institutes particularly involved) eventually withdrew. Scarcely any candidates could be found to cooperate with the Côte d'Ivoire. With no equivalent organization to take over, the centres fell into decay. Many of the country's scientists left the institution or set up their own NGOs, in order to gain the means and freedom of action. The international donors were distrustful and the funding was blocked completely.

For several years the establishments concerned were to remain inactive. In order to break the deadlock, the World Bank had to be approached. It offered help under conditions which it found difficult to enforce in less desperate situations. The Bank's idea was that the research had to support organically innovation among direct producers. Those who practised such research had to prove their spirit of enterprise. The establishments had to provide services and be largely self-funded. The state, as guarantor of continuity and the general interest, kept a right to scrutinize but not to decide. In order to concretize the scheme, the World Bank first obtained the formation of professional associations, that could represent planters and farmers. Then agricultural research as a whole was put back into one establishment which no longer had public-sector status. The state (with a minority holding of 40%), agro-industrial companies and professional agricultural associations were put in possession of the capital. The Board reflects this composition. The management has become independent of the government. As for the scientists already holding posts, they were firstly put at the disposition of the civil service. Those who wished could take part in an international call for tender, aiming to recruit a new compliment of staff cut down by a half. Selected on the basis of their knowledge and their abilities in popularizing science, they had to resign from the civil service. They benefit from short-term contracts, accompanied by financial advantages and an obligation to produce results.

Neighbouring countries watch the experiment attentively. There are many questions that bear on the ability of the professional associations to formulate what they need in research, on the quality of management and on the scientific qualities of the chosen candidates. The scheme has not been operating long enough to assess its progress. The significance is that this is the first example of a system of agricultural research that is not clothed in the characteristics of national science. And there we have something that goes well beyond a few isolated private research centres. The World Bank is trying to promote similar reforms in other countries. In Senegal, Tanzania and Uganda, the Bank has stimulated the formation of farmers' associations, which are the possible participants in such operations. Before it grants any new funding, it obliges Institutes of Agricultural Research to draw up a strategic action plan, which envisages some self-funding and more direct links with end-users of research. However, a solution akin to that of the Côte d'Ivoire is not yet very common: governments want to keep control and researchers cling on to their careers.

Senegal can be seen experimenting also, but in another area: the rehabilitation of the universities. The situation is not in fact one of the worst. Because of unions strength, the teaching staff has obtained a revised status and acceptable pay. Academics have free time outside teaching hours. History has pushed them in three directions: political involvement (often in the opposition, but sometimes as advisors at the government), archetypical academic (the scholar who looks at obtaining respect from his peers worldwide, who still conducts research without public support), or consultancy on all sorts of subjects sought by a myriad of newly arrived NGOs. Moreover, the opening up of universities to the wider population at the height of economic crisis has considerably degraded working conditions. Facilities have deteriorated, students protest and the number of students obliged to repeat academic years is swelling. Laissez-faire here would lead to a catastrophic situation.

For several years, the World Bank has been wary of student unrest and takes a new interest in the training of a new elite. It therefore proposes loans and donations to finance plans for university recovery. The scheme offered to Senegal comprises three main elements:

- renovation of all educational facilities: material, buildings, and - the most spectacular aspect - a large extension of the library, connected to immense virtual resources. In return, student numbers have to be got under control and the number of entrants and students repeating years have to be restricted.
- restoration of student facilities such as restaurants and halls of residence, prior to a full-fledged privatization of these services. In return, a significant level of tuition fees is to be introduced and grants will be replaced by loans.
- initiation of a research fund, fed by the various faculties. Faculties have to identifiy a proportion of their budget as research and the Fund would automatically be supplemented by the state with an equivalent sum. The Bank will then add the same amount.

The new buildings are under construction (the enormous library was almost finished by 2000). A real effort to control student flows has already been undertaken. However, various points are still causing dissent. The introduction of tuition fees for instance is far from being settled. The management and objectives of the Fund for research is giving rise to fierce controversy. The unions insist that its priority should be to finance scientific and sabbatical visits, promised as part of their conditions to every member of the teaching staff without restriction on the choice of subject (this entails the accomplishment of personal work necessary for promotion to higher academic grades). The academic authorities are in agreement, but want to keep the right to judge the quality of the research. The department that oversees research at the Ministry of Higher Education would prefer to confer this role on national committees and to orient research towards sectors more directly linked to employment. Finally, the Senegalese side of the World Bank's Programme, a parallel administration functioning at Prime Minister's office level, intends to spend money through a tendering system, which would restrict subjects and take this opportunity to structure academic research into laboratories, which are currently missing. The extended nature of these debates has up to now paralysed the Fund's launch. It remains that the model for university rehabilitation, first tried in Nigeria where the teachers' political hostility had sunk the idea, here in Senegal revealed its principles: universities of high quality, highly professionalized, reserved for hard-working students, with a requirement for teaching staff to do research, linked if possible to real economic needs and capable of yielding resources for the institution. Other countries are currently being urged to take a similar path: those in East Africa (whose universities are considered to yield too few results) and Madagascar (whose university has just made an exceptional effort to control student flows).

3.3 New National Policies

In the case of Agricultural institutes in the Côte d'Ivoire, just as in that of universities in Senegal, it can be seen that donors are asking the state to resume its involvement, albeit to a limited extent, in research funding and in devising the aims of research. A number of countries already do this (or even have never stopped doing so). We will pay special attention to the case of two large countries, whose determination is tracing the outlines of new national policies. They are South Africa, the continent's main productive force in almost all disciplines; and Morocco which in a few years has risen to third place in the whole continent and which typifies the dynamism of the Maghreb countries.

South Africa is doted with Research centres and universities accustomed to working under contract with already developed industries. The "private sector" contributes 50% of the funding for research and accounts for 50% of its performance. The two worlds have long been acquainted. However, a number of new problems have emerged. The most prestigious universities are called on to make intensive efforts to integrate black students. They have to accept not only to curb their inscription fees but also to make reasonable increases to their teaching load: even if it means cutting into research time. In order that the latter remain active, a mechanism of profit-sharing has been brought in, which rewards substantially the establishments and laboratories for each article published by one of its members. Besides this, the National Research Foundation runs an incentive research fund, with substantial resources, which although not reserved exclusively for them, gives a new chance to "black" universities to acquire the culture of research which it is missing. The Foundation evaluates the proposals and the principal investigators of candidate laboratories by international panels. A prestigious array of prizes stems from this and, at the same time, an important function of strategic scientific and technical intelligence is fulfilled for the country.

The Ministry of Research has in addition submitted the national institutes to stiff auditing, involving their clients. Some exceptional advantages have been granted on senior researchers to relinquish their posts, thus liberating these, and take opportunities to set up their own research companies. This is what many have done. Precise directives are given to multiply joint research projects with industry in sectors deemed to be strategic; basic popular needs are to be answered by several original instruments managed by specialized NGOs which the state helps to organize.

In fact the government, having assessed the powerful research system bequeathed by the apartheid regime, is beginning to define more clearly the main thrust and instruments of its policy. The budget of the Ministry of Arts and Sciences is continually being increased, (an increase of 80% in constant Rands from 1995 to 2000), particularly for the Sciences sector (that doubled over the same period). The basic subsidy assigned to establishments (Institutes as well as Universities) rose by only 30% (a slight decrease in constant Rands). However, considerable resources are apportioned through the channel of "competitive" Funds. While submitting them to evaluation, the National Research Foundation continues to devote nearly half of its finances rather to basic research, initiated by highly qualified scientists and their laboratories. In contrast, two other Funds, undergoing rapid growth, are intended to support what is termed strategic research.

The Technology and Human Resources for Industry Programme (THRIP), by-half co-financed by industry, supports scientific work that seeks to develop the competitiveness of South African firms (technology learning for small and medium-sized businesses; anticipatory research, whose costs are shared by large local firms or multinationals). Its budget has increased more than tenfold between 1995 and 2000, to reach the equivalent of the total distributed by the NRF (and 20% of the subsidy to the establishments).

The National Innovation Fund (NIF) supports large-scale, long-term cooperative programmes, between research establishments, industry and non-profit organisations, in subject areas of national priority. Up to now, it has been centred around crime prevention, promotion of an information-based society and biotechnology. Its budget has quadrupled in the past three years; today it has reached 10% of the subsidies received by research establishments. The emphasis is therefore put on research deemed to be strategic (effort focused on a topic with foreseeable medium and long-term practical scope), or even for "application" (linked to a real economic need) or "participative" (implying joint action with recipient social groups).

A recent survey showed that university academics actively engaged in research themselves classified 40% of their work into categories extending from "strategic" to applied research (including the "incremental improvement of existing processes"); it also indicated that only 30% of their funds came from sources that encourage basic research as a priority, 15% from sources favouring strategic research and 55% from other funds intended to feed applied research. It has to be added that the policy displayed is not hostile to basic research (conducted notably by the universities) which are considered to guarantee quality "without which applied research would not take long to die off". It is a question rather of stimulating a new spirit of research: to initiate a new contract between scientists and society, the former finding their legitimacy to find inspiration from the interests of the latter.

In the Maghreb the vigorous development of universities in the 1975-85 period was the stimulus for a flourishing of research which was not followed up as a priority. However, some outstanding institutions (for example the Hassan II Agriculture and Veterinary Institute in Morocco) have made it an instrument of their prestige. Research ensures its place in society and explains their financial resources. Well supported by sound management structure in stable institutions, with the aid of cooperation schemes, the young teaching staff have carried on the research tradition in which they were immersed as students in the best establishments in Europe or the USA. As in the case of South Africa, the scientific profession has neither noticeably lost its richness nor been relegated in status and diverted from its missions. With little political support, production of articles and studies grew as the scientific centres themselves developed. More recently governments have begun to take an interest in the inherent potential of research. In all these North African countries, governments have created Secretariats of State for research within the Ministry of Higher Education, or outside of it. In Tunisia, for example, the Secretariat of State's office had jurisdiction over the national research centres; but also, using a tendering system, it tries to build a structure of university laboratories which was never a priority for the Higher Education ministry.

The Secretariat of State for research in Morocco is highly active. It works to channel the initiative of scientists or of certain institutions which are winning international tenders and build-up cooperation networks to match demand. It uses three devices for this: its own calls to tender; priority programmes to stimulate particular research areas; the constitution of major poles in specialized areas. The invitations to tender, open for any research subject, have two objectives. They aim to encourage talented young scientists or scientists scattered among recently founded establishments to come forward and hence be identified and to generate a culture of evaluation, which is lacking in the ministry as well as in the scientific community. Success has been achieved simply by setting up independent commissions in each subject/discipline. Also, seven priority programmes have been defined. The intention is to proceed towards an accumulation of results and to steer research effort towards themes which have been neglected. Finally, the creation of specialized centres aims to build a structure in areas judged to be strategically important in the long term. Here the principle is to draw on research strong-points and to link them up in a network. A strong establishment can for example act as a nucleus and a site where other centres can share the use of its equipment. However, the concern is also to establish a strong centre in each region and to attribute a role to each type of institution (Institutes, Engineering Schools, or Universities).

Added to these initiatives are a host of other measures, such as: prizes to reward scientific invention and technical innovation in agriculture; the creation by the industrial sector of an association for research and development, which puts out tenders for work conducted with certain companies; intensification of cooperation activities in the form of twinning with a "scientific region" in France. The state has taken on coordination of all these initiatives with the objective of transforming what is a real but fragmented potential (though well supported by robust institutions) into an effective force. There are presumed budgetary costs attached to this process. It translates a medium-term vision of economic transformation (the entry into a "large Mediterranean market") necessitating an improvement on the quality of products, brought up to official European standards, and the development of new sectors of production. It is based on the realization of the importance of innovations (including technical and technological ones) in an open economy; and on the necessity not only for highly applied research but upstream as well where the need arises, in order to facilitate quick entry into new niches.

In Tunisia, the Secretary of State's office handling of national research centres is guided by similar considerations. Such principles prompt it also to issue its own calls for tender and thus to try to unite university laboratories into a structure within the Ministry of Higher Education. Algeria, which has often neglected its research and which makes slower progress in this sector than the other Maghreb countries, nevertheless equipped itself with the means to direct it and to favour its expansion (invitations to tender).

3.4 Towards Regional Policies?

In contrast to the cases described, in countries where the State has lost interest in science, some of its professionals put some hope in regional policies. Donors are indeed interested in this level as well. Bilateral and multilateral joint schemes have been employed to consolidate inter-governmental bodies concerned with agricultural policy (such as SADEC, CORAF), which are not insensitive to the need for research. In the health sectors, a WHO initiative has placed the management of endemic disease control at the regional scale, with a coordination of local programmes, and the creation of supranational specialized research centres. The European programmes of scientific cooperation are derived from priorities assessed at the same level. Scandinavian cooperation schemes have divided their intervention into several fractions: they support "institutional building" negotiated country by country, but maintain scientific cooperation programmes shared by neighbouring countries. Periodically the idea comes up of abandoning aid to national research systems that are in an impasse and to concentrate it on trans-national centres of excellence endowed with a critical mass in a specialty, which are well equipped, offering good pay, but demanding as to the quantity and quality of work expected. Examples of these do exist: such as trans-national Schools of Civil Engineering, Water Engineering, Statistics. There are international centres also in agricultural research. (see the article *Stakes and New Prospects for North-South Scientific Cooperation Policies*).

The implementation of such solutions, however, comes up against some strong resistance. Rivalry between nations and the taste each national government has for seeing the institutes it funds serve local needs weakens the supranational establishments. Previous enterprises have been seen to fall apart, such as East African institutions (universities and research centres) in the 1970s. It was at that time also that Burkina Faso wanted to finance universities in neighbouring countries rather than found its own, but its hopes were soon dashed. Science is moreover always considered by states as an area of their sovereignty. Research, in mining for instance, and by extension in all sorts of fields, is sometimes seen at risk which exposes part of a country's resource heritage to plundering, unless they give rise to a government rent. Even if they barely provide any fund, they are on their guard against any sharing of authority over it. The abandonment of national centres would in any case face them with the formidable problem of what to do with civil servants employed there; the shedding of agricultural research, which the Côte d'Ivoire has gone along with, is a difficult bet. Furthermore, researchers and professors-researchers form a corporate body whose consent has to be won. It is true, though, that academics in French-speaking Africa are glad that their careers are managed by CAMES, an international peer commission independent of any political interests. The full-time researchers, orientated more towards engineering, now would like a similar arrangement, provided that the specific nature of what they produce is taken into account. The dual advantage lies in making career promotion independent of the vagaries of politics or bureaucracies and facilitating mobility, within the neighbouring countries. It also permits to provide scientists with valid international qualifications. Still to be solved is the problem of job security, which up to the present only national institutions could guarantee.

4. Conclusion

The upsurge of regional institutions is a reflection of two forces: pressure from donors in the industrialized countries, anxious to find a political level for discussion related to research; and the desire of many research scientists, no longer bound to state-governed science, but wishing to avoid the extreme volatility of the institutions. In the first place, the move to a market economy and globalization removed the bonds that kept researchers attached to their systems of national science. Devaluation of their profession and the radical withdrawal of certain governments have had a destructive effect. Besides a large-scale loss of numbers (owing to changes of profession and exodus abroad) the result has been to convert those who stay to market-led research. They do research sustained by a few firms and, more often, by international NGOs in their quest for information, or by certain bilateral cooperation schemes promoting work associated with engineering. Research institutions in the classical mould (universities or national centres) find themselves often disqualified and out of the running. The scientific community has been fragmented into isolated units by the chase after contracts. What still has to be guaranteed is the training to the trade of research, the internalization of professional norms to younger scientists, the definition of policies and the continuity of research activities.

It is to these aspects that governments of some countries (Maghreb, South Africa) vigorously turn their efforts today. These are countries where their professional scientists have not been decimated by the crisis of the past 15 years. Scientists themselves are convinced today that an aggressive strategy of development is necessary and that it is possible They seem to believe that, in an open economy, this can be attained by improving "quality" and therefore by way of technical innovations which have scientific content, discernible by a well-informed scientific community, linked up both with the outside world and the parties involved locally.

Other influencing factors converge today to recreate the needed institutional bodies, possibly at regional scale. These would need to invent an original three-stranded configuration that brings together clients for the research, public authorities and organizations that perform the research. There is no guarantee that these efforts will be successful, except in the case of particularly far-sighted states. But, at a time when "mode 2" science is expanding, Africa is indeed the scene where science is seeking to find its place in a relevant form of revised institutionalisation. The tension between national sciences and demand-led science is at play here in a context of institutions living in precarious situations.

Acknowledgements

Our studies on the *State of science in present Africa* have been made possible by a grant from the European Commission (DG XII: Science, INCO-DC Programme) and by a grant from the French Ministry of Foreign Affairs (Cooperation: Bureau of Research).

Bibliography

Arvanitis R. and Villavicencio D. ed. (1998). Comparative perspectives on Technological Learning, *Science, Technology and Society*, special issue, **3** (1), 1-238. [A point of view of learning from industrial firms. Articles are dealing mostly with Latin America and Asia but concepts can be applied more generally]

Bennett J. (1998). *Science and Technology Policies and Capacities in East Africa*, 138 pp., Köln, Germany: Institute of African Studies [A comparative Review]

Busch L. (2000). *The Eclipse of Morality: Science, State and Market*. 219 pp., New York, USA: Aldine de Gruyter. [A brilliant essay, rooted in the mastery of philosophical theories as well as practical knowledge of the most recent advances in agricultural biotechnologies. Deals with development and the problem of the power of technology, the expansion of property rights, the "networks of democracy". A warning to the developing countries, and much information for them to be known.]

Chatelin Y. and Arvanitis R. (1988). *Stratégies scientifiques et développement*, 143 pp., Paris, France: Orstom [Carves out interesting concepts (autocentred and cosmopolitan communities, styles of science, scientific autonomy...), and demonstrates useful methods (mapping of scientific topics). Part of the results is available in English: Arvanitis, R. and Y. Chatelin (1988). "National Scientific Strategies in Tropical Soil Sciences." *Social Studies of Science* **18**(1): 113-146.]

EisemonT.O. and Davis C. H. (1997). Kenya: Crisis in the Scientific Community, *Scientific Communities in the Developing World* (Gaillard, Krishna and Waast ed.), 105-128, New Delhi, India: Sage. [A recent survey, by experts of the field].

Enos E.L. (1995) *In Pursuit of Science and Technology in Sub-Saharan Africa*, 213 pp., London, UK: Routledge. [The Impact of Structural Adjustment Programmes].

Gaillard J. (1999). *La coopération scientifique et technique avec les pays du Sud*, 340pp. Paris, France: Karthala. [Up to date with the doctrine and policies of the main donors to S&T in developing countries: Canada, USA, Japan, and several European countries. A useful analysis of the historical background, the political implications, stakes and prospective. Large summary in English: 70pp.]

Gaillard J, Krishna V.V. and Waast R. (ed.) (1997). *Scientific Communities in the Developing World*, 398 pp., New Delhi, India: Sage. [A reference book. History, scientific communities and professionalization, links with politics, social and cultural inscription of science. Rich and recent evidence on scientific institutions and capacities. A panorama of 12 countries by their own leading specialists, among which 6 African countries: Algeria, Egypt, Senegal, Nigeria, Kenya, South Africa.]

Gaillard J. and Waast R. (1992). The uphill emergence of scientific communities in Africa, *Journal of African and Asian studies*, vol. **27** (1-2): 41-67. [Contains a review of the notion of a scientific community. Good data about the 1970-90 period.]

Gaillard J. and Waast R. (2000). L'aide à la recherche en Afrique: comment sortir de la dépendance?, *Autre Part*, Vol. **13**, 71-89. [Science in Senegal and Tanzania to-day]

Gibbons M., Limoges C., Nowotny H, Schwartzman S., Scott P. and Trow M. (1996). *The New Production of Knowledge*, 179 pp., London, UK: Sage. [A reference book, giving rise up to now to endless controversy (see *Science, Technology and Society*, **5** (2)). A rehabilitation of the political approach to science, against epistemology and ethnology. Stimulating and prospective (is there a new mode of knowledge production, steered by industry, and destined to substitute the academic regulation of science?). The interesting paradox is that this "mode 2" seems to spread quicker than elsewhere in Africa, where industry is lacking.]

KrishnaV.V., Waast R. and Gaillard J. (2000). The Changing Structure of Science in developing Countries, *Science, Technology and Society*, **5** (2), 209-224. [Globalization and the reorganization of science in developing countries.]

Lebeau Y. and Ogunsanya M. (ed.) (2 000). *The Dilemma of Post-Colonial Universities* 334 pp., Ibadan, Nigeria: IFRA/ABB. [A collection of essays, heartfelt and documented, on the scientific profession and institutions in present Nigeria and Senegal. And a detailed account of the national policies and the World Bank strategy]

Mouton J. and Boshoff S.C. (2001). *Science in Transition*, 89 pp., University of Stellenbosch, South Africa: CENIS. [A synthesis about S&T in South Africa today].

Ndiaye Falilou (2000). La condition des Universitaires Sénégalais, *The Dilemma of Post-colonial Universities* (ed. Y. Lebeau and M. Ogunsanya), 169-207. Ibadan, Nigeria: IFRA/ABB. [The scientific profession in Senegal today]

Salomon J.J. and Lebeau A. (1993). *Mirages of Development: Science and technology in the Third Worlds*, 263 pp. London: Boulder. [A book that has left a strong mark. Without concessions, examines the strategies of R&D open to different types of Third World countries.]

Shinn T., Spaapen J. and Krishna V. ed. (1997). Science and Technology in a Developing World, *Yearbook of the Sociology of the Sciences*, vol.**19** special issue 1-411. Dordrecht: Kluwer. [A thougtful analysis of the challenges to science in developing countries: epistemological (what difference with an ethno-science, the rhetoric of progress, ethnocentricity), historical (colonial background, institutional culture, differences and imbalance with science in the North), social (hegemonic attitudes, and counterhegemonic movements in the society). Useful as well for a thought on modern science anywhere.]

Tostensen A., Nordal I. and Andersen R. (1998). *Norwegian Research Support to Developing Countries: The Cases of Uganda and Zimbabwe*, 118pp., Oslo: The Research Council of Norway. [Incidentally, a good review of the present state of institutions in those two countries]

Tostensen A, Oygard R.,,Carlsson J. and Andersen R. (1998).*Building Research Capability in Africa*, 156 pp., Oslo: The Research Council of Norway. [A Review of Norway's assistance to Regional resarch organisations]

Waast R. (ed.) (1996). *Les sciences au Sud, état des lieux*, 332 pp., Paris, France: Orstom-UNESCO. [A resource book. Mapping science. Facts and figures about the recent evolution of science in selected countries and policy matters (internationalization and privatization of science; science as a public good; technological choices). Comments on the "Occidentality" of science (indigenous knowledge, social inscription of science, protest movements against the hegemony of science).]

Waast R. (dir.) (1996-1997). *Les sciences hors d'Occident au 20° siècle/Science beyond the Metropolis*, Vol. 1 to 7, Paris, France: Orstom-UNESCO. [A broad coverage of topics: Vol 1 Keynote texts, Vol. 2: Colonial Science, Vol 3: Nature and environment, vol 4: Medecine and Health, vol 5: Science and Development, vol 6: The State of South Science, Vol 7: International scientific cooperations. Contains contributions by some of the best specialists. Half of the chapters is in English, half in French.]

Widstrand C. (1992). *Tanzania: Development of Scientific Research 1977-1991*, 154 pp., Stockolm, Sweden: SAREC Documentation. [A good review of the present state of institutions in Tanzania]

Zahlan A.B. (1997). Scientific Communities in Egypt: Emergence and Effectiveness, *Scientific Communities in the Developing World* (Gaillard, Krishna and Waast ed.), 81-104, New Delhi, India: Sage. [A good review of the present state of institutions and capacities in the country]

Biographical Sketch

Roland Waast is a senior researcher at IRD (Institut de Recherches pour le Développement, France). He holds an engineering diploma from the Ecole Polytechnique, Paris, and graduated in sociology from La Sorbonne University, Paris. As a sociologist, he spent numerous years in developing countries (particularly Madagascar and Algeria). He has been head of the Department of "Development Strategies" at IRD, and is a member of several scientific Commissions on S&T policies as well as a former member of the French High Council for Science and Technology. He is currently a member of the French Commission for UNESCO.

He has set up a research team specializing in the Sociology of the Sciences at IRD fifteen years ago, and an international network (ALFONSO) dealing with the same topics in developing countries (main nodes in India, Venezuela, Brazil, Argentina, Algeria and South Africa). He is the founder and co-director of the Journal *Science, Technology and Society*. He has authored numerous papers in French and International Journals, collaborates to the *World Science Report* (Unesco), and completed various research projects for the European Commission. He is the Series Editor of *Les sciences hors d'Occident au 20° siècle/Science beyond the Metropolis*" (IRD-UNESCO: 7 volumes). He recently published (in collaboration): *Scientific Communities in the Developing World*, and *Les Sciences au Sud: état des lieux*. During the two last years, he organized an extensive survey of the state of the sciences in Africa, and is now working out the results.

CHANGING POLICY IN SCIENCE AND TECHNOLOGY IN INDIA

Krishna V.V.
Jawaharlal Nehru University, New Delhi, India

Keywords: India, science and technology olicy, innovation policy, policy cultures, liberalization

Contents

1. Introduction
2. Four Science and Technology Policy Cultures
3. Different Phases of S&T Policy
4. Changing Trends in Science as Social Institution
5. Conclusion
Acknowledgements
Bibliography
Biographical Sketch

Summary

Four S&T policy cultures—political-bureaucratic, industry-market, academic, and civic—are defined here in order to explore the institutional growth of science and technology in India during the second half of the twentieth century. Within this perspective, the article attempts to trace different phases and trends in S&T policies. Three main phases are identified—1947–1970 (optimism in "policy for the sciences"), 1970s–1990 (from optimism to critical evaluation), and after 1991 (new economic reforms, liberalization, and globalization)—and used when exploring the growth of S&T. Personalities in science and politics who have played an important part in shaping India's S&T policies during different phases are considered. Having traced the growth of S&T policies in historical terms, the article focuses on S&T policy challenges in the present era of market reforms and globalization. How are these factors influencing the research system? What institutional changes are being introduced? What are the implications concerning "science as public good" versus "science as market good"? And what are the current challenges?

1. Introduction

In India discussion of S&T policies emerged, in terms both of scholarly and policy relevant discourse, in the latter phase of the colonial period around the beginning of the twentieth century, and has clear roots in the anti-colonial struggle. Debates on the reception of modern western science in India, on modes of industrialization during the colonial phase, on the struggle to institutionalize and professionalize the Indian scientific community, and the efforts towards establishing some key universities and scientific and technological institutions between 1900 and 1947, were all rooted in the actions and S&T policy debates of the Indian scientific élite and political leaders led by Gandhi, Subhas Bose, and Nehru. Some glimpses of this discourse in the important decade before India's independence can be found in the first-ever Indian science policy journal, *Science and Culture*, launched by the eminent Indian physicist M.N. Saha in 1938 and still published today.

This is not the place to go into the details of the genesis of these studies in India, which is covered by much scholarly writing. The present essay attempts to map out a perspective of S&T policy cultures relevant to developing countries such as India in the context of the post-colonial and post-war period. After this brief survey, it explores different phases in S&T policies in India during the period between the late 1940s and the 1990s. This will examine the key actors, agencies, and institutions that have shaped India's S&T policies, key milestones in different periods, the main agendas pursued, and other issues. We will conclude by exploring current trends in the era of economic liberalism and globalization, placing them in the context of developments in the emerging knowledge industry.

2. Four Science and Technology Policy Cultures

In India more than 75 percent of total S&T funding, including gross expenditure on research and development (GERD), comes from the government. For this reason, government policies and attitudes play a crucial role in decision making regarding the development of science and technology. Despite this, other actors, agencies, and institutions contribute to the overall structure of S&T policy. Thomas Kuhn and Ruivo have drawn our attention to "phases" or "paradigms" of science policy. Different paradigms signify different models of the utilization and regulation of S&T research systems. Heterogeneous groups of actors—including politicians, scientists and engineers, academicians, diplomats, industrialists, business representatives, and opinion leaders from civil society—influence S&T policies or goals in science, collectively or otherwise, leading to different patterns of science and policy frameworks (*Science and Technology Policy*). We can identify four distinct but overlapping policy patterns as four different S&T cultures: "political-bureaucratic," "industrial-market," "academic," and "civic," as relevant to the Indian context. As Elzinga and Jamison observe, these policy cultures:

… might be thought of … coexisting within each society, competing for resources and influence, and seeking to steer science and technology in particular directions. These cultures, which stand out as representative of the dominant voices … represent different political and social interests and draw on different institutional bases and traditions for their positions. Each policy culture has its own perceptions of policy, including doctrinal assumptions, ideological preferences, and ideals of science, and each has a different set of relationships with the holders of political and economic power.

These four policy cultures display differences in development priorities, policy instruments, ethos, and core constituencies with regard to science, technology and development issues at the national level. The categorization of these policy cultures can be taken as "universal" in the sense that they are relevant within the contexts of most individual nations. However, they are to be understood and framed here with specific reference to India. Each policy culture will be considered here briefly to this end.

2.1. Political-Bureaucratic Culture

The historical roots of this policy culture can be traced back to the centralized S&T decision-making processes established by the British colonial administration. In the post-independence period Pandit Nehru, India's first prime minister, played an important part and is credited with having forged an important alliance with the scientific élite. From the beginning, élite scientists who were heads of large science agencies like the *Atomic Energy Commission* and the *Council of Scientific and Industrial Research* were made part of the bureaucracy, as they were given positions equivalent to those of civil servants and thus came under the Public Service rules of the government.

In the post-independence period, the "tacit" alliance between this scientific élite (a form of technocracy) and the political leadership come to dominate the decision making system in India, and continues to do so to a great extent even today. This policy culture is dominated by science departments, councils, advisory bodies, committees, and science agencies, where the technocracy controls the S&T budgets and takes major decisions relating to S&T "in consultation" with the government of the day. Priorities in scientific research are set by the government and the political party in power, and the approach to decision making is generally "top-down." The core constituencies of decision making are centered around bodies such as ministries, S&T councils, and state planning regimes. Since political power ultimately rests on the democratic election process, "science as public good" may claim considerable legitimacy.

2.2. Industry-Market Culture

This policy culture is dominated by private business and market interests, and is generally represented by bodies such as the *Confederation of Indian Industry* (CII) and the *Federation of Indian Chambers and Commerce* (FICCI). It emphasizes entrepreneurship, the use of knowledge in businesses, liberal policies for technology transfer, and tariff concessions for local industrial firms. By and large market-related criteria are adopted for assigning priorities in R&D; and "science as market good" assumes considerable importance.

2.3. Academic Culture

This hardly needs much elaboration here. Much of the concern here is for maintaining autonomy and scientific excellence. The academic community and its élite play a crucial role in setting priorities in science. The academic policy culture is grounded in what has been labeled "mode 1" of knowledge production, as opposed to a "mode 2" more open to influence by day to day challenges. This policy culture emphasizes the importance of science as a profession, and of scientific communities, a disciplinary-bound science, peer evaluation in scientific decision-making, and the importance of universities.

2.4. Civic Culture

As Elzinga and Jamison point out, "civic culture articulates its position through public interest organizations as well as through campaigns and movements, and its influence is obviously determined by the relative strength of the civil society in a country's overall political culture." In the Indian context, the civic culture in S&T is represented by various groups and movements. In the environment and ecological field, movements led by Baba Amte, Medha Patkar, S. Bahuguna, and C.P. Bhatt, among others, provide good examples. Then there are large popular science movements led by organizations such as *Kerala Shastra Sahitya Parishand* (KSSP)—an all-India movement. In different ways, the historical roots of such recent civic involvement can be traced to the efforts of M.K. Gandhi and the Gandhian-based Sarvodaya Movement, with regard to the application of science and technology for development.

3. Different Phases of S&T Policy

The S&T policy-making process and its bearing on society is best understood from a historical perspective. In the Indian context, three main overlapping phases in S&T policy making can be defined. The intensity or varying influences of the different S&T policy cultures mentioned above during these different phases are shown in Table 1.

| Periods | Main science and technology cultures | | | |
	Political–bureaucratic	Industry–market	Academic	Civic
1947–1970	Very high influence	Low influence	Moderate influence	
1970s	High influence	Low influence	-	Low influence
1980s	High influence	Low influence	-	Moderate influence
1990s	Moderate influence	High influence	-	Moderate influence

Table 1. The influence of different science and policy cultures

3.1. 1947 to 1970: Phase of Optimism in "Policy for Science"

In this phase the political-bureaucratic culture exerted a dominant influence, mainly through a science–politics alliance initiated by Nehru with scientists such as Homi Bhabha, Shanti Swarup Bhatnagar, Mahalanobis, J.C. Ghosh, and D.S. Kothari, among others, who played an important role in drawing political support for building various science institutions and science agencies. The growth of Indian S&T in this initial phase cannot be understood without examining closely the relations between science and politics, particularly the close alliance referred to above. As early as 1947, when addressing the 34th Session of the Indian Science Congress, Nehru initiated the alliance with scientists by observing "that in India there is a growing realization of this fact that the politician and scientist should work in close cooperation." In contrast to Gandhi's critical stance towards modern science and technology, Nehru's modern, secular image and—most of all—his unquestioned support for science made him a "messiah" for the development of science in India. The scientific community in general, and its élite in particular, could immediately identify with his vision of science and development as they also found him a great promoter of their interests. Nehru once declared that:

It is science alone that can solve the problem of hunger and poverty, of insanitation and illiteracy, of superstition and deadening custom and tradition, of vast resources running over waste, of a rich country inhabited by starving people. I do not see any way out of our vicious circle of poverty except by utilising the new sources of power which science has placed at our disposal.

(These statements by Nehru are quite well known, but their source is rarely identified. The first is from *Science Reporter*, July–August 1, Volume 1, 1964. The second is from "The Tragic Paradox of our Age," *New York Times Magazine*, September 7, 196])

This era witnessed a great deal of optimism about science and development. The Manifesto of the Congress Party for the first national government in 1945 declared:

Science, in its instrumental fields of activity, has played an ever increasing part in influencing and moulding human life and will do so in even greater measure in the future … Industrial, agricultural and cultural advance, as well as national defence, depend on it. Scientific research is, therefore, a basic and essential activity of the state and should be organized and encouraged on the widest scale.

This period reflects a phase of "policy for science," during which the main emphasis was on creating a basic infrastructure for S&T in the country, including the expansion of the university sector to supply the necessary human resources. It was during this period that India's finest five Indian Institutes of Technology were planned. Infrastructure development in S&T also included substantial efforts towards building what may be termed the techno-industrial capacity of engineering, consulting, design, and development organizations. There were 42 such organizations by 1970 in the private sector and eight in the public sector. These institutions were to promote partnership between science and technology in the processes of capital goods industries; absorption of imported technology into areas such as power, chemicals, and metallurgy; and to complete turnkey processing, plant design and engineering, and erection and commissioning of plants in the major sectors of S&T.

Major mission-oriented science agencies such as DAE and CSIR, DRDO were either established or rapidly expanded during this phase. Pre-independence Indian science was focused on universities, but the post-independence expansion of science under the auspices of the government emphasized these science agencies. The postwar "science push" and "pipeline innovation" models triggered considerable optimism in the organization of science institutions in various sectors, from atomic energy to industrial research. Nehru and eminent scientists like Homi Bhabha, who is regarded as the father of India's atomic energy programme, were instrumental in getting the first-ever official Scientific Policy Resolution (SPR) passed in the Indian Parliament in 1958. This document is still an important landmark, since it has repeatedly been used to justify the funding and expansion of the S&T institutional base.

One of the notable features of the science–politics alliance of the Nehru era was that the growth and nature of the S&T institutions in different sectors was influenced by the interests of the élite scientists who were close to Nehru. These people included S.S. Bhatnagar in CSIR, Homi Bhabha in Atomic Energy, J.C. Ghosh and P.C. Mahalanobis in the Planning Commission, and D.S. Khothari in the Defence Related Organisation. In other words, the form adopted by the "policy for science" may be viewed as an informal science policy determined by the alliance.

Although Nehru was instrumental in setting out a scheme for planned economic development articulated through national Five Year Plans, and despite the fact that Nehru was one of the founders of India's Planning Commission, India's first ever Five Year S&T Plan (1974–9) only came into being in 1973. Close study of its origins shows that despite the presence of the advisory bodies created in this period, drawing scientists from various organizations and agencies, only a very small number of élite scientists close to the political leadership wielded real power during Nehru's era and that of Mrs Gandhi, extending into the early 1980s.

Even though Nehru consulted with a wide section of the scientific intelligentsia, the science-politics alliance of the Nehru era led S&T growth into very "specific" directions. CSIR had no laboratories worth mentioning in 1947, but by the 1950s S.S. Bhatnagar was able to establish a network of fifteen. The world-famous physicist C.V. Raman called this the "Nehru–Bhatnagar effect." This had a parallel in the Atomic Energy Agency, with Homi Bhabha as its head. Bhabha eventually convince Nehru to set up the Department of Atomic Energy headquarters in Bombay, where he (Bhaba) wanted it.

Thus, for about two decades after independence, the real expansion of S&T infrastructure took place in CSIR, DAE, and defense-related establishments. As Parthasarathi rightly pointed out in the early 1970s:

It is perhaps not surprising to find that decisions regarding the allocation of scientific resources, for example, have been taken not on the basis of the advice tendered to the political leadership by either of these bodies [the Science Advisory Body to the Cabinet and the Planning Commission], but as result of informal and tacit interactions between concerned individuals in the scientific community, the executive and the polity. Even today, decisions about defense, public health, atomic energy, industrial research and even agricultural research are apparently being taken almost independent of the formal national science policy.

With hindsight, the structure of Indian S&T institutional growth reflects the way in which agriculture and medical research were two important fields that witnessed only marginal development till the late 1960s. The close alliance between Nehru and élite scientists in industrial research and atomic energy had consequences for work in other areas. It is not surprising that the "grand old" agriculture scientist B.P. Pal lamented in 1977:

… how much the application of science to agriculture might have advanced if Nehru had been directly associated with Indian Council of Agriculture Research (ICAR) in the way in which he was associated with the CSIR and DAE. It is a pity that when these modern scientific organisations were set up, the older ICAR was not drastically reorganised on similar lines

Historically speaking, the university sector also suffered from this science–politics nexus, through a relative stagnation in the allocation of R&D funds. Though higher education witnessed considerable expansion, the locus of R&D was somehow restricted to the mission-oriented science agencies. By rough estimates the university component of R&D budget as a percentage of total R&D expenditure remained less than 10 percent from the 1960s to the 1990s. One reason for the domination of the mission-oriented science agencies (such as DAE and CSIR, among others) was that they were represented by élite scientists close to the political establishment, a tradition that continues today. The academic community did not come to play a major part in S&T policy issues.

Implicit in the "policy for science" perspective that was adopted was the view that most problems inherent in scientific development could be tackled once the infrastructure for research and development had been created, personnel trained, and a set of institutions and universities established. This phase of the policy discourse saw unbridled optimism from Nehru and élite scientists.

Furthermore, creating a base in science was seen as crucial for absorbing and eventually replacing foreign technology, as well as for generating new capacities in technological innovation for the industrial development of the country. While the government ethos reflected a "top-down" model of operation, the S&T policy adopted by the political-bureaucratic regime pushed ahead strongly with policies of import substitution and self reliance. The other three S&T policy cultures did not have any major part to play in setting science and development goal direction during this phase (Table 2).

1983	Technology Policy Statement issued
1984	Computer Policy
1985	Textile Policy
	Electronics Policy
	Setting up of Centre for the Development of Telematics
1987	Technology Information Forecasting and Assessment Council (TIFAC) created under the Department of Science and Technology
	Technology Missions launched in water, telecommunications, oil-seeds, etc.
1991	New Industrial Policy Statement issued
	New Industrial Policy for Small Scale Sector
	Liberal policies on MNCs and FDI
	Automatic permission to import technology up to R10 million
1992	National Policy on Education, 1986 (modified)
	New Fertiliser Pricing Policy
1995	New reforms in CSIR and other science agencies
1996	Setting up of the autonomous Technology Development Board to assist firms in the commercialization of technology from the national laboratories
1998	Phokran Nuclear Explosion II; launching of indigenous space

		satellites
		USA making S&T collaboration with several Indian R&D institutions
		India developing new models of super computers
		Five-year tax holiday for commercial R&D companies
		Excise duty waiver for three years on goods produced based on indigenously developed technology and patented in any of the European countries
		Income tax relief on R&D expenditure
	1999	New Patents Policy confirming to WTO
		Exclusive marketing rights for five years to companies as part of WTO
	2000	Information Technology Bill
		Creation of a new Ministry of Information Technology
		Introduction of Protection of Plant Varieties and Farmer's Rights Bill in Parliament

Table 2. Major S&T policy-related developments in India since the 1980s

Though Gandhian values and the sarvodaya model promoting the rural and agricultural sector had considerable influence in the 1940s, the death of Gandhi in 1948 did not have any major influence on developmental policies. The institutions involved in rural development which were inspired by the Gandhian values continued to function, but had no major influence on the political–bureaucratic policy regime.

3.2. 1970s to 1990: From Optimism to Critical Evaluation

In hindsight, the major benefit of the first phase of development was the rapid expansion of the S&T institutional base, including the expansion of higher education in universities, that was achieved by the 1970s. More than two decades of optimism about science and development, and considerable investment in S&T, had both positive and negative consequences. Apart from in the period 1977–80, when the Janata government under Mr.Morarji Desai held power, the Congress Party governed India throughout, largely with Mrs Gandhi and her son Rajiv Gandhi at the helm.

In this period, India began to shift towards a "science for policy" perspective, in contrast to the "policy for science" of the first phase. India entered the world nuclear and space "clubs." The science agencies of the DAE, DRDO, and Indian Space Research Organisation (ISRO) witnessed considerable growth and continued to gain both social and political legitimacy. Despite several shortcomings, India has had relative success in developing some industries, including agriculture, milk production, and chemicals and pharmaceuticals.

India's first government-announced Science and Technology Plan (1974–9) made explicit reference to technology absorption and assimilation, and to the development of indigenous capacity. Developing the local research and development base in science agencies was still regarded as crucial to the generation of technological capacities, as well as for averting technological dependence on foreign sources including transnational companies. In an effort to protect the local R&D base, these policies of self-reliance and import substitution were strengthened during the 1970s and 1980s. Unlike the so-called East Asian "dragons" (South Korea, Singapore, Hong Kong, and Taiwan), Indian policies did not emphasize the export-oriented model during this phase. The continuing emphasis on technology absorption and innovation during the 1980s led to the announcement of the Technology Policy Statement in 1983, which underlined yet further the existing protective measures.

By the early 1980s India, as well as becoming involved in nuclear and space development, experienced some success in the agrarian "Green Revolution" and in the "White Revolution" in milk cooperatives, as well as in the development of industry. Some crucial technologies were transferred successfully into the food, chemicals, pharmaceuticals, agricultural machinery, and other industries from national laboratories. These development gave credibility to the concept of "science for policy."

The dominance of political–bureaucratic culture continued under Mrs Gandhi into the 1980s, with M.G.K. Menon, an élite scientist who was a close associate of Homi Bhabha, to the fore. During the period of Mrs Gandhi's second government, Menon assumed an important position in the political-bureaucratic regime. He was made Member Planning Commission in charge of S&T, and Chairman of the Science Advisory Body to the Cabinet. Later he also went on to become the Science Minister of India.

As the role of S&T came to be questioned more and more from the viewpoint of the basic needs of society, the government begun to steer S&T towards development and socio-economic objectives. By the mid-1980s, the government of Rajiv Gandhi announced a new policy on "Technology Missions." A good example of this new policy is provided by Sam Pitroda, a non-resident Indian and eminent technologist said to have at least 70 patents to his credit, who returned to India during this period to usher in the telecommunications revolution. He was a key person behind the creation of an excellent R&D institution in telecommunications, the Centre for the Development of Telematics, which developed India's first rural digital exchange. Pitroda is also known for catalyzing the concept of Technology Missions in several sectors, although milk was looked after by the father of India's milk revolution, Dr.V. Kurien.

During these times, strategies intended to tackle basic needs through redirection of S&T inputs in water, immunization, oil-seeds, telecommunications, literacy, and other economic domains such as the leather sector, were seen as crucial for employment generation. On the whole, despite some problems, the previous optimism regarding the use of science was not completely forsaken, but a whole new paradigm of S&T policy developed.

Nonetheless, the optimistic assumptions about science and development underlying the perspective of "policy for science" from the first phase began to erode since the early 1970s. The oil crisis in 1973, along with the rise of alternative and appropriate technology movements, raised questions about science and technology for development. The wave of interest in appropriate technology which began in the mid-1970s gained momentum during the period of Janata Party government, but soon lost it in the following decade. For the first time since India's independence, a party other than Congress was in power for about two years between 1977 and 1979. The new Janata administration included several Gandhians who emphasized small-scale technologies and rural-based development programs. While the new administration was not in power for long, Gandhian perspectives on development gained increased currency in this period and gave strength to the "civic culture" movement

A major issue for this movement was that of alternatives to production and development and the choice of technology in the process of development. By contrast, the question of economic returns from science and technology investments came to assume greater significance during the initial phase of this period. As concerns over technology absorption and assimilation came to the fore, the earlier "science push" was of less interest. This phase also saw the emergence of a critical stream of thinking and writings pointing out the failures of S&T efforts over the previous three decades, seen in the increasing disparities between the urban rich and rural poor as a result of S&T-based developments such as the Green Revolution. India, which had been a leading country in the Green Revolution, also came to nurture some of its strongest critics, who insisted on developing the ecological and socio-economic preconditions that were supposed to apply to it. High yields in agricultural crops, it was said, could only be obtained under certain ideal conditions: irrigation, intense use of fertilizers, strict monocultures, the use of rich soils, and other factors were all invoked.

A consensus developed that the "trickle-down" approach behind the optimistic assumptions of the earlier phase of policy-making had failed. The disenchantment over the role of modern science and technology in society was made clear by the emergence of Peoples Science and Alternative Science movements (including environmental and ecology groups); these came to the center stage of politics and decision making in the early 1980s, signaling the increasing importance of civic culture. The Chipko movement led by Sunderlal Bahuguna and Chandi Prasad Bhatt, which was very successful in mobilizing people against the felling of trees in the Garhwal region, is an example of one such movement. Later this movement, led by Bahuguna, was also successful in stopping the Tehri Dam project on ecological grounds. Similarly the Narmada Bachao Andolan (NBA), led by Medha Patkar and Baba Amte and dedicated to combating World Bank-funded dam projects, offers another example of civic involvement in S&T policy issues coming to the fore.

The Bhopal gas tragedy, which claimed more than 5,000 human lives as a result of a poisonous gas leak from a Union Carbide factory in Bhopal, in the Central Indian state of Madhya Pradesh, exposed the possible dangers of not paying attention to an R&D risk assessment, particularly since India had seen massive inward transfer of foreign technology during the preceding three decades. At the intellectual level, both as a result of various events in India and under the influence of a growing western discourse on counter-science movements, a critical stream of thinking emerged. In various forms this underpinned the emergence of popular science and alternative science movements in India from the early 1980s.

The 1980s continued to pose challenges, and this decade saw Indian S&T in a double bind. As noted earlier, on one hand the issue of basic needs posed a challenge and led to a critical evaluation of the policies followed. On the other, the rise and importance of new technologies (including micro-electronics, information technologies, and biotechnologies) and their institutionalization led to problems with absorption, development, and diffusion. As a consequence, the government had to lift various restrictions that had been applied to international technology transfer. In addition, India saw considerable political instability between the mid-1980s and 1990; in five years there were two elections and three changes of government. No new policies were announced. On the whole the major emphasis lay on evolving technology policies, and the utilization of S&T for policy purposes.

On the whole, this phase was saw critical examination, scrutiny, and assessment of official S&T policies by various science movements and interest groups in society. Though the political–bureaucratic regime continued to dominate S&T policy it was certainly weakened by the rise of movements critical of science and of other stakeholders, while civic culture and concerns became more significant.

3.3. After 1991: New Economic Reforms, Liberalization, and Globalization

By comparison with China, which began liberalization immediately after Chairman Mao's death in the 1970s, India made a late start (See *Science and Technology Policy in China*). The new government of P.V. Narasimha Rao, elected in June 1991, catalyzed this transition dramatically. It embarked upon two major programmes— a stabilization programme for tackling the immediate economic crisis the country was facing, and a structural adjustment programme for the medium term. What is of interest here is that the second programme introduced new economic reforms and policies of liberalization—indeed a paradigmatic change—which have long-term implications for S&T.

The main thrust of the new economic reforms was the introduction of the New Industrial Policy (NIP) in 1991. A major departure from the earlier protectionist policies, this was an "inward looking" import-substitution regime. The major elements of the New Industrial Policy of 1991 are set out below. All of them remain valid today, in an even more liberalized form.

- Industrial licensing for setting up new industries and expanding existing ones has been abolished, except for a small number of strategic significant and hazardous industries. Few industries reserved for the small-scale sector were given protection, however.
- The new policy allows and encourages direct foreign investment with few restrictions within a defined list of 34 industry groupings, subject to a limit of 51 percent in any given firm. Foreign firms seeking higher levels of participation were also provided with a window through the Foreign Investment Promotion Board.
- Import of technology for use by the above list of industries would be automatically approved, subject to a limit of Rs10 million for lump sum payments or 8 percent of sales for royalty payments. This makes the earlier protective measures of import substitution and self-reliance in technology more or less redundant.
- No permission would be required to hire foreign technical personnel.

- The government allowed unrestricted imports of capital goods, technical knowledge, and managerial expertise across a range of industrial sectors and infrastructure sectors over and above the Rs10 million ceiling set out earlier, on condition that these sums were managed in hard convertible foreign exchange as a result of the foreign investment.
- The reforms opened up the public sector enterprises (which are owned by the government) in power, telecommunications, transportation, railways, electronics, mineral exploration and production, fossil fuels, and parts of the defense and atomic energy sectors for private participation. For the first time the government placed strong emphasis on export promotion and global competition in its economic policies.

In a significant departure from the earlier phases, the industry–market culture has come to assume a dominant influence in S&T policy-making. As shown in Table 1, the 1990s witnessed a corresponding decline in the influence of a political-bureaucratic culture in the S&T policy regime. The liberal policies were associated with expansion in the numbers of private market leaders. The Confederation of Indian Industry (CII), the Federation of Indian Chambers of Commerce and Industry (FICCI), and other private business groups, including non-resident Indians, have come to play a major part in the policy-making process. Their role has far-reaching implications for national S&T policies.

Unlike in the era of centralized S&T policy regimes, the emphasis in the post-liberalization phase has been on sectoral S&T policy regimes, which are constituted by various individual ministries. Here the private industrial and commercial representatives, through specially created appropriate consultative bodies, have come to assume a greater role. This is clearly evident in areas such as telecommunications, information technology, heavy industrial sectors, railways, and energy, among other sectors.

In contrast with the earlier phases (at least prior to 1989)—when élite scientists close to the political leadership wielded control over national S&T policy making—the late 1990s saw dramatic change. The political–bureaucratic culture of the Nehru era continued but was confined to strategic and defense-related S&T policy issues (atomic energy, defense, and space research). On the other hand, the industry–market culture and its various actors have come to assume a greater role in civilian R&D sectors, where S&T policies are constituted in a sectoral manner at the ministerial and science agency level. In a significant departure from the earlier phases, these sectoral S&T policies pick up signals from the overall economic policies of liberalization and globalization.

CSIR is one representative science agency that instituted several liberal economic policies relevant to S&T in the 1990s. The "new" policy discourse initiated by R.A. Mashelkar, Head of CSIR, through a "Vision Document", spells out several of these new measures. The first impact of reforms forced the agency to generate 30 percent of its budget from beyond CSIR sources. Since 1991 it has aimed to generate 50 percent of its total budget and 100 percent of its operational budget from non-governmental sources through industry partnerships and by selling its technologies. Giving the significant turn in the direction of research output, CSIR is giving top priority to patents and intellectual property rights, above sending research results for publications. CSIR had set a target of 500 patents by 2001, from the earlier level of 50 patents in 1991. In a departure from the earlier era, since the early 1990s scientists have been allowed to obtain royalties from the commercialization of patents developed by them in the agency's laboratories. This new policy is being applied across the board to other science agencies. By creating business development groups both at the national laboratories and the headquarters of CSIR, the agency is set to foster commercialization of technologies, and enter into new partnerships with the national and international multinationals. For the first time, over the last four years, CSIR has been able systematically to set up joint R&D programmes with multinational companies such as Du Pont, Abbot Labs, Parke Davis, SmithKline Beecham, Akzo, and General Electric in the chemical and pharmaceutical areas of research. From \$2 million earnings in 1990, the agency is set to earn about \$50 million by 2001. The impact of these policies is also evident in other sectors like agriculture, however. As one important functionary of the Indian Council of Agriculture Research commented in the early 1990s:

Market force has started giving signals and demanding advanced technology, particularly for the commodities considered important for international trade. In commercial sectors such as horticulture and fishery, the organised segment of producers as well as the commercial sectors (processing, export and marketing) are taking a proactive role. They are demanding knowledge from the research system and as a result, the research agenda is changing.

Two S&T fields that are likely to play important economic roles in coming decades, and which deserve some attention, are biotechnology (BT) and information technology (IT) research. A major difference between the two fields is that the S&T policy perspective of "science for policy" in the second phase catalyzed the growth of a professional community and an institutional base in biotechnology, whereas the emergence of IT is the result of the industry–market-oriented policies of the last five years in this third phase of policy-making.

3.3.1. Biotechnology

In the mid-1980s the government created a separate Department of Biotechnology (DBT) to develop this new field with a separate budget. From a budget of Rs413 million in 1990, DBT was allocated around Rs1 billion in 1999. The major achievement of DBT is its contribution to the creation of postgraduate and doctoral programmes in biotechnology in about 35 universities, and the creation of new chairs and scholarships; above all, over the last fifteen years the Department has either expanded or created six new world class national laboratories of molecular biology and biotechnology research. Currently postgraduate programmes in biotechnology are offered by 51 universities in India. As early as 1991, leading biologist G. Padmanabhan observed that "for a government outfit, DBT has been extraordinarily active, vibrant and forward looking."

Though India's biotechnology industry is only emerging at present its major strength lies in the government and academic sector, where there is a prominent community of specialists. Its main contribution so far has been in the development of world-class infrastructure, research programmes at doctoral level and, above all, development of hepatitis and anti-leprosy vaccines (including tests in malaria vaccines). DNA fingerprinting analysis has contributed to developments in legal and social fields; a variety of genes have been cloned; there has been development work on genetic drugs and antigen-antibody-DNA based diagnostics. As the world market for drugs developed from indigenous or traditional knowledge is worth $46 billion, the development of the S&T base in biotechnology is expected to pay off in the coming decade.

3.3.2. Information Technology

The second field to emerge in recent years has been the information technology industry. The number of educational institutions teaching IT courses has grown tremendously and now includes seven National Institutes and 41 Engineering Colleges. In 1998 there were 338 degree colleges and 806 diploma colleges teaching IT courses. In the private sector there were 2300 institutes offering courses in IT. The growth of software professionals has also been rapid, rising from 6800 persons in 1985 to 200 000 in 1998. The newly created IT ministry is set to catalyze the IT industry, which is dominated by the private firms (Table 3).

Rank	Companies	Total revenue, million Rs. (export + domestic)
1	Tata Consultancy Services	16 522.73
2	Wipro Limited	7 572.40
3	3. NIIT Limited	5 428.50
4	Pentafour Software & Exports Ltd	5 256.80
5	Infosys Technologies Limited	5 088.90
6	Satyam Computer Services Ltd	3 781.23
7	IBM Global Services India Pvt Ltd	3 544.10
8	Cognizant Technology Solutions	2 900.00
9	Tata Infotech Limited	2 498.00
10	DSQ Software Limited	2 233.00
11	Patni Computer Systems Limited	2 199.00
12	HCL Technologies India Pvt Ltd	2 083.20
13	Mahindra-British Telecom Ltd	1 725.50
14	L&T Information Technology Ltd	1 550.40
15	Siemens Information Systems Ltd	1 516.90
16	International Computers India Ltd	1 429.05
17	IMR Global Limited	1 398.40
18	Birlasoft Ltd	1 385.00
19	Citicorp Information Technology Ind. Ltd	1 378.90
20	Mastek Limited	1 344.50

Source: NASSCOM (National Association of Software Service Companies). *India: An Overview of IT Software and Service Industry.* New Delhi, 2000.

Table 3. Twenty top Indian software companies

In a country where about 50 percent of the population is illiterate and an equal number of people live in poverty, the major challenge facing IT policy development is that of harnessing the IT revolution while keeping social justice and equity in the forefront of the agenda. While the export-led IT industry is governed by the industry–market culture and global economic trends, the agenda of government IT policies is likely to be dominated by socio-economic concerns, and to focus on areas (for example, revenue, tax collection, transportation and railways, government administration, land and real estate, and others) where the information revolution can promote efficiency and good governance.

The coming of the new millennium saw the emergence of the IT industry, particularly in cities such as Bangalore, Hyderabad, Delhi, Pune, Calcutta, Ahmedabad, Mumbai, and Chennai (with an urban middle class population of about 70 million). This phenomenon became known as the "new Silicon Valley." Strictly speaking, the IT cluster in Bangalore first came to prominence as India's "Silicon Valley" in the 1990s. The IT software export industry started in the late 1970s in the Santa Cruz Electronics Export Processing Zone (SEEPZ), but the shortage of skilled human resources led the industry to shift to Bangalore in the mid-1980s.

The IT industry has generated tremendous excitement in the country and it has become the first choice of career for students during the last few years. There are around 1800 publicly funded institutions producing 67 785 professionals (in hardware and software, including Ph.D. holders, graduates and undergraduates, and diploma holders); 2300 recognized private training institutes trained 10 000 software professionals in 1998 alone. There are currently more than 946 IT companies in India engaged in the development and export of software. The top 25 companies in 1998–9 accounted for 63 percent of the exports. The major international players in the IT industry, including IBM, Microsoft, Novell, Sun systems, and others, have set up joint collaborative ventures or wholly-owned R&D centers in Indian cities and at major universities such as the Indian Institutes of Technology.

According to a Nasscom-McKinsey study report of 1999, India is set to achieve annual revenues of $87 billion in the IT software and services sector by 2008. From a current $3.9 billion export value for software India is expected to export to the tune of $50 billion, account for 35 percent of total exports, employ 2.2 million professionals, and contribute 7.5 percent of national GDP by 2008. In 1998–9 India's domestic software market was estimated at Rs49.5 billion ($1.25 billion), a figure that does not include in-house development by end users. The domestic software industry has shown an annual compound growth rate of 46.05 percent. It is expected that India's software domestic market would be worth $37 billion in 2008. (Data from NASSCOM (National Association of Software Service Companies), *India: An Overview of IT Software and Service Industry,* New Delhi, 2000; other information collected from Indian newspaper *The Economic Times*, January–May 2000.)

Although the private sector has assumed a dominant role in the recent years, the government played a catalytic role since the mid-1980s. The government had established the concept of software technology parks (STPs) since 1986, and the Department of Electronics developed export processing zones in software since early 1990s.

Currently there are seventeen STPs in different locations. STPs not only provide the infrastructure necessary for exports but also assistance with project approvals, imports, bonding and export certification, and other matters. For new entrants it also provides access to start-up infrastructure and high-speed international gateways. Acknowledging the importance of the IT industry to the coming decades, the government created a Ministry of Information Technology in early 2000.

4. Changing Trends in Science as Social Institution

In the later 1990s the industry-market culture of developing science policy through the processes of liberalization and globalization influenced the work of the research institutions, and is set to transform the social significance of science in a fundamental way. Unlike in earlier phases of S&T development, it has brought about the development of a new economic framework which is set to transform science's social role and the research system as a whole. The systematic advancement of knowledge, the importance attached to full publication of results, the high premium placed on professional rewards, and the make-up of reference panels from the discipline-based scientific élite all remain significant, but are undergoing change. In other words, the traditional *Mertonian* "ethos of science" is becoming less important in the context of the emerging situation. As Professor John Ziman recently spelled out, "Academic science is undergoing a cultural revolution. It is giving way to "post-academic" science, which may be so different sociologically and philosophically that it will produce a different type of knowledge." Within the confines of this paper we can only plot the most important contours of this changing structure.

4.1. Wealth from Knowledge

There has been an important ideological shift in the orientation of scientific communities, from one devoted to "advancing knowledge" to increased emphasis on the "creation of wealth." There has also been a corresponding shift of emphasis from "basic research" towards "technological innovation"; in the 1990s much of this inspiration spread across the developing world as a result of the experiences of East Asia. East Asian economic success stories, particularly post-war Japan and the Asian "dragons" of the 1980s and 1990s, were underpinned by learning in S&T, and by creating wealth out of knowledge. As the vision statement of CSIR underlines: "CSIR thus has no alternative but to view R&D as a business, a transformation process which adds value to inputs, efficiently and profitably. Each laboratory would thus be encouraged to evolve a five year business plan, and not just R&D plans, with external professional inputs and assistance."

This is not intended to suggest that countries like India have given up doing basic research, or that basic research is unimportant. However, when it comes to choice of research problems one can see a shift in the goal direction of research, which is becoming increasingly investment-oriented. Since the ideal of advancing knowledge is being enveloped, slowly but steadily, by the pragmatic value attached to creating wealth, there is increasing pressure to withhold critical elements of knowledge from open publications. In the current phase of globalization, international regimes such as Intellectual Property Rights under GATT and later WTO, or the biodiversity and environmental treaties, which are strongly supported by the multinational companies, are catalyzing this shift in the orientation of scientific communities. (See *Technology and the Environmental Market: Is Sustainability Bound to the Old World Order?*)

4.2. Withering Boundaries and Hybrid Communities

The ongoing penetration of commercial and industrial interests into academic research environments has not only resulted in the loss of autonomy, but has catalyzed the dismantling of conventional "cultural boundaries" between different work settings. This development has enabled different work cultures, styles of research, behaviors, and goal orientations to coexist, not necessarily in single physical environment but on single research programmes dispersed across a wide range of networked organizations. Whilst the established scientific disciplines, specialties, and research areas continue to provide intellectual support and a sense of community, the "dual commitment" of practitioners to the goals of multidisciplinary research programmes is a clearly visible new development. These multidisciplinary programmes had led to what may be called hybrid communities—these share a commitment to common goals which often relate to results that are significant in different market, trade, technological, and scientific spheres.

4.3. Incorporating Interests, Accountability, and a Reward Structure

The loss of research autonomy is related to the new importance of external interests to scientific communities' research programmes, coming from government, industry, political parties, and various social groups and movements. Globalization has catalyzed the influence of industrial and trade-related interest groups in the S&T decision making process at the top level of government, which was previously dominated by élite scientists and technocrats. However, the new agencies have also come to the center stage of politics and society in the last decade. Their research agendas are influenced by social interest groups, and by social movements in ecology and environment. The increasing prominence of these interests is one factor at the heart of transforming the social institution of science in India.

One important consequence of these shifts is that scientists are now subject to scrutiny by—and are accountable to—a number of interests groups in addition to their scientific peers. With the increasing importance of intellectual property rights, there is a trend towards patenting, designing, software development, and similar activities, in contrast to the earlier emphasis on open publication. Full time research and academic institutions are beginning to open up and revise existing reward structures in line with changes in social and market interests and circumstances.

Liberal economic policies under globalization, which led to the expansion of the private sector in education and high and new technology, have fostered discrete "cultures" of "private science" and "public science," with distinct work cultures, salary levels, incentives, and other staff benefits, leading to institutional and management problems for research institutions. As R.A. Mashelkar, the present head of the CSIR in India (which maintains over 40 national laboratories and where over 10 000 scientific and technical personnel work), commented recently:

The emergence of R&D centres in India by multinationals needs to be viewed in a perspective also. With super attractive remunerations, world class facilities and cutting edge and challenging problems many top brains from the S&T community will be attracted by these centres. It is absolutely essential that the Indian R&D institutes urgently create an intellectually stimulating environment, where young minds could be challenged and performers could be awarded handsomely.

Lala Karam Chand Thapar Centenary Memorial Lecture, March 4 1995, New Delhi

4.4. Management of R&D and Entrepreneurial Activity

Science agencies and universities in India are undergoing transformation in order to incorporate an entrepreneurial mode of work and culture. For instance, as the vision document of CSIR observes, in the new environment, "the entrepreneur in a scientist would be awakened, equipped and motivated to venture out in knowledge marketplaces." Different manifestations of this new trend are discernible in the promotion of industrial and private consultancy, corporate industrial investment, aligning universities closer to industrial needs and demands, and the creation of new firms by professionals from both university and non-academic research laboratories. We can now speak of a new "community" of "professional entrepreneurs" in India who have started companies in new and high technology fields such as software, computing and information technologies, telecommunications, biotechnologies, and modern horticulture and floriculture.

5. Conclusion

In the latter decades of the twentieth century, India built up a relatively substantial S&T infrastructure compared with that of other developing countries. Some quantitative studies indicate that India accounts for one-half of Third World scientific output, measured in terms of production of research papers. The growth of S&T institutions— including national journals, academies and scientific societies, R&D centers and higher educational institutions, and (most importantly) appropriate political and social regard for science—has fostered the growth of a large and professional scientific community, comparable to those of many developed countries.

India, unlike several other developing countries, no longer suffers the problems of "isolation of scientists," and dependence on foreign money for carrying out research in certain areas of national and local importance. However, while India is not a "metropolitan" scientific power the status of Indian science is by no means "peripheral." The achievements of Indian scientists in nuclear, space, agricultural, and to a lesser extent industrial, research defy any such suggestions in the scientometric literature. Despite criticisms of the Green Revolution, India was able to meet growing demands of food grains and avert dependence on imports. Although several critics called this achievement a "grain revolution," this was indeed a major milestone of the 1970s and 1980s. Some experts argue that a second Green Revolution is needed. Professor M.S. Swaminathan—the head of a large research foundation in Chennai, and one of the leading agricultural scientists responsible for the 1970s Green Revolution—has repeatedly called for this second Green Revolution to have a basis in sustainability.

Swaminathan has advocated the use of new information technologies at a village level, and his team of researchers are involved in innovations at the rural farm level, as suggested by some experts. This is not to say that all is well with Indian science.

S&T development efforts to date have enabled India to attain self-reliance in production for local consumption in several important areas, such as fertilizers, steel, manufacturing industries, mining, and the railway industry. India's five decades of effort in higher education has given it benefits in "human capital" for areas such as information technology industries. The same cannot be said of export promotion and economic competitiveness in "high" and "advanced" technology based industries, however, with some exceptions such as software. Protective government policies coupled with the long gestation of import-substitution policies did not help technological innovation in several industries, particularly in "new technologies" to aid technological competitiveness of Indian R&D. Not long ago, as Nathan Rosenberg observed while reviewing various models of industrialization:

India represents what appears to be a case of low payoffs from a relatively well-developed and extensive scientific and technological infrastructure. Specifically, it is widely accepted that by comparison with her agriculture research, which enabled India to approach self-sufficiency in food grain production in the late 1970s and early 1980s, industrial research in India has been distinctively disappointing. I believe that this has a lot to do with the extremely tenuous links between the various public and private institutions that are involved in the process.

… research activities (of CSIR) were at times effectively insulated from information about the needs of the public and private sector firms that would be the ultimate users of their output … Studies showed that most projects tended to be initiated by scientists themselves and that users of technologies generated by CSIR labs tended to be confined to firms situated in close geographical proximity. A related problem was that work on these technologies was terminated at the prototype stage.

Rosenberg's observations (1990) are still valid in large measure, and are confirmed by several other studies. One study by the National Institute for Science, Technology and Development Studies (NISTADS) surveyed 2744 scientists in twelve leading Indian S&T institutions and found only twenty collaborations and joint projects with industry and only twelve patent applications. Out of these 2744 scientists, only 54 (or 1.9 percent) visited industry for research or consultation in the year 1988.

In the present era of liberalization and globalization there are new challenges, and at the same time "serious concerns" about undermining indigenous R&D capability. In an interview in the *Times of India,* India's leading national daily, Dr. Paul Ratnaswamy, director of the National Chemical Laboratories in Pune, Maharasatra and an eminent CSIR scientist, recently commented (November 14 1999):

R&D endeavours are indeed falling short of conducive motivations as the management controls can be seen passing into foreign hands. But multinational companies cannot be expected to possess national fidelity. Hence indigenous technological development is going to be a less and less motivating force than technological development per se.

It should be noted that Dr. Ratnaswamy spoke from a position of knowledge: he holds 148 patents and is responsible for an equal number of scientific publications, and is one of the leading scientists of CSIR.

Compared with China, India has attracted less foreign direct investment, yet in the later 1990s it emerged as one of the nations favored by the world's big multinational companies when opening up R&D centers. Many of these centers have developed collaborative links with national S&T institutions. One of the main reasons for this trend is the availability of "cheap" intellectual labor, expanding urban middle class markets which are huge compared to those in other countries, and a reasonably developed R&D institutional set-up.

To what extent India will be able to benefit from this new inflow of foreign R&D tie-ups and centers remains to be seen. Authors observe that the multinational companies are most heavily involved in "adaptive" rather than "creative" and "advanced" technologies. The present era demands the permanent upgrading of industrial companies, and a profound reorientation of the industrial structure would require efforts such as in-house R&D, expanding the information bases of firms, and enhancing quality assurance and support, as well as providing appropriate policy and institutional support and assistance for firms technological learning. Some analysts have noted that Indian S&T structure is being transformed and made more responsive to industrial needs. There is also growing government support for firms, through a venture capital industry and technology development mechanisms, but much of this is confined to the "modern" industrial sector. The small and medium enterprises with connections to 60 percent of India's rural population have yet to receive the attention they deserve from the formal S&T and R&D institutions. In contrast with earlier concern to develop small and medium-scale enterprises, the current phase of globalization demands a new "rural innovation system" that would place the knowledge institutions (that is, all educational and S&T bodies) at the center stage of rural industrialization at a district level.

The most dynamic sector, with considerable linkages between the different actors responsible for innovation, has been that of defense and strategic-related R&D systems, including the space and atomic energy sectors. Elsewhere, when compared to countries and blocs such as Japan, South Korea, and OECD, India has yet to evolve an innovation policy that will bring together actors in science and technology, on one hand, and the market place, on the other. Of crucial importance here is the recognition of the role of "organizational" and "institutional" innovations in the overall innovation policy. R&D, although very important, is only one among several factors responsible for success in innovation. Nodal agencies such as DST, Planning Commission, and advisory bodies in S&T have yet to take a pro-active role in the formulation of an explicit "innovation policy." The last formal Technology Policy Statement was in 1983. The goals pursued in S&T agencies and institutions may still be said, to a great extent, to be still based on "signals" from other key ministries and the political regime in power.

One problem faced by Indian S&T today is declining government support as a proportion of GNP: this fell to 0.66 percent in 1996–7 from 0.89 percent a decade before. Although the R&D budget has been increasing, it has not kept pace with GNP. The government still invests over 81 percent of total R&D expenditure and has not been very effective in inducing private enterprises to invest in it. The private sector accounted for around 19 percent of total R&D expenditure in 1996: in South Korea, however, it provides nearly 75 percent of total R&D funding. This is a problem for the Indian S&T policy regime, and reflects its underdeveloped mechanisms (Table 4).

	R&D expenditures (millions Rs)			Number of published articles
	1985/86	1994/95	1996/1997	1997
Council of Scientific and Industrial Researc	1 622	3 564	4 440	1 535
Defence Res. & Dev. Org.	4 519	12 460	14 620	143
Dept. of Atomic Energy	1 427	4 451	5 298	1 363
Dept. of Biotechnology	-	660	742	57
Dept. of Electronics	47	331	424	14
Min. of Non-Conventional Energy Sources	209	91	45	-
Dept. Of Ocean Development	104	462	502	-
Dept. of S&T	900	2 038	2 495	703
Dept. of Space	2 125	7 590	10 624	200
Indian Council of Agricultural Research	1 422	4 349	4 618	215
Indian Council of Medical Research	410	482	535	157
Ministry of the Environment and Forests	555	2 398	535	43
Universities	NA	NA	NA	7 105

Source: Department of Science and Technology, R&D Statistics, 1999; NISTADS, 1999. Total number of articles published in 1997 was 14 355.

Table 4. Major scientific agencies and R&D expenditure

"Science as public good" versus "science as market good"—these two conceptualizations of science are based on two different logics. One emphasizes disclosure, open knowledge, and free circulation of information; the other intellectual property rights and knowledge as "commodity" and private property, and thus retention of information leading to concealment of research results. The concept of science as market good has been encouraged by the ongoing process of globalization. This new view of science has challenged the prevailing mode of science as public good. This is a problem in developing countries like India where R&D funds come mainly from public sources. As the operating mechanisms of the market-driven commercial interests of "private science" are increasingly applied to regulate research in publicly funded science agencies, there are signs of research expenditure cutbacks or stagnating public R&D funding in welfare, education, health, risk-related and other small-scale sector research which enjoys considerable status under the ideal of "science as public good."

India is increasingly entering a "double bind" situation in seeking to respond to market forces, while at the same time sustaining research activities directed toward the public good. Wholesale endorsement of market-oriented science is likely to have dangerous consequences. If the East Asian experience is of any relevance, then the message is very clear. The state must shoulder a major responsibility and intervene to hold the balance between two sets of policies, sustaining science as public good until society is able to absorb the shocks generated by market forces.

Natural resource endowments and cheap labor are unlikely to give comparative advantages to developing countries like India in the way they have done in the past. There is little doubt that value creation through technological change and knowledge will play a central role in the creation of wealth and in the developmental processes in the future. In other words, Indian society is transforming from an agricultural and industry-based economy to a knowledge-based one. There is a policy discourse around taking India into the new millennium through creating a knowledge society, mainly with the assistance of the IT industry. However, in a country where 50 percent of the population is still illiterate and an equal number of people still lack proper sanitation and access to safe drinking water—and when the Indian average of child mortality is around 72 percent—the challenges posed by education and literacy hardly need underlining. The situation calls for doubling of national education expenditure as percentage of GNP in the coming years. Most importantly, human and social development indicators must be a central policy concern in any discourse on creating a knowledge society.

Acknowledgments

This paper was first presented at the Development Policy Research Unit, University of Cape Town, South Africa, May 12 1998. The section on changing trends in science as social institution is based on Krishna et al. (1998). The author is grateful to Maison des Sciences de l'Homme and Institut de Recherche pour le Développement, Paris, for their invitation as Visiting Professor in June–July 2000, when an earlier version of the paper was completed. I wish to thank Professor J.J. Salomon, Dr. Roland Waast, Dr. Ashok Jain, Professor Aant Elzinga, and several other colleagues for fruitful discussions over the years.

Bibliography

Abrol D. (1999). Science & technology policy. *The Indian Economy 1998–99: An Alternative Survey*, pp. 183–189. New Delhi: Delhi Science Forum. [Examines collaborations between industry and S&T institutions.]

Ahmad A. (1985). Politics of science policy making in India. *Science and Public Policy* **12**(3), 234–240

Babar Z. (1996). *The Science of Empire—Scientific Knowledge, Civilization, and Colonial Rule in India.* New York: State University of New York. [A historical analysis of science in the colonial period in India.]

CSIR (1996). *CSIR 2001: Vision and Strategy.* New Delhi: Council of Scientific and Industrial Research.

Elzinga A. and Jaminson A. (1994). Changing policy agendas in science and technology. *Handbook of Science and Technology Studies* (ed. S. Jasanoff et al..), pp. 572–597. Newbury Park, CA: Sage.

Etzkowitz H. (1983). Entrepreneurial scientists and entrepreneurial universities in American academic science. *Minerva* **21**, 198–233.

Gibbons M., Limoges C., Nowotny H., Schwartzman S., Scott P., and Trow M. (1994). *The New Production of Knowledge: The Dynamics of Science and Research in Contemporary Societies.* London: Sage. [This book explores the notions of "mode 1" and "mode 2" with regard to knowledge production, which have triggered fierce debates in policy-making.]

Glaser B. (1989). *The Green Revolution Revisited: Critiques and Alternatives.* London: Unwin Hyman. [A review of critiques of the Green Revolution, which have been particularly strong in India.]

Jain A. (1989). Signals and regimes for science policy in India: some observations. *Science Policies in International Perspective: The Experience of India and Netherlands* (ed. P.J. Lavakare and J.G. Waardenberg). UK: Pinter.

Krishna V.V. (1991). Colonial "model" and the emergence of national science in India, 1876–1920. *Sciences and Empires* (ed. P. Petitjean et al.). Netherlands: Kluwer Academic. [A historical review of the emergence of the scientific community in India.]

Krishna V.V. (1997). Science, technology and counter hegemony: some reflections on the contemporary science movements in India. *Science and Technology in a Developing World, Sociology of the Sciences Year Book 1995* (ed. T. Shinn, J. Spaapen, and V.V. Krishna), pp. 375–411. Netherlands: Kluwer Academic. [The impacts of science movements in India and their effects upon science policy-making.]

Kumar D. (1995). *Science and The Raj: 1857–1905.* New Delhi: Oxford University Press. [A classic on Indian science in the colonial period.]

Kumar N. (1998). Technology generation and technology transfers in the world economy: recent trends and implications for developing countries. *Science, Technology & Society* **3**(2), 265–306. [An analysis of R&D activities in recent years, and particularly of the role of multinational companies.]

Osborne M. and Kumar D. (1999). Special issue: social history of science. *Science, Technology & Society* **4**(2). [Focuses on many aspects pertaining to the social history of science in developing countries.]

Padmanabhan G. (1991). Government funding and support for the DBT. *Current Science*, **60**(9–10), 510–513.

Pal B.P. (1977). Science and agriculture. *Science and Technology* (ed. B.R. Nanda), pp. 43–54. New Delhi: Vikas.

Parthasarathi A. (1974). Appearance and reality in two decades of science policy. *Science Policy Studies* (ed. A. Rahman and K.D. Sharma). Bombay: Somaiya.

Rosenberg N. (1990). Science and technology policy for the Asian NICs: lessons from economic history. *Science and Technology Lessons for Development Policy* (ed. R.E. Evenson and G. Ranis), pp. 149–150. London: Intermediate Technology Publications. [Examines the main lessons to be drawn from the development of the East Asian countries.]

Ruivo B. (1994). Phases and paradigms of science policy? *Science and Public Policy* **21**(3), 157–164. [An attempt at identifying the differing paradigms that govern the making of a science policy.]

Salomon J.J. (1977). Science policy studies and the development of science policy. *Science, Technology and Society: A Cross-Disciplinary Perspective* (ed. I. Spiegel-Rosing and D. de Solla Price). London and Beverly Hills, CA: Sage.

Sethi H. (1988). The great technology run. *Economic and Political Weekly* (May 14), 999–1002. [A critical appraisal of technology missions in the 1980s.]

Sitaramayya P. (1969). *History of the Indian National Congress*, Vol. 2. Delhi: S. Chand and Co.

Ziman J. (1996). Is science losing its objectivity? *Nature* **382**, 751–754.

Biographical Sketch

Dr. V.V. Krishna is Associate Professor in Science Policy at the Centre for Studies in Science Policy, School of Social Sciences, Jawaharlal Nehru University, New Delhi. He initiated the programme on sociology of science and history of science groups at the National Institute for Science, Technology and

Development Studies (NISTADS), New Delhi, in the 1980s. After serving for over twenty years in this institute, he was invited by the Jawaharlal Nehru University in 1997 to rejuvenate the Science Policy Centre. This is the first center in South Asia to offer M.Phil./Ph.D. programmes in Science Policy Studies. Its main areas of teaching and research are sociology, politics, economics, history of science and technology, S&T policy analysis, technical change, and innovation studies.

Dr. Krishna holds a Ph.D. in Sociology of Science from the University of Wollongong, New South Wales, Australia. He has over 24 years' research experience in the sociology of science, science and technology policy studies, and social history of science. He has published 30 research papers and four books including *Scientific Communities in the Developing Countries* (1997, New Delhi: Sage) and *Science and Technology in a Developing World* (1997: The Netherlands: Kluwer). He is the founder-editor of *Science, Technology & Society—An International Journal Devoted to the Developing World*, published through Sage Publications. Dr. Krishna is a consultant to UNESCO, Paris, for its programmes on electronic publishing in the developing countries and its World Science Reports 1998 and 2000. He is a Council Member of the Society for Social Studies of Science (4S), USA and a member of the International Council for Science Policy Studies, ICSU, UNESCO, Paris.

SCIENCE AND TECHNOLOGY POLICY IN CHINA

Shulin Gu

TsingHua University, Beijing, China and UNU/INTECH, Maastricht, The Netherlands

Keywords : China, science and technology policy, innovation policy, economic development and catching-up, market reform, learning, capability building, national innovation systems, learning models in the Asian NIEs, sustainable development, technological infrastructure, policy capacity

Contents

1. Introduction
2. Science and Technology Policy from the 1950s to the 1970s
3. Science and Technology Policy in the 1980s and 1990s: Market Reform and the Transformation of the Innovation System
4. Conclusion
Glossary
Bibliography
Biographical Sketch

Summary

This chapter examines the case of China in science and technology policy. It analyzes the role of science and technology in the economic development of China, in relation to the formulation and implementation of the policy that assisted the establishment and thereafter adjustment of the science and technology system, and served to co-ordinate scientific and technological activities. The time span covers fifty years from the 1950s to the 1990s. This is the time when China achieved record development performance in her modern history. The article emphasizes the importance of institutional settings on the development of science and technology. Incentive structure and co-ordination mode are therefore analyzed as critical factors responsible for the functioning of science and technology, to link it with technological change and hence productivity growth. Learning and capability-building are also given analytical emphasis, as indicators of the dynamic process in which, as an industrial latecomer, China improved her capacity in innovation and in policy-making. The chapter examines the latest developed schools of thought on technological innovation and development economics in the context of China. It provides in-depth, but succinct, explanations for the development of China in the centrally-planned (1950s to 1970s) and market reform (1980s and 1990s) periods and achievements, from the point of view of learning and capability building, and their institutional bearing. The chapter ends by pointing out the new challenges China will have to address in her science, technology and innovation policy for the twenty-first century, if she is to sustain and accelerate her modernization.

1. Introduction

This article analyzes the science and technology policy in China. Before entering into the detail of the subject, however, let us emphasize that the article assumes that the science and technology policy is aimed at social and, mainly, economic development. Although this focus might appear as a partial point of view, since other values such as the health of the scientific community itself might be pertinent, it nevertheless relies on the fact that, in the long run, development performance is the result of the effectiveness of innovation, one of the major goals of the science and technology policy. The capacity to innovate embraces the capacity of the economy to create new products and services, new markets and institutions. In this sense, innovation policy, technology policy, and science and technology policy are used as synonyms in this chapter. The analysis is organized around institutional aspects—the organizational structure, the incentives and the co-ordination mode of the innovation system— in order to explore why innovation performance was what it was in different periods of development in China. The strength and pattern in generation and application of scientific and technological knowledge of an economy are largely supported by its institutional setting. The innovation system is always "country-specific". In the case of China, the specificities of the innovation system are not only the result of historical, political and cultural factors, but also, or perhaps more so, of economic regimes.

The modern history of China is a history of a traditional society struggling to respond to the challenges from the frontier of science and human civilizations. During the hundred years since the Opium War of 1840—an event that divides the history of China between ancient and contemporary— little action was taken in a consistent and effective way in order to answer them. The conservatives were powerful; they beat down the reform initiatives at the end of the nineteenth century, and then the court of the final imperial Qing Dynasty terminated altogether in the 1911 revolution. In the following several decades, China lived in social turmoil and wars; old institutions were discredited and destroyed but new ones had not been tested out. It was only in the middle of the twentieth century, by the establishment of the People's Republic in 1949, that China got onto the track where it was able to impose internal order, accelerate economic development and restore national dignity.

The establishment of the People's Republic was, as Maddison terms it, "the Chinese equivalent to the 1868 Meiji revolution in Japan". Figures 1 and 2 give an overview of how China passed through the years from the nineteenth to the twentieth century in international comparative terms.

In 1820, China had a GDP per capita which accounted for around 40 percent of the United States' GDP per capita. Since then, this figure declined to its lowest point in 1950, which was around 5 percent of the USA standard. Labor productivity must have taken a similar declining curve but the data is not available. China had been falling severely behind in terms of productivity as well as in terms of GDP throughout the nineteenth century and the first half of the twentieth century. From 1950 onwards, the indicators stopped declining, with slow improvement or relative stagnation for the first period between 1950 and the 1970s, and rapid improvement since the end of the 1970s, corresponding to the respective centrally-planned and market reform periods.

Economists call this process "catching-up" in which the gap with the economic forerunner is narrowed. China has been in the track of catching-up since 1950.

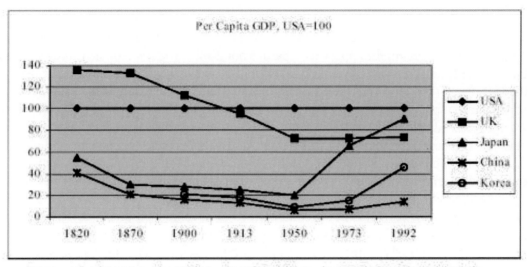

Source: Author reproduced based on Maddison A. 1995: 23-24, Table 1-3.

Figure 1: GDP per Capita: selected countries, 1820-1992

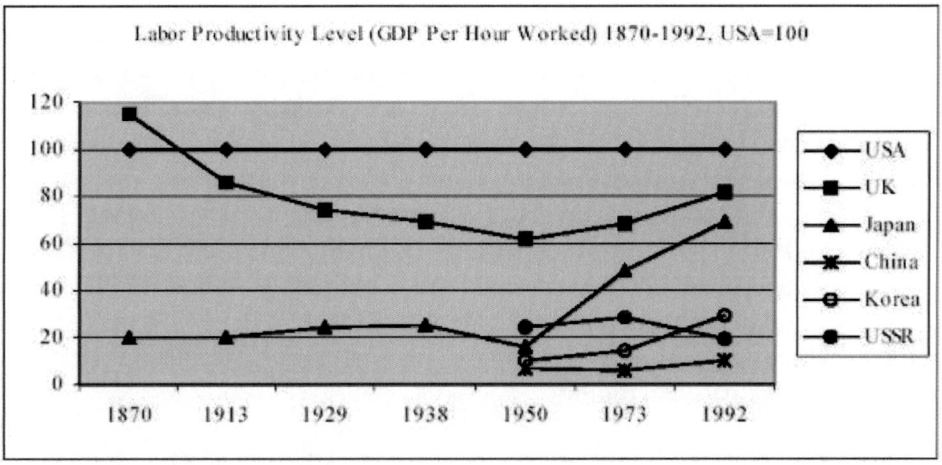

Source: Author reproduced based on Maddison A. 1995: 47 Tables 2-7 (a) and (b)

Figure 2: Labor productivity level: selected countries, 1870-1992

The data concerning education in the same period demonstrate that China had been experiencing a sweeping destruction of social life and a difficult transformation of its educational system over the past centuries. In 1820, Chinese people were not much less educated than American people were, in terms of years of education of ages 15 to 64. Up to 80 percent of Chinese men used to be able to read and write in traditional China. However, in 1950 China was the poorest among those that are listed in the table; about 80 percent of Chinese adults were illiterate. The data on education show the importance of education in the successful catching-up economies in Asia--Japan, South Korea and

Taiwan (China). These countries experienced faster educational developments and a quick upsurge of economic developments. In comparison, although China has invested importantly in education since 1950, progress in education is yet to be facilitated further.

	1820	1870	1913	1950	1973	1992
USA	1.75	3.92	7.86	11.27	14.58	18.04
UK	2.00	4.44	8.82	10.60	11.66	14.09
Germany	-	-	8.37	10.40	11.55	12.17
Brazil	-	-	-	2.05	3.77	6.41
Japan	1.50	1.50	5.36	9.11	12.09	14.87
China	1.50	1.50	1.50	1.20	3.40	7.60
South Korea	-	-	-	3.36	6.82	13.35
Taiwan	-	-	-	3.62	7.35	13.83

Source: Maddison A. 1995: 37, 77 for all countries except China. The author's calculation, based on data in Statistical Yearbook of China 1993: Tables 3-4, 3-5, 18-4, the figures for China in 1973 and 1992; and estimated for 1820, 1870. As a reference, Maddison provides the data for China which are 1.70, 5.33, and 8.93 for 1952, 1978 and 1995 (Madisson A. 1998: 61).

Table 1: Years of education: selected countries, 1820-1992 (Per Person Aged 15-64)

The two periods of post-war development in China are divided according to the political-economic regimes that China has undergone, which laid down an overall environment in which the science and technology policy was formulated and implemented. In the first period, from 1950 to the end of the 1970s, China underwent a centrally-planned regime; the second period, since the end of the 1970s, is a period of market reform.

The two periods differ considerably in development performance, although both periods achieved positive catching-up. In labor productivity growth and per capita GDP increase, the second period shows an accelerating pace compared with the first period. Thus, the average annual growth rate of per capita GDP was 6.0 percent for 1978 to 1995, versus 2.3 percent for 1952 to 1978 (Figure 1). This means that the living standard of Chinese people improved rather rapidly in the second period, whereas living conditions had changed only modestly between 1950 and 1978. This indicator did not improve at all in the previous period between 1890 and 1952. Likewise, labor productivity (Figure 2) increased faster in the second period, but it is still very low compared to the productivity leader. To offer a reference point for comparing the two periods, investment levels were similar, both at around 30 percent of GDP. Hence, there must be reasons other than investment levels which are responsible for the difference. We shall explore it from the perspective of innovation and innovation policy.

2. Science and Technology Policy from the 1950s to the 1970s

There were two lines of development strategy in the post-war time: a strategy based on comparative advantages, which advises developing countries to start with labor-intensive and low capital industries, and a strategy which insists on developing capital goods industry as a priority. China followed the second strategy, also called "self-sufficient", in this period. With the capacity of capital goods technology, it was believed the economy would be able to reproduce itself by the domestic provision of means of production or intermediate inputs for the development of various productive sectors. The implementation of the strategy in China was carried out in a centrally planned political and economic regime.

In retrospect, both of the two strategies were partial. Both were "static" in conceiving the essence of economic development. The first strategy emphasizes the comparative advantages derived from the allocation of factor endowments of a developing country; it pays little attention to the creation of new comparative advantages, and does not give any importance to science, technology and innovation. The second strategy, although aware of the importance of technology that is embodied in capital goods and often accompanied with high investment in science and technology, views the acquisition of the technology as a matter of gaining it once and for all. However, the development process is essentially a highly dynamic one. Learning is the most important feature for successful and sustainable development, where capital investment, technological capability-building and institutional restructuring are crucial.

Apart from the political alliances that China had in the 1950s, the centrally planned approach to science, technology and economic production is related to its historical experience. Nationalization of important industries and the large institutional setting needed for centrally-planned economies has a long tradition in China; it can be traced back as far as two thousand years ago, when important industries such as salt and iron began to be nationalized. China was a great contributor to pre-modern science and technology. Inventions that originated in China, such as the compass, gunpowder and printing techniques, were transmitted directly or indirectly and became important inputs for the rise of modern Europe. In ancient China, science and technology were largely government-organized too, especially astronomy, hydraulic engineering, and agricultural and medical science.

2.1. Investment in Science and Technology and Institutionalization of the Innovation System

The self-sufficient strategy involves high investments in science and technology. The "linear" model that supported this strategy was popular in the 1950s and was accepted in China as well. It assumes that scientific breakthroughs lead to technological development, and the economic value of science is obtained by applying these scientific breakthroughs. In other words, the linear model assumes that R&D has direct economic effects. In his keynote speech to the most important S&T policy conference held in 1956, Mr. Zhou Enlai, then Prime Minister, announced the massive investment plan, and, associated with it, the plan for institutional construction. The Chinese Academy of Sciences was assigned to be the national center for natural and engineering science, and simultaneously industrial R&D institutions were to work for science and technology in specific sectors: "the most excellent scientific manpower and the best university graduates should be organized to work in scientific research in the Chinese Academy of Science". Industrial sectors "should establish and strengthen their research institutions, sharing responsibilities with the Chinese Academy of Sciences, to develop the most advanced technology for the sectors" (extracts from Zhou Enlai's policy statement known under the title "On the Issue of Intellectuals" announced in 1956).

China showed a remarkable capacity of resource mobilization for the establishment of a modern science and technology system, including physical capital and S&T personnel. By the end of the 1950s, the S&T system in China had developed a comprehensive structure, embracing the Chinese Academy of Sciences, the Chinese Academy of Agricultural Science, the Chinese Academy of Medical Sciences, and a large number of institutions for industrial technologies. In the 1960s and 1970s, the system expanded and elaborated with some disturbances, while the whole framework remained well-established in the 1950s.

Year	Percentage of R&D Expenditure Based on National Income	Year	Percentage of R&D Expenditure Based on GDP
1953	0.1	1978	1.5 (1.8 of national income)
1954	0.2	1979	1.5
1955	0.3	1980	1.5
1956	0.6	1981	1.3
1957	0.6	1982	1.3
1958	1.0	1983	1.4
1959	1.6	1984	1.4
1960	2.8	1985	1.2
1961	2.0	1986	1.3
1962	1.5	1987	1.0
1963	1.9	1988	0.8
1964	2.1	1989	0.8
1965	2.0	1990	0.8
1966	1.6	1991	0.8
1967	1.0	1992	0.7
1968	1.0	1993	0.7
1969	1.5	1994	0.7
1970	1.6	1995	0.6
1971	1.8	1996	0.6
1972	1.7	1997	0.6
1973	1.5	1998	0.7
1974	1.5	1999	0.8
1975	1.6		

1976	1.6		
1977	1.6		
1978	1.8 (1.5 of GDP)		

Sources: China Statistical Yearbook on Science and Technology various issues; also at http://168.160.78.249/kjnew/maintitle/MainMod.asp?Mainq=2&Subq=4

Table 2: R&D expenditure in China (1953-1999)

Table 2 outlines the R&D expenditure in China from 1953 to 1999. Because of the limitation of data, the percentage of R&D expenditure for the years before 1978 are provided based on national income, and the data for the years after 1978 are based on GDP. During 1958-1978, China spent about 1.75 percent of the national income (roughly 1.5 percent of GDP) on average for R&D, an expenditure that is higher than Japan and all NIEs in that period. This is a record which was never reached by any developing country with a similar level of per capita income.

The most prominent character of the system was its entire hierarchical structure. This was made under a centrally planned regime in parallel to the industry sectors, which were, too, organized as entirely integrated to the planning system. Table 3 shows a picture of the hierarchy. At the central level of the structure, there was the Chinese Academy of Science embracing 122 R&D institutes and 32 thousand scientists and engineers. At the central level were also 622 R&D institutes, belonging to central ministries and commissions and specialized for their respective sectors which accounted for 93 thousand scientists and engineers. The rest of the system included provincial, municipal and countywide levels, and accounted for 3,946 R&D institutes and 106 thousand scientists and engineers.

Affiliation Relation with the Planning Body	Number of Institute	Scientists and Engineers Recruited (thousand)
Central Government * Chinese Academy of Sciences	122	32
* Central Ministries and Commissions	622	93
Local Governments, Total, including * Provincial	3,946	106
* Municipal		
* County level		

Source: SSTC 'White Paper' 1987: 232

Table 3: Distribution of S&T activities in terms of performing institutions and their relation to the government hierarchy

From the systemic point of view, science and technology is a sub-system of the overall socio-economic structure. Institutional "match" between subsystems is a necessary condition in order for the system to be operational. Under the planned regime, highly or totally "specialized" sub-systems in R&D, education and production were established through importation of the Soviet Model; they "matched" to the centrally planned operational rules and "matched" to each other under central coordination. Industrial factories were committed solely to production; universities had the division of labor solely for teaching; and R&D was almost exclusively the commitment to independent R&D institutes. This is a social organizational pattern that is in sharp contrast to those developed in market economies. Within the S&T system, R&D institutes were also specialized. For example, most of the important institutes for industrial technology were precisely assigned with the objective to develop one type of product or process technology. Above these functional units was the planning apparatus, whose role was to co-ordinate supply-demand links. It is interesting to remark that although such an organizational pattern commonly occurred in formerly centrally planned systems, there were some specific features in the Chinese version. The Chinese system was relatively more decentralized; local levels of administration possessed a substantial command of S&T activities. But at the same time, the Chinese system might have been more homogeneous among hierarchical layers with regard to its functioning routines, a feature that however cannot be demonstrated with statistical data.

Another characteristic of the system was that it strongly focused on technology. It is worthwhile to note that, based upon the same linear model, Latin America developed elite scientific societies and their R&D systems were biased to basic science, while China developed a technology-aimed R&D system, and scientists and engineers in China were regarded more like ordinary workers. In Figure 3, R&D activities under the heading of "industry" in which the majority of the central and local industrial technology R&D institutes were engaged, were aiming at industrial technological development; R&D activities under the term of "comprehensive S&E" which were mainly the job of a large part of the Chinese Academy of Sciences, were in fact for industrial technology as well. Besides, R&D for agriculture, medicine, environment and public utility were also rather technology-oriented. Very likely, the traditional concern of Chinese culture about secular utilization of science as Needham observed, and the co-ordination capacity of central planning may have contributed to a scientific endeavor oriented toward application.

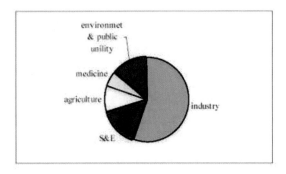

Note: The figure is produced based upon the data for 1990. To illustrate the overall structure developed in the 1950s to 1980s, the data is still meaningful because the traditional portion of the system shows the legacy of the previous decades (Gu S. 1999a: 64).

Figure 3: Distribution of S&T activity in terms of social and economic objectives

2.2.   Innovation Performance: Why Was It Low?

What were the achievements in the period? They were very mixed. The bright side was in areas such as resource survey and exploration, hydraulic and railway engineering, agricultural and medical science and technology, as well as weapon armament technology. These are areas in which centralized organization can work reasonably well. However, disappointing records were obtained in important modern industries, prominently exemplified in the case of the machinery industry. A pattern of technological change was imposed in the machinery industry, which can be labeled as one favoring "general technology". This means that it was capable of producing incremental changes for general-purposed machinery at a slow pace. However, it was insensitive to the development of "specific" technologies, suitable for particular machining functions or particular operational conditions. Consequently, this industry ranked low in international competitiveness. But the impact of such technological change was even deeper, since, as we know from the work of Rosenberg, for example, machinery industry is one of the industries that serve as "innovation centres" for an economy as a whole. The low efficiency of the whole of China's economy in the period can be to a large extent attributed to the fact that it relied on inefficient, machine-embodied techniques.

Three reasons might be put forth as possible explanations: there were low incentives for innovation and learning, low intensity in interactions between technology producers and users, and low international inflows of knowledge and technology. We will focus on the first two aspects which are intimately linked to the institutional setting. The third one, which we leave for Section 3, was to an extent a result of the strategic choice of the self-sufficient approach.

2.2.1. Low Incentives

Low incentives for innovation and learning resulted from the fact that both production factories and R&D institutes were excluded from decision-making, which was assumed to belong to the planning body. The objectives of production factories were set basically in quantitative outputs, so that factory managers did not care much about improvement in terms of quality, variety, or cost of their products and services. This was reinforced by the context of "seller's market"; namely, goods and services were often in short supply, consumers did not have much choice but to accept whatever was provided. According to the author's interviews, knowledge in many cases did accumulate on the shop floor of factories as to how to improve the technology in use. However, institutional restrictions dulled their attention. Procedures for making change, if a factory manager or engineer might want to do so, were slow and costly, since every initiative had to be reported to, and be approved by, the planning system.

2.2.2. Blocked Information Flows between Producers and Users of Technology

We can take the machinery industry as an example. Machinery technology relies greatly on particular machining requirements raised by user industries, and this technology has been largely a user-needs tailored technology. Feedback from users makes up critical input for continuous improvement of the technology. Furthermore, as highlighted by the technology life cycle theory, specific machinery developments come through product diversification and process specialization in the more mature stages of the technology. Specific machines bring about productivity gains to both the machine producers and the machine-using sectors. Since the 1980s this division between "general" and "specific" technology has become vague because CNC machine tools emerged by merging electronic technology with mechanical technology, which provided new generations of flexible and integrated equipment. China lagged behind in CNC-integrated machines for similar reasons. The Chinese industry experienced little development of specific machines although it possessed the necessary capacity in testing, design, and product diversification as early as the 1950s. Among developing countries only a very few have acquired such capacity—South Korea, Brazil and India -- and it appears China was less capable of improving this technological capacity when compared with her counterparts.

The "general-purpose" trajectory of machinery technology is an intrinsic feature of centrally planned economies. Similar observations were reported with the former Soviet Union as well. This is related to the fact that all the units that were associated with the technology, such as the factories and R&D institutes, were operating under a rule where only "vertical" information flows were institutionally legitimated to support the top-down decision-making norms. Autonomous horizontal information flows between them were largely banned. With limited capacity of management of the planning body and lack of specific information on particular users or functions, the technology could only develop following general categories. Furthermore, the firms were largely organized as vertically integrated entities—i.e., they produced themselves all necessary parts and they possessed internally in themselves all kinds of process means, which impeded the development of specific technologies for processes and parts among producer factories.

It has been shown that intensive user-producer interactions are indispensable for modern industrial technologies (see Learning, competencies and networks for innovation and other chapters in the Topic on *The Innovation System and S&T Policies*). Knowledge, along with these interactions and information flows mediated through market institutions, is created and accumulated faster, specialization is accelerating, and returns to investment tend to increase rather than decrease in the economy. The dynamics for innovation is also motivated by higher incentives and wider opportunities when ordinary people are able to access information and other resources more "democratically". Compared to the market-based institutional settings, rigid and centralized co-ordination by government authority is far from effective.

3. Science and Technology Policy in the 1980s and 1990s: Market Reform and the Transformation of the Innovation System

The market reform, which began in 1978, by the chance of changing political leadership, was a step in moving out of the disappointing experiences of the previous decades. In a strong "neighborhood effect", the success stories of the Asian "Tigers" or NIEs may have played an important part in inspiring the reform. The primary target of the reform was to change China from a *command economy* to a *market-based* one. That was accompanied by "opening the door" of the economy to international exchange in goods, services and technology. Strategic targets of development were revised in a more balanced manner to include not only heavy industries but also agriculture as well as light industries. The improvement of the people's living standard received more attention. It is not an exaggeration to state that the reform was a second revolution in China's long march towards modernization.

With the reform, China moved out from the unrealistic pursuit of growth relying on heavy industry and high-tech. It took comparative advantages as an important reference for choice of simple and labor-intensive technologies, while also active in acquiring relatively sophisticated technologies. China therefore was implicitly imitating the Asian NIEs. However, the institutional basis that China developed was rather different. As technological development depends crucially upon the institutional transformations, and redeployment of accumulated scientific and technological strengths required restructuring of established institutions, we review these aspects of the Chinese experience in comparison with that of the other Asian NIEs.

Science and technology policy in this period was aimed at lifting the obstacles to closer interactions between producers and users of technology. But no one knew how to do that, since there was no preceding experience of this kind. Hence, a "trial and error" process characterizes the development of the policy, which is discussed in Section 3.1. Section 3.2 explores the policy process and the institutional and structural change of China's innovation system. Section 3.3 examines the incentives and resources for learning and innovation, a process of "innovative recombination" achieved in this transitional period.

3.1. From "Technology Market" to Organizational Restructuring

At the beginning of the S&T system reform, the program was simply a straightforward copy of the notion of commodity market. The idea of creating a "technology market" was formally declared in the 1985 "Decision on Science and Technology System Reform". Two other measures complemented the creation of this "technology market": the reduction of government funding and the assignment of decision-making autonomy to the R&D institutes. At the same time, a Patent Law and a Law of Technological Contracts were promulgated and enacted. More than 2,000 industrial technology R&D institutes were the major target of this reform, for which operational funds were to be reduced gradually, and it was supposed that such funds would be totally eliminated by 1990. Although these measures were primarily stipulated for industrial technology R&D institutes, nearly all R&D establishments likewise acquired an autonomous decision-making capacity; they were pushed to some extent to go to the technology market as well. Several other measures were additionally decided: a competitive bidding procedure for government R&D funding, the restoration of peer review, based on academic excellence, for R&D performance evaluation and R&D management, and measures to support and legitimate the mobility of scientific and technological personnel. These measures were important, considering particularly that the political intervention in academic lives in the time of "Cultural Revolution" (1966-1976) continued having serious impact on the academic community.

The scenario of 1985 in which China began S&T system reform was the following: the rural economic reform had been making good progress since the end of the 1970s, and urban economic reforms were initiated just one year earlier. It was apparently necessary to re-formulate the role of the R&D institutes as providers of technology in this new context where the number of rural (township and village) enterprises was growing, FDI-related firms were rapidly being created, and state-owned firms were becoming autonomous in making decisions on technological acquisition.

The technology market-centered measures assumed that producers and users of technology could easily meet in the market place once the transactions of technology were legitimized; there was little insight over the need to modify the existing organizational structures. Very soon after the enactment of these measures, frustrated sellers and buyers of technology began making complaints. Transactions of technology appeared to be difficult. Both sellers and buyers were discovering the tacitness of the knowledge that is not easy to be conveyed in the transactions. They were recognizing as well the uncertainty of technological change that was supposed to accelerate upon technology transfer realized at the market (see *International Technology Transfers* for a similar argument on the complexity of technology transfers). Contracts could not be written accurately and transactions could not be concluded in a once-and-for-all manner. The reform program then responded in 1987 by an additional initiative to encourage mergers of R&D institutes into industrial factories, under the rationale that this would permit the internalization of the uncertainty into united organizational territories. This was, again, only partly functional. Hence, a further step was taken in 1988 with the launching of the Torch Program which aimed at creating, supporting, and regulating "spin-off" enterprises spun from existing R&D institutions and universities. In the meanwhile, the policy evolved more measures in the late 1980s and 1990s in order to transform R&D institutes into technology-intensive and profit-earning enterprises. All the latter policy measures were made towards offering guidance to organizational restructuring.

The approach followed in these years might be labeled as a "positive pragmatist" approach. The reform program could not be managed as a once-and-for-all process. Instead, it was open to the necessity of modification, developed through a sequence. Each step forward was taken by looking at the information and experience gained from the preceding step. A positive pragmatist approach also accepted plural solutions. It was not restricted to a single mode for reforming an R&D institute but it eventually developed the sum of many solutions, including transactions of technology in the market place, organizational merger; spinning-off new enterprises, and transforming an R&D institute to be profit-earning producer of technology and engineering services. All these made the complex transformation of China's system possible, even though some of the solutions might be seen as unorthodox from the perspective of established theories. Perhaps the positive pragmatism has a root in the cultural heritage of East Asia. The NIEs and Japan may have benefited from such an approach. Japan, for example, was observed by Hayami and Ruttan rather unique in the modernization of the agricultural sector in the Meiji era.

3.2. Technology Market and Transaction of Mature and Complementary Technologies

It is interesting to see what happened in the technology market that was created for the transactions of technology. Table 4 offers a view as to how and where the industrial

SCIENCE AND TECHNOLOGY POLICY - Vol. II – *Science and Technology Policy in China* - Shulin Gu

technology R&D institutes moved in response to the pro-technology market policy thrust.

Group of R&D Institutes	Market earnings	Composition of market earnings				
		Technology development	*Technology transfer*	*Technological consultancy & technological services*	*Trial production*	*Other production and sales*
All industrial technology R&D institutes	10.59 billion yuan (78% of overall income)	20%	5%	12%	19%	44%
Centrally affiliated machinery technology R&D institutes	1.502 billion yuan (84% of overall income)	38%	12%	15%	22%	13%

Notes:
1. "All industrial technology R&D institutes" include those that were in 1993 affiliated to central industrial ministries and those that were affiliated to local industrial bureaus at provincial and municipal levels. Total 1804 institutes are in this group. The coverage is not accurate, restricted by data availability.
2. "Key machinery technology R&D institutes" include 64 institutes affiliated to the central Ministry of Machinery Industry.

Sources: *China Statistical Yearbook on Science and Technology*, calculated by the author (Gu S. 1999a: Tables 15.3 and 15.6); The ministry of science and technology of China, at http://168.160.78.249/kjnew (visited on 15 January 2000)

Table 4: Activities of industrial technology R&D institutes on the technology market (1993 current price)

First, as shown in the table, industrial technology R&D institutes had, by the first half of the 1990s, moved substantially to the marketplace. In 1993, market earnings became the major source of their incomes, 78 percent for them as a whole and 84 percent for the centrally affiliated machinery technology R&D institutes. However the institutes relied more on the commodity market, rather than the technology market, for their incomes. Market earnings composition is shown in the right part of the table. The first three categories of market earnings ("technology development", "technology transfer" and "technological consultancy and technological services") represent the degree of *technology market* participation, and the latter two categories ("trial production" and

©Encyclopedia of Life Support Systems (EOLSS)   118

"other production and services") represent the actual involvement in the *commodity market*. Industrial technology R&D institutes as a whole had 63 percent of their market earnings from selling products in the commodity market (from "trial production" and "other production and services"). On the opposite, the group of the machinery technology R&D institutes had main earnings from the technology market (65 percent), since they serve as major providers of various tools and equipments for machine-use sectors.

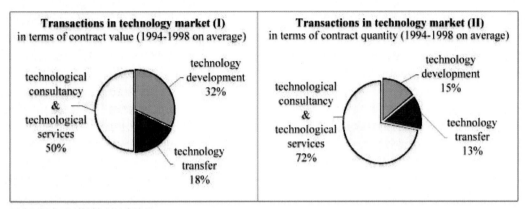

Sources: same as Table 4 above

Figure 4: Transactions of the R&D institutes on the technology market

When looking at the technology market itself, a first message is that the market was growing in spite of the difficulties that were involved in transactions of technology. The sales value at the market increased to 43 billions Yuan in 1998, from 2.3 billions Yuan in 1985, the first year when the market was created. In comparison, the total expenditure for R&D was 55 billions Yuan in 1998. Not only industrial technology R&D institutes, but also universities and large firms went to the technology market, to sell (and buy) technologies. A second message is that mature and complementary technologies dominated the transactions in the technology market. Transactions under the headings of "technological consultancy & technological services" and "technology transfer" are more likely to involve complementary and mature technologies. They represented 85 percent of the deals, a figure far higher than the transactions for "technology development" which are more likely to be related to tacit knowledge and entail a higher degree of uncertainty (See Figure 4).

Altogether, the technology market approach was not as successful as it was hoped to be. But neither was it a failure. It appeared to be one of the many ways of organizing and coordinating innovation activities. It also provided information and incentives to the participants to adjust their internal operations and external relations. The technology market has hence been growing along with the restructuring of its participants of R&D institutes and industrial firms.

3.3. The Transformation of the Innovation System in China

Profound organizational change had been going on throughout these years. Industrial technology R&D institutes had continuously been re-organizing their structure and the management of their internal resources. They had been trying out market niches and expanding or re-creating their users' links. Some of them had developed a consolidated position either as suppliers of technology-intensive goods, or as providers of engineering and information services. Merger with industrial enterprises was an approach initially difficult to implement because of institutional and cultural gaps. For some R&D institutes merger became an acceptable choice later on, either because the gaps narrowed or simply because there was no other alternative left. Mobility of individual professionals and new spin-offs from R&D institutions helped to release scientific and technological expertise to business organizations. At the same time, large firms were developing in-house innovation centers, in order to be able to improve technologies internally, and to absorb and to adapt technologies from external sources. As both firms and R&D institutes were revising their organizational boundaries and adjusting their innovation activities, cross-organizational information flows were intensified. Knowledge activities that involve a high degree of tacitness and uncertainty were more internalized in various organizations. Accordingly, the innovation system in China was fundamentally transformed.

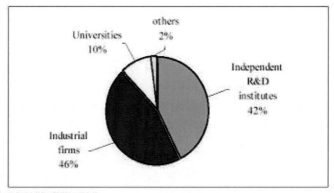

(In terms of expenditure in current price)

Source: The ministry of Science and Technology of China at http://168.160.78.249/kjnew (visited on 15 January 2000)

Figure 5: R&D performers in China's innovation system: 1998

Figure 5 illustrates the innovation system in China as it appeared by the end of 1990s, in terms of R&D performers and their relative positions. In 1998, three major performers of R&D were industrial firms, independent R&D institutes, and universities. Their shares were 46, 42 and 10 percent respectively with the total R&D expenditure being 55 billion Yuan, or 0.67 percent of GDP. Industrial firms became for the first time in history the number one R&D performers, no longer the merely passive receivers of technology as they were some twenty years ago. Universities increased their role in R&D while, before, they dedicated almost exclusively in teaching. Independent R&D institutes became close associates to firms and held no more monopoly over the R&D enterprise.

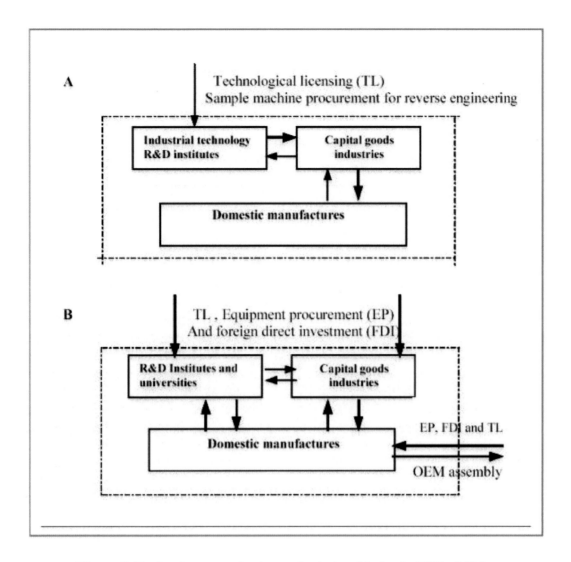

Figure 6: Technology supply-demand relationship in the NIE of China

A slightly different perspective of the system's transformation, the perspective of technology supply-demand relationship, is depicted in Figure 6. Before the market reform, as shown in the upper "A" half, industrial technology R&D institutes were virtually the sole suppliers of technology. Capital goods manufacturers employed outcomes from the institutes as their inputs in production, and they in turn produced inputs in the forms of machines for the production of the final products. Feedback from capital goods production to R&D was weak, as was that from end-users to capital goods producers. The flow of foreign technology was low and mainly went to the R&D institutes for their reverse engineering and diversification. By the end of the 1990s, as shown in the lower "B" half of the figure, the system turned out to have more intensive domestic and international sources of technology. Domestically, independent R&D institutes, large firms and universities all engaged in innovative activities with some division of labor.

Internationally, technology came in through technological licensing (TL), equipment procurement (EP), and foreign direct investment (FDI), among other channels.

International flows of technology became accessible to all of these groups of innovation performers. Figure 7 illustrates international flows of technology, approximated by FDI. In 1998, FDI into China amounted to 45 billion $US, the total utilization of foreign capital being 58 billion $US. In comparison, the sum of FDI from1979 to 1984 was 1.8 billion $US, and before the reform FDI was restricted at minimum level.

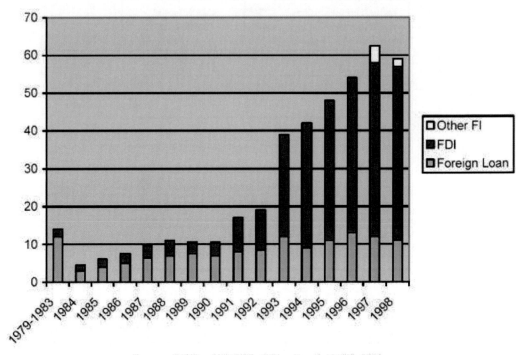

Source: China Statistical Yearbook 1999: 594

Figure 7: Utilization of foreign capital

Up to the year 2000, the fifteen-year approach to the S&T system reform seemed to be coming to a close. It is announced that industrial technology R&D institutes should eventually end their affiliation relationship to the government body as previously assigned. The majority are to be officially legitimated in their transformation to become firms or parts of firms. A few institutes, the most important ones, maintained their position as state-owned. In terms of innovation activity distribution, we expect that industrial firms will show an even larger proportion in statistics since many of the independent R&D are now included in the category of industrial firms.

3.4. "Innovative Recombination" and Learning in Economic Transition in China

How were the resources and incentives geared to learning and better innovation performance? What characterized the learning and acquisition of technology in this transitional time? In addition, what role did the accumulated scientific and technological capacity play?

China enjoyed impressive deployment performance during the 1980s and 1990s, as has been mentioned. In retrospect, good luck helped, such as the re-allocation of production by Asian NIEs to China, which brought in massive FDI. The accumulated science, technology, and manufacturing capabilities were important as sources of endogenous inputs. Economic policymaking was pragmatic and responsive; it adopted a "trial and error" process which responded to opportunities quite effectively.

A two-pronged development strategy came into shape by the second half of the 1980s. It was, on the one hand, aimed at labor-intensive manufacturing, which was focused for the development of coastal areas. It developed, on the other hand, the technology and knowledge-intensive sectors, reflected typically in the launch of the Torch Program in 1988. The essence of the strategy was "to rely on two kinds of resources (domestic and international) to meet the demands from two markets (domestic and international)".

Incentive structure was revised and sharpened. It was based on decentralization of the decision-making authority by delegating power to lower levels of the government system. Local governments, at provincial and lower levels, as well as managers of state-owned enterprises and R&D institutes, acquired a great deal of autonomy, whereas state ownership was not modified immediately. With the delegated autonomy of decision-making, local governments as well as R&D institutes and large enterprises were empowered and activated with higher incentives. Local governments became enthusiastic in the construction of infrastructure and the attraction of FDI. Meanwhile, local governments became involved rather directly in capital investments. This model, labeled as an M-form planning, gained a greater openness to international trade and to new domestic entries into economic activities, although it still had limits in improving investment and innovation efficiencies. Its negative aspects were that it encouraged duplicated investments among provinces and regions, and it allowed corruptive, rent-seeking behaviors. Nevertheless, in the 1980s and 1990s, the system supported the mobilization of capital, labor and skills.

International inflows of technology came in mainly in tradable terms. All the international procurement of producer goods, the technological know-how licenses and the foreign investment brought in technology, which was either embodied in the purchased equipment or incorporated in the codes of technical specifications, patents and investment programs. These technologies were developed in commercially profitable forms. Chinese companies, with this knowledge diffusion from international trade, altered their attitude to international technology, understanding that doing everything domestically is much more expensive. The self-sufficient approach had wasted many resources by "re-inventing the wheel".

International inflows of technology intensified and deepened over the period. In the 1980s, FDI came almost exclusively from Hong Kong, Taiwan, Singapore and, to a lesser extent, Japan. During the first half of the 1990s western multinational companies became increasingly important. Nevertheless, by the end of the 1990s, 60 percent of the FDI stock that China had accumulated was from Asian NIEs, and 90 percent of the total FDI went to coastal provinces.

China appeared to be taking diverse approaches to international sources of technology. For the eastern coastal provinces, FDI was the major means that brought in relatively advanced mature technology used in labor-intensive manufacturing, and that came largely from the Asian NIEs who were attempting to re-allocate labor-intensive manufacturing to lower waged countries. China adopted active policies to attract these investments. Geographical adjacency and Chinese ethnic relations were additional factors in favor of these investment flows into China. On the other hand, technological licensing, capital goods procurement, and some contractual OEM were more important for inland areas. The sources of these technologies came from more advanced industrial economies, and these technologies were used for the renewal of the obsolete technological basis of established industries.

Compared with other Asian NIEs, China was distinguished in acquiring diverse forms of international technology, while other Asian NIEs relied more on single means for the purpose. More importantly, China acquired international technology not only for the demands of the international markets but also for the demands of the national market. This was so partly because of the size of the economy and partly because the consumption of daily necessities and durable household electronics had been depressed in previous years, so that the national market had a great demand potential. In comparison, Asian NIEs had the international markets as their major target. The experience in China may help to keep a balanced view regarding the role of international markets and domestic market in providing incentives and resources for development.

3.5. Successful Mobilization of Endogenous Capabilities and "Recombination Learning"

3.5.1. The case of the TV Sector

The development of the TV sector exemplifies learning in a relatively mature sector. Foreign investment played a direct role in the upgrading of final products of the sector. To attract FDI, "Specific Economic Zones" were created in coastal provinces. These Zones developed consumer electronics including TVs among a number of labor-intensive sectors, based on FDI related technology.

Endogenous capabilities were mobilized to meet the opportunities from foreign investment associated technology. In order to channel engineering and management expertise from inland areas, many inland factories and research institutes, some of which had previously engaged in military electronics, were encouraged to open their "window" offices or branch companies in these Zones. Skilled persons were also hired from around the country. The 1985 Decision stipulated measures to encourage the mobility of scientific and technological persons that helped the migration of well-trained people. It was popular among young university graduates "to go to the development zones of the south" throughout the 1980s and 1990s. The presence of supportive industries played a part as well. Technological acquisition for components and parts, which were simultaneously absorbed in inland factories, did give support to the final TV sets production in coastal areas. R&D work such as that for CAD, remote control technology, digital technology and testing equipment, all in relation to TV, continued. With foreign capital involved, learning was enhanced not only in technological terms, but equally or more importantly in production management and marketing through which managers learned advanced managerial skills for running the industry under competition, skills that were missing in the pre-reform time. Consequently, the competitiveness of China's consumer electronics sector improved considerably.

This sector's situation has dramatically changed in fifteen or twenty years. In the early 1980s domestic products were distrusted and the market was dominated by imported and smuggled TVs By the end of the 1990s, this sector had firmly established its position in the domestic market with its own brands, and exports were increasing.

3.5.2. The Case of the Computer Software Sector.

The computer software sector highlights learning and capability building in an emerging and dynamic sector. It resulted from the S&T system reform, with little FDI involvement in the early stages of development. In the mid-1980s, a new kind of enterprise, the spin-offs from the R&D system, appeared. Initially these were created by individual scientists, engineers and R&D institutes who actively sought to commercialize their laboratory achievements and to gain advantage from the onset of the PC market boom of that time. The Torch Program endowed these initiatives in 1988 to promote this kind of commercial activity.

"Development Zones for New Technology Industries" was a major policy measure devised to serve as institutional and geographical basis for the development. By the early 1990s, 3,000 such companies had been established in the Beijing Zone alone; most of them were relatively small businesses. Although the Torch Program had aimed at promoting a wide range of new technology industries, about two thirds of the new businesses were working on computer-related software development and services in the 1980s and early 1990s. They constituted the most important players of the young industry, evidenced in the fact that, of the "biggest 20" computer software and service companies in China in the mid-1990s, 17 were of this kind. Most companies began with the adaptation of computer language from English to Chinese and simple application services, from which some of the companies grew up to be competitive in PC assembly, PC parts production, Chinese text processing and Chinese publication systems, application software particularly in finance and education, and Internet services. The emergence of the software industry changed the structure of the electronics industry in China, which was, prior to the reform, centered in hardware manufacturing and weak in applications and services. With the enhancement in applications and services, the consumption of personal computers increased to 6 or 7 million units in 2000, and Internet connection increased to about 20 million. Growth of PC sales and Internet connections was exponential. at between 20 to 50 percent a year in the second half of the 1990s.

For the development of the industry, design, testing and other engineering capabilities accumulated from pre-reform time were crucially useful. At the same time, the presence of internationally accessible technology either embodied in parts, components, architecture and interface codes of computer systems, or in the provision of international engineering services, were supportive to the speedy learning of the sector. Hence, Chinese developers were able to embark on the path in which they identified the niches and improved their core technologies for the identified niches compatible to the state-of-the-art computer structures. For example, some of the most successful Chinese developers were thereupon capable of upgrading their core businesses by the use of internationally available design and fabrication services. The opening to international trade enabled the industry to take benefit from the international sources of technology.

3.5.3. "Innovative Recombination" of Capabilities.

Based on the evidence, we have termed the learning and innovation process carried out in the transitional period of China as a process of "innovative recombination". Until now, the learning models of developing countries are mainly based on the experience of the neighboring Asian NIEs. Their process is usually summarized as a sequence in terms of technological sophistication. They went step-by-step from simple assembling of mature technology, to the mastering of process improvement, to minor design diversification, and eventually to brand creation.

In contrast, learning in the course of "innovative recombination" embraces many more factors --technological, institutional, managerial, and political -- at work simultaneously. In the macro-economic environment, market reform and trade liberalization produced new incentives for re-allocation of innovative capabilities and activities. Then capabilities in production, design, R&D and testing accumulated in previous periods were actively used and re-organized in novel and productive ways, so that they met the new challenges and opportunities. Intensive technological learning was devoted to identifying and filling major gaps left from the previous system. Relatively "advanced" capabilities and more simple skills were complemented and re-combined, which was not as linear a sequence from simple to sophisticated technologies as the NIEs experienced. Finally, enormous efforts were directed at institutional restructuring supportive to innovative recombination learning. Of course, not every sector did equally well in the recombination learning. Learning at state-owned enterprises was slower. Difficulties were caused by, among other reasons, institutional inertia that impeded the pace of the firms.

4. Conclusion

At the end of the 1990s, the conditions of success of the 1980s and 1990s were changing or eroding. The second half of the 1990s saw China slow down in its growth rates. The price indexes for consumer goods were declining and domestic investment was reducing even after the strong fiscal and financial policies that the Chinese government took from 1997. Certainly, the Asian financial crisis exacerbated the problem; but deep reasons came from inside the system.

The development approach of China and the underpinning innovation system, although much more open to international technology acquisition and more "market-friendly", had some serious biases and weaknesses. They had favored those who live in urban areas, and rural people had been at a disadvantage although they constitute 70 percent of the population. They had benefited less from the reform and development, and even their migration to urban areas, although massive in effect, remained not legally free. Small initiatives and private entry were still less encouraged or discriminated against. The widening of income differentials between urban and rural residents, and the unequal and limited opportunities depressed demands for both consumption and investment. In addition, environmental degradation reached the point where it was both unacceptable and unsustainable.

All of these call for renewal of the development strategy, which is actually the key issue being debated in China. It is agreed that the essential aspect of the new strategy should be balanced (in urban and rural modernization, between large and SMEs, and among various sectors) and environment-friendly development. It is also agreed that further development in China must be based on knowledge and innovation rather than mainly or solely on physical investment to expand production capacity, something which seems now necessary and feasible; the legacy of the planned economy -- constant shortage of supply -- had become a matter of the past since the mid-1990s.

The revision of the development strategy implies, however, a fundamental departure from both the experiences of China herself and conventional thoughts about development. Conventionally, manufacturing sectors were considered central, and for this purpose, large firms were assumed to be important. Gerschenkron's proposition in his famous book on the development of late-comers' economies, for example, states that the more backward a country's economy, the more pronounced is the need for its industrialization based on big plant and enterprises, and the intervention of the government. Propositions such as this have long been influential. What China is faced with is in a sense making a departure from this convention.

Table 5 summarizes the characteristics of social-economic development in the past and for the future, in terms of economic regime and strategic targets for development, integration with the world economy, sector priority of investment, pattern of growth, income distribution, urban-rural relationship, and environmental effect.

	1950-1978	*1979-1999*	*2000 and onwards*
Economic regime and strategic target of development	Centrally planned; Quantitative leap-frogging of heavy industry	In transition to market regime; Quantitative expansion of both light and heavy industry	To move to maturity of market regime; Quality of growth; People-centered development
Integration with the world economy	At low level (5-6% of GDP); Self-sufficiency achieved by planned import-substitution	At higher level (20%-30% of GDP); Export of labor-intensive manufacturing, import of technology and capital goods	To be at higher level; Deepening in international specialization
Sector priority of investment	Capital goods industry; Military industry; Centralized and high R&D	Consumer goods industry; Infrastructure development; Relatively low R&D	Balanced sector structure and structural deepening; Human capital development; high R&D and intensive learning and innovation

Pattern of growth (indicated by major contribution factors to growth)	Capital investment; Through planning	Labor and capital investment; In response to international and domestic demands	Increasingly largely from human capital and technical progress; In response to domestic and international demands; Active coordination by innovation policy
Income distribution	Absolute equalitarian distribution	Concentrated to those engaged in market activities successfully and those privileged in rent-seeking; Gap-widening	Fairly equal
Urban-rural relationship	Rigid separation between industrial urban and populated agricultural rural	Temporary migration of rural population; Relative income distribution worsening	Convergence through acceleration of agricultural and rural modernization
Environmental effect	High and wasteful consumption of energy	Severe environmental deterioration	Sustainable and environment-friendly development

Sources: The author's summary in reference to discussions at http://forum.cei.gov.cn/Forum50 and http://forum.cei.gov.cn/UnionForum especially by HU Angang and LIN Yifu.

Table 5: Characteristics of social-economic development in different historical periods in China

Science and technology policy in China will have to address at least three critical issues if the innovation system of China is to support a sustainable and innovation-based development. The first one concerns education. To achieve universal second education and expand the enrolments of tertiary education will be very necessary. Training quality is more important than merely quantitative considerations. Basic research will need to be improved in universities because China's universities might have overdone the championing of the commercialization of R&D, and the quality of R&D and training there might have been in stagnation.

The second issue involves the development of technological infrastructure, for it to become effective, extensive and accessible. Technological infrastructure serves knowledge and information and necessarily supports the broad participation of small firms and private initiatives. Services by technological infrastructure are to an extent public goods. The notion of technological infrastructure is new for China, since it was not important or was even meaningless in the centrally-planned system where everything was internalized in the planning co-ordination. However, in a market economy and as required by the rules of the World Trade Organization, the development of technological infrastructure, together with the development of communications infrastructure and educational infrastructure, will be priorities in public policy. Innovation in products and services should be in principle left for the initiatives of firms and individuals.

The third issue concerns policy capacity. In a market economy, heterogeneity and diversity characterize innovation activities. It is indispensable to have a "specific" policy capacity. At the national level, innovation policy needs to be specialized and handle many aspects such as financing, trading and taxation in relation to innovation. More importantly, specific policies are to be developed to promote collective efforts for learning and innovation at less aggregated levels. Thus specific innovation policies need to be developed for regions, for sectors (such as agriculture-based and resource-based sectors), and for clusters (that are constituted by a number of agents tied up in a value chain, a missing perspective in the past). A single and homogenous set of national innovation policies, as was the case in previous times, would not work effectively today. This makes up a challenge to the capacity of the policy system itself in China.

Glossary

CAD:	computer-added design
CNC machine tools:	computer numerically controlled machine tools
EP:	equipment procurement
FDI:	foreign direct investment
GDP:	gross domestic production
NIEs:	newly industrializing economies; in this paper the term embraces South Korea, Taiwan (China), Hong Kong, and Singapore, which are also called the "first tier" of Asian NIEs, having emerged earliest as successful Asian developing economies in the post-war period.
OEM:	original equipment manufacture
PC:	personal computer
R&D:	research and development
SMEs:	small and medium enterprises
S&T:	science and technology
TL:	technological licensing
TV	television sets
WTO:	the World Trade Organization

Bibliography

Chudnovsky D. and Nagao M. with the collaboration of Jacobsson S. (1983). *Capital Goods Production in the Third World*, London: Frances Pinter. [This book provides a comparison of machinery industries in a few developing countries including China, Brazil, and South Korea among others which had developed the industry relatively substantially by the end of the 1970s.]

CPC and State Council of China (1999). *Decision on Encouraging Technological Innovation, Developing High-tech, and Realizing Commercialization of New Technologies*. Beijing: CPC and State Council of China. [A policy statement of the Chinese government for innovation-based development in the future.]

Fei J. C. and Ranis G. (1997). *Growth and Development from an Evolutionary Perspective*, Malden MA USA and Oxford UK: Blackwell Publishers, Malden USA and Oxford UK. [This book widens the evolutionary view of economic development to combine technological innovation and institutional change to development economics.]

Gerschenkron A. (1962). *A Book of Essays*, Cambridge, Massachusetts, and London, England: The Belknap Press of Harvard University Press. [Gerschenkron's book deals with development approaches of backward countries. His contribution is based on the study of Russia and other backward European countries in the late nineteenth century and the early twentieth century.]

Granick D. (1967). *Soviet Metal-Fabricating and Economic Development, Practice versus Policy*, Madison, Milwaukee, and London: The University of Wisconsin Press. [This book gives a detailed analysis of the machinery industry in Soviet Union in terms of planning process, firm's structure, and the ways in which technological change was organized.]

Gu S. (1996). The Emergence of New Technology Enterprises in China: A Study of Endogenous Capability Building via Restructuring, *The Journal of Development Studies*, Vol. 32, No. 4, 475-505, London and Portland USA: Frank Cass Journals. [This paper analyzes the emergence of the new technology enterprises, which gave birth to the software industry in China resulting from S&T reform policy and particularly the Torch Program.]

Gu S. (1999a). *China's Industrial Technology: Market Reform and Organizational Change*, London and Now York: Routledge in association with UNU Press. [This book contains a detailed survey on science and technology policy in both centrally planned and market reform periods of China. It analyzes the innovation system and its transformation in the machinery and electronics industries, and links it with development performance in China. It develops a learning model as "innovative recombination" as manifested during the market reform.]

Hanson P. and Pavitt K. (1987). *The Comparative Economics of Research Development and Innovation in East and West: a Survey*, Harwood Academic Publishers. [This book makes a major contribution in the comparison between East and West--centrally planned and market economies, in terms of institutional structure and innovation performance.]

Hayami Y. (1997). *Development Economics, from the Poverty to the Wealth of Nations*, Oxford: Oxford University Press. [A well-written book containing latest insights of induced innovation and thoughts about economic development, which consider development incorporated with learning to innovate.]

Hayami Y. and Ruttan V. W. (1985). *Agricultural Development, An International Perspective*, Baltimore and London: The John Hopkins University Press. [This book analyzes agricultural developments in United States and Japan based on the author's substantial studies, and advises for developing countries. It develops and uses an induced innovation model to explain agricultural development.]

Hobday M. (1995). *Innovation in East Asia, The Challenge to Japan*, Aldershot, England and Brookfield, USA: Edward Elgar. [The book explores technological development in the Asian NIEs—South Korea, Taiwan, Singapore and Hong Kong in their export-oriented developments. A learning model is thereupon summarized as from OEM to ODM (own-design and manufacture) to OBM (own-brand manufacture).]

Jian X (2000). FDI Inflow and Technology Progress in a Transition Economy: Chinese Experience, a paper presented in the Ministry of Science and Technology of China and United Nations University (organized): *the International High-Level Seminar on Technological Innovation*, Beijing, China. [It provides statistical data and analysis of the effect of FDI on innovation in China.]

Kim L. (1997). *Imitation to Innovation: The Dynamics of Korea's Technological Learning*, Boston, Massachusetts: Harvard Business School Press. [One of the best studies on learning process carried out in South Korean firms. It summarizes the learning model as moving in a reverse order to the life cycle of a technology, namely from simple assembling to mastering minor processes and product improvements, up to the creation of major innovation, from "imitation to innovation".]

Kornai J. (1980). *Economics of Shortage,* Amsterdam, New York, Oxford: North-Holland Publishing Company. [This classic work describes the operation of centrally-planned economies.]

Maddison A (1995). *Monitoring the World Economy 1820-1992.* Paris: OECD [This book measures economic developments of Western Europe, Western Offshoots, South Europe, Eastern Europe, Latin America, Asia and Africa from 1820 to 1992, and explores the causal influence and phases of the developments.]

Maddison A. (1998). *Chinese Economic Performance in the Long Run.* Paris: OECD. [This work analyzes economic development in China from 1820 to 1995 with a substantial reference to the earlier history of economy and technology of China.]

Needham J. (1979). *The Grand Titration, Science and Society in East and West,* London, Boston and Sydney: George Allen & Unwin Ltd. [The book studies science and technology in Ancient China and their interactions with the Chinese society, from an East-West comparative perspective.]

Nelson R. R. ed. (1993). *National Innovation Systems: A comparative analysis,* New York and Oxford: Oxford University Press. [An important work. It develops the national innovation systems approach by providing a number of country cases. In addition to advanced economies it includes some developing economies such as South Korea, Taiwan and Brazil.]

OECD (1999). *Managing National Innovation Systems,* Paris: OECD. [This is a summary of findings from the OECD "Project National Innovation Systems". The book contributes on country-specific innovation policies as well as in the strengths and weakness of national innovation systems.]

Ostry S. and Nelson R. R. (1995). *Techno-Nationalism and Techno-Globalization, Conflicts and Cooperation,* Washington, D.C.: The Brookings Institution. [This book concerns the world-wide dissemination of technological and market information. Techno-nationalism is accordingly developed in the context of techno-globalization.]

Qian Y. and Xu C. (1993). Why China's Economic Reforms Differ: The M-Form Hierarchy and Entry/ Expansion of the Non-State Sector, *The Economics of Transition* 1, 135-170. [This paper analyzes the decentralized administration structure of China and explains the entry and expansion of the non-state sector during the economic reform because of the openness of the so-called M-Form hierarchy.]

Rosenberg N. (1963). Technological Change in the Machine Tool Industry, 1840–1910, *Journal of Economic History,* 23, 1963, reprinted in: *Perspectives on Technology* (Rosenberg N.), 9–31, Armonk and London: M.E. Sharpe. [This is a classic work concerning technological change in the machinery industry. It develops the notion that the machine tools industry serves as one of the innovation centers of an economy.]

SSTC "White Paper" (1987). State Science and Technology Commission (SSTC): *zhongguo kexue jishu zhengce zhinan 1986, kexue jishu baipishu di"yihao (Guide to China's Science and Technology Policy (1986), "White Paper on Science and Technology No. 1),* Beijing: SSTC. [A government policy book with statistical data on science and technology in China.]

Wade R. (1990). *Economic Theory and the Role of Government in East Asian Industrialization.* Princeton, New Jersey: Princeton University Press. [This book analyzes the role of the Taiwanese government in the successful economic development in Taiwan.]

World Bank (1999). *Knowledge for Development, World Development Report 1998/1999,* World Bank. [A mile-stone publication of the World Bank on knowledge–based development to advise developing countries. But it is still primitive in its theoretical rationalization and policy recommendations.]

Zhou E. (1956). guanyu zhishi fenzi wenti de baogao (On the Issue of Intellectuals), 14 Jan. 1956, in *Zhou Enlai wenxuan (Selected Works of Zhou Enlai),* Beijing: renmin chubanshe (People's Publishing House), 185. [This is one of the most important policy statements made by the Chinese government since the establishment of the People's Republic of China. It outlines the major principles and measures of science and technology policy under the centrally-planned system. These were policies actually implemented in the 1950s to 1970s except for the times when China was in political movements such as 1958-1960 and 1966-1976.]

Biographical Sketch

Shulin Gu was trained as a natural scientist in Chemical Physics. From the late 1960s to the 1970s, she worked in several distinguished Chinese institutes. In the Institute of Mechanics of the Chinese Academy of Sciences, she was Assistant to the Director and worked in the field of chemical dynamics; and in the Central Research Institute of the Beijing Petrochemical Corporation she took a leading role in several areas of petro-chemical catalyst processes. Since the 1980s Shulin Gu has been engaged in science, technology and innovation policy studies. She worked in the Institute of Policy and Management of the Chinese Academy of Sciences in the 1980s, and took the leadership of the Science and Technology Policy Department. From 1992 to 2000 she was Senior Research Fellow at UNU/INTECH, Maastricht, where she completed several INTECH projects in science and technology policy and transformation of the innovation system in China, and contributed to the research on national innovation systems in developing countries. Her work is now based on China, and she is Visiting Professor of TsingHua University and Adjunct Senior Research Fellow of UNU/INTECH. Shulin Gu is a member of the Advisory Board of several academic journals. She is also involved in activities organized by UNDP and UNU. Of her wide publications in both Chinese and English, the book *China's Industrial Technology, Market Reform and Organizational Change* (Routledge 1999) is highly appraised in book reviews (*China Quarterly*, 1999 December, *Research Policy* 2001 August) as an important work on S&T policy and the reform of the S&T system in China. Her research interest is in technological innovation and innovation policy in developing countries, and clustering and regional development strategies.

EUROPEAN SCIENCE AND TECHNOLOGY POLICY

Regina Gusmao
Technical Advisor, FAPESP, Brasil.

Keywords: European Union, science and technology policy, R&D projects, international research collaborations, scientific and technological networks, EU Framework Programmes, R&D Funding.

Contents

1. Introduction
2. The International Context
3. Formalization and Implementation of the EU Research Policy
4. The EU Research and Development Programs
5. Strengths and weaknesses of the EU intervention: a general overview
6. Conclusion
Glossary
Bibliography
Biographical Sketch

Summary

The article shows the principal components of the European policy in research and technological development. The first part exposes some aspects of the international context. The second part examines the processes of formalization and implementation of the European research programs with a brief overview of other multilateral cooperation programs which exist in Europe. The third part details the "Framework Programmes" (FWP), their funding mechanism and the structures it has created. The article reviews the thematic changes that have been underway since the 1980s. In the fourth part a general balance is proposed of strength and weaknesses of the European research action. Major changes and perspectives are the subject of the conclusion.

1. Introduction

The wide range policy action of the European Union (EU) in research and technological development (RTD) is rather recent. It dates back to the 1980s. Nonetheless, the history of community research, as well as the constitution of the European Community, is a large process of growth and maturation where new domains are introduced, along with the progressive growth of allocated resources.

The implementation of the 1987 Single European Act was an essential step in this process. It legitimated the European dimension of science and technology, making RTD an area of new competence for the EU, along with other sectorial policies. The major objective in the policy action was "reinforcement of the scientific and technological basis of European industry in order to develop its international competitivity".

Although it still occupies a small fraction of the total EU budget (around 4%), the RTD budget has constantly been growing in recent years in absolute terms.

Compared to member countries, resources for research are still rather low in the European budget: it only represents 5% of the total civilian research expenditure of all member states. The primary basis for research expenditure is still national rather than European. The total sum of support for different European international ventures in R&D (be they community or inter-governmental) does not exceed 17% of total national expenditure on R&D.

Support of the EU towards research might appear as modest, but it is necessary to keep in mind that the support given to projects by the EU is in the form of "incentive credits", i.e. sums of money that do not imply heavy equipment funding and represent additional funding to already existing structures. EU funds are not "subsidies" to research organizations and companies, and may only be used for carefully described work or research developments. Compared to other types of public investment, these funds are immediately used on research activities, whereas most public expenditures are investment in fixed costs and are used for salaries and infrastructure. Moreover the EU funds around 50% of the total cost of projects ("shared-cost projects"). This means that the volume of R&D generated by these actions is at least twice as large as the amounts listed in the community budgets.

In most industrialized countries, since the mid 1980s, the slower growth of public and private funds allocated to R&D has been compensated by the growth of external funds. Thus the support brought by the EU has had an important impact on national research and development.

Most evaluations done in recent years tend to show a globally positive image of the last 15 years of EU action in this area: community programs have an important impact on research in the continent and create a specific added value, namely the "Europeanization" of research. To put it differently, there has been a large development of transnational networks in RTD which have opened national scientific communities and have re-inforced common projects between academics and the industrial world, which were previously poorly developed in Europe.

Apart from the insufficient resources, the most evident gaps concern the absence of an industrial strategy, the administrative methodologies and the complexity and length of legal and institutional procedures. These limits point to the main difficulties of European research: its insufficient capacity to translate research and technological advance into viable industrial projects and the very feeble coordination of actions between the national and community areas.

The European Union is going to be enlarged to 25 or 30 countries in the twenty-first century. This demands more appropriate methods that the ones used up to now. New measures and modifications have been introduced in the pluriannual programmes directed towards this enlargement.

2. The international context

In a context of growing costs for research, governments have more difficulty in maintaining their funding levels for R&D. On a world scale, public funds have stabilized or diminished in real terms since the early 1990s.

In most OECD countries, the 1980s were characterized by a slowdown of funding for R&D—a trend, which has been accentuated in the 1990s along with the appearance of economic recession. National S&T policies have been geared towards economic competitiveness and growth, by focusing on industrial research and by reinforcing interfaces between industry and universities.

One of the principal trends of the last decade has been the growing participation of international funds in R&D expenditure (foreign companies as well as foreign states and international organizations). International cooperation assumes great importance in public R&D budgets, notably so for European Union member states. Even industrial R&D in the 1980s received a growing amount of foreign funds. This overall growth of external sources has partly compensated for the loss of funds provided by firms.

Most national policies want to encourage the internationalization of their national scientific and technological potential. In the EU these measures range from the simple exchange of researchers to the more complex establishment of multilateral cooperation agreements, including cases such as the creation of international research institutions (e.g. CERN, the European Laboratory for Particle Physics).

The promotion of "common interests" in the scientific and technological domain has been associated with the fear of a European decline relative to other economic potencies. This has been the principal engine behind most decisive community measures in favor of RTD, which were legitimized by a consciousness in the European industrial world of growing gaps with USA and Japan.

In the 1980s, most cooperation activities in the European Union have been done in the promotion of industrial competitivity. In high technology sectors, the competitive position of Europe as compared to other members of the "Triad" (European Union, USA and Japan) has been deteriorating. The USA is ahead of Europe in the majority of the so-called "critical" technologies, both in terms of results and potential evolution. Moreover, in some sectors (electronics and semi-conductors), Europe is behind Japan.

As far as scientific research is concerned, indicators show that basic research is growing as compared to other types of research. But industrial research is slowing down in some major economic sectors for Europe. High technology products in the EU represent only 30% of exports whereas this percentage is more than 50% for Japan and USA. R&D expenditure by private companies is higher in Japan. Additionally industrial R&D in Japan is almost not supported by public funds, whereas in USA, 28% of industrial research is linked to public contracts (mainly military) and the equivalent figure is more than 18% in the EU.

All global indicators in Europe show a gap between basic research and technological development and between innovation performances and competitiveness. They are the result of three main deficiencies of the European research and technological development system:

- Insufficient transformation of research results into applied commercial results (new products, new processes and services). Europe seems to have a comparatively limited capacity to transform its scientific and technological discoveries into industrial and commercial successes.
- Fragmentation of effort and lack of coordination of research actions and programmes, at both European and national level. A "Unique Europe" is still not a process that is completed, and the structural support that exists in Japan and USA is still lacking in Europe since what could serve as a "national" framework is still not totally defined.
- Insufficient R&D efforts, mainly in the domain of education and research training, and in knowledge transfer and dissemination. This is particularly true with enterprises where efforts are still limited.

Insufficient investment in research and development plays an important role in this context. Europe's efforts are lower than its competitors. The R&D expenditure of the European Union is lower than the USA in absolute terms. It is also lower than Japan's when considered relative to Gross Domestic Product (GDP). The mean research effort of Europe is only 1.94% of its GDP against 2.80 for USA and 2.98% for Japan (as estimated by EC-Eurostat for 2001).

Nonetheless, insufficient funding is not the only aspect of Europe's decline in this domain. Many other factors have to be added. Among the more obvious ones we have to consider:
- poor management of commercial issues;
- inadequate organizational structures and training in companies;
- isolation of research teams and lack of coordination, and
- feeble cross-border exchanges between European countries.

Europe has difficulty in integrating R&D and innovation in a global strategy. More efficient management of research and technology in European firms is considered a necessity by the European Commission.

The idea of more systematic European cooperation in research and technology has been progressively accepted by scientists, industrialists and political leaders. Since the launching of the first large European programme with industrial aims (the *ESPRIT* programme in 1983, in the information technologies) a long-term European policy has been forged. Up to the 1980s, nearly all research and technology development actions were funded by community funds but executed nationally. After the 1984/85 framework, multi-partnership technological programmes, transnational cooperation, and inter-firm collaborations were encouraged. This has now developed into a multinational framework.

3. Formalization and implementation of the EU research policy

3.1. Evolution and legislative framework

The first aspect that distinguished European RTD from similar multilateral systems was its position in the complex institutional setting that defined by the Treaties establishing the European Communities and the European Union.

For a long time, apart from the ECSC Treaty (establishing the European Coal and Steel Community) and the EAEC Treaty (establishing the European Atomic Energy Community), most action of the communities in the scientific and technological domains was only referred to in one single article of the Treaty of Rome (March 1957) that established the European Economic Community (EEC). With the exception of agricultural and fisheries research, the EEC Treaty had no provisions for common research activities. Thus, until the 1980s, most community action in this domain was taken without any clear legal authority and had very limited power.

The institutional framework on RTD has been defined by two major moments: the Single European Act (1987) and the European Union Treaty (or Maastricht Treaty, 1993). Both treaties define general guidelines and principles for S&T actions, and introduce S&T action in the European political context.

The Single European Act provided great motivation and added impulse to the launching of a common research policy since it aimed at the creation of an "internal European market". It extended the competencies of the Community and introduced significant changes in the functioning of institutions. The Single European Act explicitly legitimized the regional dimension of S&T cooperation in Europe. It gave RTD a formal area of competence through the community institutions. A new chapter was integrated into the Rome treaty, on "Research and Technological Development", in which objectives and means were defined.

With the Maastricht Treaty (which became effective in November 1993), community research received a more political dimension. The treaty opened new possibilities and obligations aiming at enlarging the objectives of research. Added to the industrial finality, research was called to accompany "collective" objectives and integration with other policies, for all countries in the Union.

3.2. Principle of regulation

Community support for research is not comparable to the formerly existing national forms of support. The community programmes represent many advantages, both short-term and long-term, as compared with national or other multinational forms of support. In order to attain its objectives, the European Commission bases its action on a fundamental principle: the "principle of subsidiarity".

This general principle was formally introduced in the Maastricht Treaty, although it was already largely applied in practice. It states that "activities whose execution through European cooperation presents an evident advantage, and that will generate more beneficial effects, should be transferred at the community level". In other words, this principle supposes that activities will be effected at the European level when it is more rational, less costly or more efficient at that level, either for reasons of cost, complementarity or competence, or because of the very nature of the problem to be solved.

Among the numerous possible cases where the principle applies, we could mention the following examples:

- research which needs larger amounts of financial, technical and human resources than any member state could provide and which has expectation of results of benefit to the whole Union;
- research directly linked to the construction of the European unified market because it results in eliminating internal barriers to movements between member countries, such as unified norms and standards (sometimes called "pre-normative" research);
- research in areas where problems are naturally set at a European scale and go beyond national frontiers (such as environmental, security and health issues).

In these cases "subsidiarity" is wholly applicable. Nonetheless there exist domains of research where the "principle of subsidiarity" applies in less obvious ways. Its applicability needs then to be decided on a case by case basis.

3.3. Instruments for the Implementation and Funding of R&D Projects

The formulation of a global research strategy took place in the context of the Single European Act with two support mechanisms: a) the adoption of a "Framework Programme" (FWP) for research and technological development where the main directions, instruments and number of actions are decided for a five year period; b) the establishment of "Specific Programmes" for many technological areas. The "Specific Programmes" share three attributes: they concern the so-called "precompetitive" research; they are multidisciplinary and promote collaboration between universities, research centers and companies; they are conducted as transnational projects. These attributes define what might be called "European research"

Until the current Sixth Framework Programme (for the period 2002-2006), these multi-annual Framework Programmes and Specific Programmes relied on three types of funding or types of action: (a) the Community's "own research", carried out at the Joint Research Centre (JRC), (b) the "concerted actions", and (c) "shared-cost actions", which are the major form of EU intervention in the RTD area. Additional mechanisms have been designed in order to better respond to the challenges posed by the changing context and the changing nature of research and technological activities (see below, Strengths and Weaknesses of the Community Action).

The activities of the Joint Research Centre (JRC) were initially limited to nuclear fission. Now they extend to many domains such as environmental protection, industrial risk, nuclear security and advanced materials. It always concerns actions funded by the community. The JRC comprises nearly 2000 persons in nine institutes located in five different sites in Belgium, Germany, Italy, the Netherlands and Spain.

"*Concerted actions*" represent an alternative form of intervention where the EU covers only the cost of previous concertation of research, but the actual research is executed at a national level. In this case, the Commission helps to establish a close coordination and covers only those costs linked to coordination (information exchange, publications, meetings, travel costs, and so on). They currently only represent a small fraction of the EU financing.

In "*shared-cost actions*" community credits fund research projects that are executed in a collaborative international way. These actions are becoming the principal means of action of the EU in RTD. Most sectorial programmes are designed under this scheme; more than 75% of financial resources of the Framework Programme are allocated in this way.

"*Shared-cost contractual research*" is done on the basis of contacts that are given after a call for tenders has taken place. Universities, research centers and companies of the member states and in some cases of associated states participate in these call for proposals. The community funding covers up to 50% of total costs of research, the rest being found by the participants themselves or by other contracting institutions. Universities, colleges and organisations that do not have cost accounting making it possible to reveal total costs receive up to 100% of additional costs.

Funds can be distributed also under other forms, such as preparatory, accompanying and support measures. These forms are complementary to the above mentioned modalities. The sixth FWP has extended the instruments for funding (see http://www.cordis.lu/fp6/instruments.htm). They rely on the idea of promoting a new European Research Area (ERA).

3.4. Links with other European S&T cooperation structures

Not all European S&T cooperation schemes operate under the auspices of the European Union. Member States engage in bilateral and multilateral arrangements between themselves, and sometimes extend to other European or non-European nations. In some of these arrangements the European Commission is a partner, and in some cases, it also provides funding and the secretariat for the management committees.

The research action of the community goes beyond the limits of the Framework Programmes. It also lies in the many cooperation structures that have developed in Europe over the years.

In order to avoid the fragmentation of research action in the continent, the last EU Framework Programmes have specific measures designed for collaboration between community structures and other European structures. European Commission leaders state that these measures are intended to provide a framework in order to discuss of the respective roles of the Union and of these other structures. Initiatives of this sort, by creating common science and technology policies and real condition for political concertation between these structures, are geared at creating a real "European research space".

3.4.1. European Scientific Facilities and Organisations

In recent decades a series of international scientific and technological cooperation research agencies have been created both within and without the community structures. Their origins go back to the end of the Second World War, when multilateral organizations were set up around some major facilities of scientific equipment. In most cases the objective was the pursuit of basic research (mainly in particle physics, and nuclear fusion), followed by space, aeronautics and information technologies.

Nonetheless, according to many authors, it was not the research itself but the research tools that added momentum to this "Europeanization", independently for the construction of the European Communities. This process materialises in special organizations for a specific basic research area. CERN is the first and best example of such an institution.

The model inspired many more entities through the 1960s and 1970s, such as ESF (European Scientific Foundation), ESA (European Space Agency), EMBO (European Molecular Biology Organization), EMBL (European Molecular Biology Laboratory), ESO (European Southern Observatory), ESFR (European Synchrotron Radiation Facility), and ILL (Institute Max von Laüe-Paul Langevin).

Today, the part of the scientific community most engaged in European cooperation is that concerned with basic research. When appropriate, it regards its action as a "European endeavour". This European endeavour takes particular forms. All the above mentioned cooperation has common characteristics:
- They are highly specialized in areas where the scientific community has been able to structure itself at a European level.
- They are based on a mutual dependency between the created European level entity and national research teams that use the implemented means.
- They are, from their origin, external to the European Union, usually under some intergovernmental agreement that, in each case, rules the cooperation; in this way these cooperative agreements are unaffected by the process of consolidation of the European Union itself.
- At the same, they constitute a limited scale success because they involve a small number of disciplines where the European Union has already accumulated a very strong knowledge tradition in order to face competition.

In fact, in order to create a real "scientific and technological European co-fraternity" major equipment installed in Europe might not suffice. It also necessary to create networks of relations on the most dynamic themes. All applied research requires other solutions that can help confront problems of industrial competition. Multilateral cooperation is accomplished by other means, namely the multidisciplinary "Technological Programmes".

3.4.2. Other Multilateral Programmes

Way down basic research, European S&T cooperation has been oriented toward Technological Programmes. Among those that cross the European Union's borders, the best known are COST and EUREKA..

Although they have specificity (mainly in their operational details), these two programmes share common aspects: they are based on multiple partnerships variable in size and composition, within a large array of technological fields. Their fundamental logic is to "incite" public research institutions, universities, technical research centers, and industrial enterprises to work together on themes judged "strategic" by the governments. Contrary to the community programmes for basic research outlined above, they are a flexible, rather loose framework, that uses operational concertation/coordination procedures and not associations around a common entity.

Two basic principals differentiate these programmes from the EU Framework Programmes: a) the implementation is a *bottom-up* process, i.e. actions are **not** linked within a pre-defined framework as they are initiated directly from the research interests of participants; b) as a general rule, research is funded directly at the national level.

The *COST* program *(European Co-operation in the field of Scientific and Technical Research)* is a large European cooperation action that has existed for thirty years. The Union plays an important role in its operation: the COST secretariat is linked to the Council of Ministers and technical management is done by officials of the Commission. Instituted officially in 1971 by an inter-ministerial conference that gathered 19 countries, the COST programme has settled the basis of a new form of European scientific and technological cooperation. Moreover, it was the first programme to include other countries (not belonging to the European Union). Since 1989, some actions have been open to non-EU member countries, in particular those in Eastern Europe.

EUREKA (European Research Cooperation Agency), launched in 1985, is a cooperation programme between enterprises and European research institutes in the area of high technologies. It comprises twenty five participants—the European Commission and 24 countries (these are the fifteen member countries of the EU plus Norway, Switzerland, Iceland, Turkey and Hungary, joined more recently by Russia, Slovenia, the Czech Republic and Poland). The principal objective of EUREKA is to "stimulate and support trans-border technological cooperation that will promote innovations that support the competitiveness of the European industry in world markets". In practice, precise projects have to be defined that produce short-term commercialized industrial applications. EUREKA's domain thus excludes basic research; its action is geared toward goods and services demanded by the market.

It is important to underline that the EUREKA initiative, since its creation, can be developed in multiple ways. First, the Union can participate as a partner in a EUREKA project through its own research facilities (the JRC—Joint Research Centre), or through financial instruments. The Commission funds directly or indirectly many EUREKA projects that present a strong community interest.

On the general plan of application, these two programmes represent two successive stages. Some EUREKA projects follow research work undertaken under a European community programme. In practice the line between these instances can be difficult to draw. For several years, the lack of articulation and transparency between these two systems has been repeatedly underlined. It is often a matter of debate whether action should be under a European community programme or under a EUREKA project. In these conditions, some coordination is called in and authorities have been stimulating such a reflection."

4. The EU Research and Development Programmes

4.1. The Framework Programme conception: structure and general characteristics

The Framework Programmes, adopted by the Council of Ministers, in co-decision with the European Parliament, state the scientific and technological objectives of community support, the main priorities, the overall expenses and modalities of funding of the EU in all research programs, and the distribution of funds among different actions. Implementation is the responsibility of the Commission, in collaboration with "Programmes Committees" that include representatives of member states.

Decision-making processes inside the European Union are very complex. The adoption of research programmes is the result of a process in which the Council of Ministers, the European Commission, the European Parliament and the Economic and Social Committee are involved. Decisions are taken according the rules of these institutions. As already mentioned, the Maastricht Treaty introduced important changes in these rules.

The pluriannual "Framework Programme" as it has been institutionalized by the Single European Act, is the basis and the instrument of medium-term programming of the EU research policy. By determining the priorities and financial weight of its intervention, it constitutes a "guide" for decisions taken in the period it covers. It makes visible for other European research actors (scientific institutions, universities, companies) the actions that the Union intends to undertake and the medium-range research possibilities.

Since the mid 1980s, six Framework Programmes (FWP) have been implemented by the European Commission. A progressive growth of funding can be observed (in millions of ECU):

1st FWP (1984-1987):   3,750 MECU
2nd FWP (1987-1991) :   5,396 MECU
3rd FWP (1990-1994) :   6,600 MECU
4th FWP (1994-1998) :   13,100 MECU
5th FWP (1998-2002) :   16,300 MECU
6th FWP (2002-2006) :   17,500 MECU

One has to note the strong progression of the budget since the 4th FWP, as a result of changes introduced by the Maastricht Treaty. It has enlarged the domains of action covered by the Framework Programmes: FWPs now cover all research, technological development and demonstration activities under EU responsibility, whatever their form or orientation. Research and training in nuclear energy aspects (under European Atomic Energy Community) contributed to this sudden growth. Actually, the FWP's budget represents around 4% of the EU's overall budget and 5.4% of all public (non-military) research spending in Europe (as estimated by EC-Eurostat for 2001).

The Framework Programme includes four types of actions through one or many "Specific Programmes". They can be grouped in the following manner:

- **Research, technological development and demonstration:** this is the principal type of activities funded by the EU. Projects are mainly executed by international consortia that associate different research actors in the member states. Executed mainly as "shared-cost actions" (where EU funding corresponds to 50% of the total cost of the project), these programmes represent 80 to 85% of the total Framework Programme budget.
- **International Research Cooperation:** projects grouping third countries and or international organizations with European partners. These specific programmes correspond mainly to collaboration with neighboring countries (Eastern European and Mediterranean) and with developing countries, as well as some collaboration with advanced countries that are not in the EU.
- **Diffusion and exploitation of research results:** actions that support the creation of networks for the transfer of technology and innovation, or the support of exemplary practices in the management of research and technology, consultancy, etc.
- **Training and mobility of researchers:** the European funding aims at grants (pre-doctoral and post-doctoral) and training by research networks, as well as support for accessing the large research infrastructures ("European scientific facilities").

"Shared-cost actions" are the main instrument for implementing the R&D projects in the context of the FWP "specific programmes". These actions are put into effect through multinational consortia made up of companies (including small and medium-sized enterprises), research centers and universities from different EU member states. The trans-boundary consortia are created around projects that gather a mean number of five different teams coming from three different countries.

Global indicators show that RTD actions funded by the EU are relatively stable over time. The mean size of consortia, the number of member states, and the mean amount of funding granted by the Commission per project have not greatly changed over the years in the last FWPs. But these indicators hide a large variety of cases and a complex intervention of the community, among areas, disciplines, types of institutions and countries.

Thematic programmes have shown a high priority to information and communication, either in terms of participants or in terms of mean funding per project. Larger and wider consortia are implied, in more than four different countries. On the other hand, programs related to international cooperation and exchanges between researchers (training and mobility) rely mainly on bilateral cooperation, of reduced size, gathering at most two to three teams from two different countries.

The amounts implied also vary greatly between projects. Those with strong industrial potential, with larger firms, are usually of higher amounts than projects with public institutions. In the so-called "diffusing technologies" (such as telecommunications) amounts of funding are four times larger than Life S&T and Environment. This is common to all FWPs, although we note a tendency for the gap to reduce, since the Fourth FWP.

In pluriannual programmes, according to European Commission sources, higher education establishments and research centers (public or private) represent around 55% of total participation, against 36% for enterprises. Small and medium enterprises (SME) appear as the principal beneficiaries of changes that have been introduced in recent Framework Programmes. Their part, either in participation or funding, has been growing steadily. There has been a "progressive sliding" of research actions to the benefit of SMEs rather than larger firms.

The principal beneficiaries and participants of the Framework Programmes are the United Kingdom, France and Germany. Apart from their economic and industrial domination, they concentrate the largest portion of research capacity that is available in Europe. Along with Italy, these countries represent 60% of total participation and consequently they generate the larger number of "collaboration links" within European research consortia.

4.2. Thematic evolution

The partial overlapping of Framework Programmes and the changing names of programmes as well as of specific content, are not helpful in tracking the European action. Nonetheless, some homogenization effort has been made by the Commission to facilitate examining priorities in the first four FWP. The main result of this work is summarized in Table 1, showing the principal results of RTD activities since the mid 1980. Let us examine the main trends in this period.

Groups of Specific Programmes	1984-87 (1st FWP)	1987-91 (2nd FWP)	1990-94 (3rd FWP)	1994-98 (4th FWP)
Information and Communications Technologies	25	42	38	28
Industrial and Materials Technologies	11	16	15	16
Environment	7	6	9	9
Life Sciences and Technologies	5	7	10	13
Energy	50	22	16	18
Transport	0	0	0	2
Socio-Economic Research	0	0	0	1
International Cooperation	0	2	2	4
Dissemination and Exploitation of Results	0	1	1	3
Human Capital and Mobility	2	4	9	6

TOTAL	100	100	100	100
Total Amount (MECU)	**3,750**	**5,396**	**6,600**	**13,100**

Source : *EC, Second European Report on S&T Indicators 1997*

Table 1. Changes in R&D Priorities in EU Framework Programmes (%)

First of all we have to observe the large emphasis on information and communication technologies, notably so in the 2nd and 3rd FWP, always more than 40% of total funding, a percentage way higher than the mean figures in other sectors. After rapid growth in the 1980s, their portion in the pluriannual programmes has been diminishing steadily in the 1990s. But this fall (as low as 28% in the 4th FWP) should not hide the growth in absolute value of funds in this strategic domain (from 2.3 to 3.3 billion ECU). The 4th FWP sets a change of orientation and translates a new tendency: rather than reinforcing the international competitiveness of the European industry, the FPW aims at scientific and technologic support to the construction of European information and communication infrastructure.

Secondly, we observe a slowing down of the Energy sector. It amounted to half of the budget of the 1st FWP, and it represented less than a fifth of the 4th FWP. In absolute value, its budget fell from 1770 millions ECU to 814 millions ECU. Community funding here concentrates in three areas:

- "Controlled thermonuclear fusion", including the construction of a large reactor that is one of the more ambitious initiatives of the EU. In the 3rd FWP, financial resources diminished by 25% from the preceding FWP.
- Research on "nuclear fission ", that is limited overwhelmingly to security aspects.
- The large and multiple domain of "non-nuclear" energies, where, since 1975, many shared-cost actions have been successively launched in practically the whole range of technologies for production, transformation and use of renewable energies.

The growth of the share of the energy domain (18%) in the 4th FWP (which nevertheless corresponds to a doubling of funding in absolute terms) results from a strengthening of non-nuclear energy.

In the two other large domains of research, unlike energy research, "Industrial and Materials Technologies" and "Life Sciences and Technologies", we observe a regular growth of European funds, in absolute and relative value. Their portion grew from 11% to 16% and from 5% to 13% respectively between the First FWP and the Fourth FWP. The growth in materials technologies is associated with the renewal of the European manufacturing industry, and more precisely to the introduction of SMEs in research activities. In the Life Sciences and Technologies, the growth is the result of the importance given to research in sectorial policies such as health.

Environmental issues have also been growing in importance. In absolute value, funds from the EU to the Environmental program have multiplied by more than four from the

First FWP to the Fourth FWP. Moreover, the environmental dimension is now central in different specific thematic programs. As reflected in the Fifth FWP and Sixth FWP, the environmental concern is more than a specific research programme—it is now a general dimension of the research effort of the EU.

In the context of rapid evolution of the role of S&T activities, introduced by the Maastricht Treaty, the Fourth FWP notably reinforced international cooperation with Eastern and Central European countries, the ex-Soviet Union countries, the Mediterranean countries and the developing countries. It has also reinforced activities aimed at diffusion and exploitation of research results. Their respective parts in the total budget have doubled against the Third FWP. This reflects the integration in the Framework Programmes of activities that were formerly funded outside the FWP framework. The Fourth FWP introduced also a specific action on "targeted socio-economic research", and an area devoted to evaluation of scientific and technological choices ("technology assessment"). Among decision-makers, it is important to identify the needs of society and understand the effects of research in the workplace, the economy, in education and in cultural life, as well as understanding the evolution and diversity of the components of European society. In order to better structure and reinforce this new domain of action, the Fifth FWP (1998-2002) introduced a socio-economic aspect to all its sectorial programmes inside the FWP.

The Fifth Framework Programme, which ended in 2002, represents a cut as opposed to the preceding FWPs. It was defined in a new environment and initiated a new approach for European community research. It was stated that for the European Union, the Fifth FWP should be "the opportunity to implement a new approach for community research policy, adapted to the contemporary challenges and aspirations of European citizens". The Fifth FWP was based on two ideas: "concentration" and "flexibility". By preserving already acquired experience, this programme aimed at more precise socio-economic objectives rather than mere technological ones. Moreover, the implementation of the actions was effectively more flexible in its use of resources in order to allow for more adaptation to future evolution. This general approach is confirmed regarding the current Sixth FWP's structure and content.

5. Strengths and weaknesses of the EU intervention: a general overview

Since the First Framework Programme, in 1984, the European Union has funded, at least partially, more than 17 000 research projects, supporting tens of thousands of laboratories and institutions. The results obtained allow an overall balance of these actions. Many evaluations of these actions have been done, internally (annual reports, five years assessment reports) and also externally. In effect, many evaluations have been ordered on the execution and effects of the sectoral programmes. These evaluations aim at appreciating the objectives, priorities and financial means.

In an overall evaluation of thematic programmes of the Second FWP, the Commission presented a general reflection on its policy and its strategy after the year 1992. After that a "European Report" was drafted on indicators of science and technology in 1994, in which one can find, among other aspects, indicators on funding, participation and cooperation in the Second and Third FWP. Subsequently, a second "European Report" was issued, in 1997, and a third one has just been published in mid 2003. These and some more specific evaluation reports on thematic porgrammes allow us to extract the following conclusions on strengths and weaknesses of the EU action in RTD.

"The network effect"

Among the strong aspects, it is quite common now to underline that the principal result has been the promotion of collaborations at a European scale between many actors of research, uniting actors that until then had few common experiences. The principle of cooperation in so-called "pre-competitive" research permitted trans-border partnerships between competing companies, between public and private research institutes, and between large and small and medium enterprises. As the Commission states, "the principal merit of the Framework Programmes has been to demonstrate the *de facto* solidarity between different worlds by revealing the economy of scale of European collaborations in RTD". Community actions result, beyond the specific projects, in learning how to deal with partnerships of a new type: "European partnerships". Moreover, the international cooperation has allowed the emergence of long-standing networks, where teams are working jointly even after the end of the projects funded under the FWPs.

A global " structuring effect"

Added to the networking effect, there is a "structuring effect" on the scientific and technological universe of European countries. In few years, the community programmes became references inside the scientific community and the industrial world in Europe and elsewhere. This high attraction power can be explained by the importance of resources given to research laboratories, mainly when national resources for R&D were souring. Moreover, the participation in a community-funded research project opens up the door to information and facilitates the making of new collaborative projects, even beyond research and development. There is also a "prestige" effect linked to participation in European projects. The selection of projects is made on the basis of its "innovative character" and "excellence level". By way of consequence, participation in a research project funded by the EU is recognized as a quality label. Finally, beyond the mere scientific production, the effect on research practices is often noted. European projects stimulate the training of personnel in research institutions and participating firms.

Satisfying technological and normative results

As for the results of research activities, there is a strong controversy (see *Evaluation Practices in a Modern Context for Research: A Review)*. The low impact of EU support on commercialization and industrial initiatives is regularly denounced. Nonetheless, technological results and results in the area of norms fixation have generally met their "technological objectives". The Commission estimates it has set a relevant "strategic framework". Economic evaluation of gains of competitiveness due to the transformation of scientific and technological results into economic advantages is a rather complex process and requires a longer period of study. In addition, the evaluation of commercial success or failure of a technological innovation is due to factors far beyond the areas of responsibility of the community programs.

Normative advances (that is, fixing norms for future developments or production), particularly in telecommunications (definition of unified systems of telecoms), are quite obvious. "Broad band" integrated telecommunications has been associated to a large process of normalization, with the adoption of more than 1000 international norms and many more functional common systems. The normative mission of the *"Bureau Communautaire de Référence"* (BCR) shows more positive elements in harmonization of processes and in national metrology laboratories.

Large mobilization of SMEs

Although there is some controversy, the data show that SMEs participate to a great extent in European programs, mainly in industrial and materials technology programs. Specific measures designed in the early 1980s permitted the tripling of SMEs in these programmes, and the opening of programs to traditional industrial sectors like the textiles and construction industry. The most important measure has been CRAFT, which stands for "Cooperative Research Action for Technology", launched in 1991, inside the specific programme "Industrial Technologies and Materials" of the Third FWP. It is estimated that half of the entreprises had never been involved in a research consortium before their participation in these programs, and had no experience of multilateral research collaboration at the European level.

Unfortunately, however, obstacles still exist to the participation of SMEs. The cost of collaboration in the European framework is relatively high for these enterprises as compared to other participants. Moreover, the amount and duration of the projects is often too high for them. In order to overcome these difficulties and stimulate participation of SMEs in these projects, specific agencies have been created at the European level; they become the main interface for SMEs in the Commission.

The promotion of European cohesion

Less favored regions in Europe have benefited from Framework Programmes. Apart the financial support, which has played a crucial role in developing infrastructure in these less favoured regions of Europe, the Framework Programmes have permitted inclusion of these regions in "European networks". It is clear that a **new** enlargement of the EU can accentuate difficulties that already exist inside the European Commission, mainly with co-ordination between European and national policies. The Commission has been conscious that these difficulties could lead to a complete "administrative paralysis". Major and intense challenges have to be faced. At the beginning of the twenty-first century, the European Union enters a difficult period with many political, financial and social problems. These regions see their research activities limited for a number of reasons (e.g. isolation, insufficient highly trained manpower, insufficient financial and infrastructure). Collaborative projects have largely stimulated their research activities.

Weaknesses of the EU support

Weaknesses fall into three main categories: 1) at the level of choices and concrete orientations of the research programmes; 2) at the level of procedures and administrative methodologies; 3) at the level of legislative and institutional procedures.

As far as the way choices and objective are chosen, many evaluation reports point to an insufficient consideration of technological priorities. The rather small number of real "priority projects" hides a large dispersion of efforts in an ever growing number of research themes which imply also a large array of applications. The Commission says it is "insufficiently aware" of problems linked to this lack of selectivity of funding, and the resulting tendency in dispersion. In order to face these problems, one of the new aspects has been the launching of the Targeted Socio-Economic Research program in the Fourth FWP. As mentioned earlier, this program included a program on technology assessment and evaluation of science and technology policy options. These initiatives are geared at a better understanding of society's capacity to deal with the impacts of technologies and anticipate tomorrow's priorities.

Many obstacles have been identified in implementing administrative methodologies for the Framework Programmes. Most relate to the heavy and rigid mechanisms of the European Commission. For SMEs, the complexity of the procedures, in writing a proposal and in negotiating a contract, implies heavy additional costs. Decisions and selection of projects are done on a multiple step process that can last weeks and even months, depending on the response rates to the calls for proposals. As underlined by most evaluation reports, the experience acquired in the last fifteen years on the elaboration, decision-making and implementation is no longer adapted to the challenges of today's context (see *The New Forms of Knowledge Production and S&T Policy*). New procedures need to be defined, more transparent and efficient, for the selection and attribution of contracts as well as for the management and diffusion of results.

The complexity of the legislative procedures introduced by the Single European Act, and partially modified by the Maastricht Treaty, has been driving permanent inter-institutional conflicts. The interminable exchanges between the European Commission, the Council of Ministers and the European Parliament are slowing down the whole decision-making process concerning the pluriannual programmes. Some people estimate periods of up to eight years between the formulation of a call for a proposal and the final exploitation of research results. As some industrial partners have mentioned, the needs expressed during the phase of writing the proposal are often lost when such a long time delay ensues, due to the very rapid and profound technological advances in some key sectors.

6. Conclusion

Despite the difficulties, the constitution of "European partnerships" has been the main and most important result of community funding. Collaborative research is particularly strategic for many reasons which need to be taken into account in the analysis of impacts of the community research programmes. Partnership is no longer an additional ingredient but a very essential and central aspect of the research of many private and public laboratories, increasing possibilities of further development of innovation and technologies (see *Strategic Innovation Alliances*). Many surveys of participants in European programmes mention this strategic importance: European programs have a place of choice in their own strategy and contribute to extending their research networks.

The research policy of the European Union does not limit itself to programs devoted to its fifteen member states. International cooperation has been growing, a number of links have been built with third parties, notably so in Central and Eastern European countries, in their effort to apply science and technology to the resolution of their socio-economic development problems. Since the Fifth Framework Programme, the countries that are candidate Member states to the EU have the possibility of participating in European programs under exactly the same conditions as the member states. The objective here is to consolidate research capacity and their management capabilities. Moreover, the introduction of researchers from candidate member states into the European scientific community can only help these countries in their preparation for entry to the Union.

It is clear that the enlargement of the EU can accentuate difficulties that already exist inside the European Commission, mainly with co-ordination between European and national policies. The Commission has been conscious that these difficulties could lead to a complete "administrative paralysis" if not modified before the enlargement of the EU. Major and intense challenges have to be faced by the European Union, e.g. the growing unemployment, new demands of the public, globalization of economic activities (markets and knowledge), the emergence of new economic powers (in Asia and Latin America) which will sooner or later become technological powers. At the beginning of the new century, the European Union enters a decisive new phase of its history. An intense effort has been made to increase socio-economic research, and social sciences play an important role (see *Social Sciences, Science Policy Studies, Science Policy-Making*). Fundamentally, the European Union needs to pass from a research orientation toward technological performance to a research orientation toward its citizens, and responding to society's growing economic and social problems.

Glossary

Community:	belonging to the European Community, or to the European Union.
FWP:	Framework Programmes. Large pluriannual programs that are defining the objectives, guidelines and instruments of actions of the European Union in RTD.
JRC:	Joint Research Center. A group of research centers directly funded and belonging to the European Commission.
RTD:	A general term covering all types of research and technological development activities.
SME:	Small and medium enterprises.

Bibliography

BARRE R. and PAPON P. (1993). *Economie et politique de la science et de la technologie*, 400 pp. Paris: Hachette, Collection Pluriel. [This work provides some elements of analysis of the civil and military programs implemented in Europe, within an economic and political "geography" of public actions resulting from research and innovation].

CANNELL W. (1999). Le cinquième programme-cadre de research et de technological developmentde l'Union Européenne. *OCDE - STI Revue* n° 23, 271-301. [The author presents and discusses the main features of the 5th EU Framework Programme (1998-2002) and the new strategies proposed for Community research and development efforts].

CARACOSTAS P. and MULDUR U. (1998). *Society, the endless frontier*, 202 pp. European Commission, Report EUR 17655. Luxembourg: Office for Official Publications of the European Communities. [This work introduces and discuss the "European vision" of research and innovation policies for the twenty-first century, based on a deeper interaction between society, innovation and Europe].

CORDIS. Detailed information on the 6th Framework Programme (2002-2006) can be found on the site of the European Commission on Research and Development, CORDIS: http://www.cordis.lu/fp6/home.html The CORDIS site has also information on results of past programs, as well as detailed information on how to respond to calls of proposals.

EC (1992). *Research after Maastricht: an assessment, a strategy*, 50 pp. Communication from the Commission to the Council and the European Parliament. Bulletin of the EU Supplement 2/92. Luxembourg: Office for Official Publications of the European Communities. [A synthetic and global assessment report of Community R&D past actions, and the Commission purposes for the future programs implemented in the EU new context].

EC (1994a) *First European Report on S&T Indicators 1994*, 338 pp. Luxembourg: Office for Official Publications of the European Communities. [Among others S&T indicators, this report presents in Part IV the Community R&D objectives, structure and impacts, and provides the main results of the 2nd FWP and 3rd FWP in terms of funding, participants and collaborative links between member countries].

EC (1994b) *Livre blanc, croissance, competitivité emploi: les défis et les pistes pour entrer dans le XXIe siècle*, 176 pp. Luxembourg: Office for Official Publications of the European Communities. [This Commission study provides a portrait of the situation of the European R&D activities and underlines their positive features and main weaknesses].

EC (1994c) *Soutien communautaire à la research et au development technologique*, 91 pp. Luxembourg: Office for Official Publications of the European Communities. [This presents a general idea of the Commission methods and procedures concerning the research proposals' selection and European contracts negotiation].

EC (1997a) *Second European Report on S&T Indicators 1997*, 729 pp. Luxembourg: Office for Official Publications of the European Communities. [Succeeding the First European Report, this Second Report presents in Part IV a synthetic retrospective of the EU S&T policy and provides several indicators and interpretations respecting the 3rd FWP and 4th FWP].

EC (1997b) *Towards the 5th Framework Programme. The scientific and technological objectives*, 46 pp. EUR 17531 EC/DG-XIII. Luxembourg: Office for Official Publications of the European Communities. [This presents the official proposal, adopted by the Community institutions, respecting the new structure and orientations of the current EU framework programme].

EC (2000). *Towards a European research area*, 44 pp. Communication from the Commission to the Council, the European Parliament, the Economic and Social Committee and the Committee of the Regions. Luxembourg: Office for Official Publications of the European Communities. [Starting from a picture of the situation and objectives of R&D activities in Europe, this document introduces and discusses the "European research area" notion, and the conditions for action in the near future].

EC (2003) *Third European Report on Science & Technology Indicators 2003*, 451 pp. Luxembourg: Office for Official Publications of the European Communities. [Succeeding the Second European Report, this Third Report provides in Part II indicators and interpretations regarding performance in knowledge production, exploitation and commercialization in the European Union Member states].

GUSMAO R. (1997) *L'engagement français dans l'Europe de la recherche*, 292 pp. Paris: Ed. Economica. [This study examines EU's research policy's implementation and formalization process since the 1980s; from French experience, it focuses on Member States research systems' participation in EU Framework Programmes].

OECD (1994). *Science and Technology Policy. Review and Outlook,* 273 pp. Paris: Ed. OECD. [This report analyses the trends and perspectives of the OECD members' S&T policy; it offers a detailed presentation for each country and the recent governmental measures].

PETERSON J. (1993). Assessing the performance of European collaborative R&D policy: the case of EUREKA. *Research Policy* 22, 243-264. [This paper compares and synthesizes data from two attitudinal surveys of EUREKA participants; it underlines some conflicting aspects in regard to the EU Framework Programmes].

RUBERTI A. and ANDRE M. (1995). *Un espace européen de la science*, 186 pp. Paris: PUF. [The authors examine the evolution and the contemporary situation of the European research and innovation system, and compares the mechanisms adopted since the end of the Second World War to promote European scientific cooperation].

Biographical Sketch

Regina Gusmão has a PhD in "Science, Technology and Society" from the Conservatoire National des Arts et Métiers (Paris, France). Her thesis on the participation of Member States in European Research Programmes was carried out and published within the French Observatory of Science and Technology (OST), where she was in charge of projects for seven years. At OST she developed special competences in production and diffusion of a large sample of indicators concerning international S&T cooperation and multilateral research programs.

She is currently an adviser to the Presidency of FAPESP - The São Paulo State Research Foundation (a government agency which supports R&D projects in this Brazilian state) where she is involved in research program assessment activities and setting new S&T indicators, as important tools for the decision making process for regional research policy.

THE NORTH AMERICAN "INNOVATION SPACE": A WORK IN PROGRESS?

Paul R. Dufour
International Science Policy Foundation, Ottawa, Canada

Keywords: innovation space, S&T structure, research diaspora, regional research space, North-American Free Trade Agreement (NAFTA).

Contents

1. Introduction
2. The Characteristics of the S&T structure in North America
3. A North American Research Diaspora?
4. What are Some Key Elements of a Regional Research Space?
5. Emerging Signs of North American Collaboration in S&T
6. Reaching Beyond the North American Sphere
7. An Agenda for a North American Innovation Approach
Glossary
Bibliography
Biographical Sketches

Summary

The article presents the evolution of the North American science and technology 'innovation space' which is under construction within the framework of the North-American Free Trade Agreement (NAFTA) linking Canada, the United States of America and Mexico. The article reviews rapidly the main characteristics of the S&T system of the three countries, and the emerging signs of a more intense collaboration among them. It examines the main aspects of their research collaboration bilaterally as well as in multilateral negotiation arenas. The article comments on possible futures of this emerging regional S&T space.

1. Introduction

At this event several years ago, Richard Lipsey, a noted Canadian economist working with the Canadian Institute for Advanced Research on economic growth and public policy issues, delivered a strong message for the scholars and decision-makers on the importance of innovation to growth. His keynote address was particularly instructive for the research and technology communities. In it he speculated about the model that Mexico might adopt in support of science, technology and innovation. Lipsey argued that there are two major world views developing; one that has government providing the broad background conditions for innovation, but leaving the market free to generate innovation and diffusion that powers growth; and a second vision, a neo-liberal one, that has governments taking a more active role in the national innovation system (while accepting the importance of free markets).

His talk triggered a lively debate during the week-long summer institute devoted to this emerging policy issue. At this week-long event sponsored by Simon Fraser University in British Colombia, Mexican, Canadian and USA scholars met to discuss the potential ways in which to respond to the challenges of globalization and competitiveness and their interrelationships in the new North American region shaped by the North American Free Trade Agreement (NAFTA).

The issues identified by Lipsey over six years ago remain today and will likely help mould how respective governments in North America support science, technology and innovation in the coming decades. For several reasons described below, efforts to coordinate and integrate this field of activities at a regional level have met, and will continue to meet, with mixed results.

To a considerable extent, the decision-making levers of the three North American governments have become more restricted not only with the NAFTA conventions on subsidies, intellectual property and investment, but also with the evolving rules of the game outlined by the global trading system represented through multilateral institutions such as the OECD and the World Trade Organisation.

How will these issues impact on the so-called 'invisible and global college' of science and knowledge? Certainly the advances in technology and electronic communication have privileged scientists and researchers worldwide, and are assisting in the diffusion of knowledge through the Internet and growing global scholarship.

But there is more in store for the North American region. North American governments, by recognizing that economic growth and development are conditioned by knowledge, have become more assertive in promoting and investing in these knowledge assets. It is no coincidence that the President Clinton/Vice President Gore science policy statement of 1993 is called Science in the National Interest. In Canada, the Federal Government's 1996 S&T for a New Century argues a similar path with a view to better managing these strategic investments. In Mexico, efforts in building on its five-year plans for national development of S&T, not only focus on capacity-building for science and technology, but also stress the importance of increased competition in the private sector in fostering a stronger receptor capability for S&T.

The cultural specificity of national policies for innovation and research has also been influenced in each country by the respective instruments in place and the view each has with respect to international partnerships. Undoubtedly, the key and unique factor in this arrangement is one where two countries (one developing and another mid-sized) are the spokes of the 'hub' that is the world's only superpower – the USA.

2. The Characteristics of the S&T structure in North America

With its strong emphasis on maintaining global leadership and technological superiority in civilian and defense matters, the US approach continues to rely on the strength of its university research system for new discoveries and commercialization of research, as well as the increasing entrepreneurship and R&D investments of its firms, both global and domestic. In Canada, with long-standing weaknesses in the private sector industrial

R&D (partly a result of the significant foreign ownership of its economic base) recent efforts have tended towards improving the university research base and its links to commercial activity (as a proxy for the lack of private sector R&D), including strengthening the skills base in critical technology areas. In Mexico, much attention has been devoted to building the S&T capacity through infrastructure and education and training. Renewed efforts to build bridges between the previously separate public and private research activities are major priorities. The convergence in several of these national initiatives poses new challenges for thinking about a North American innovation agenda. Let us briefly examine the specific characteristics of the S&T institutional structures and support for each country.

2.1 The United States and Technological Leadership

While still the dominant source of the world's pool of R&D, the US has seen its technological and research lead diminish in the 90s as a result of large investments in innovation from other countries. For instance, while the US still produces over one third of all scientific publications, the number of scientific articles from European and Asian countries has increased more rapidly than the number of articles by US authors.

Considerable debate in the US has centered around how to transform a research system that was heavily dominated by defense R&D investments into one that responds to today's globalized, civilian economy. Large increases in health and environmental R&D have been the current trend to keep pace with issues surrounding improved quality of life. Private sector industrial R&D is on the upswing as more and more US-based firms invest heavily in R&D for competitiveness and maintenance of global advantage, especially in information technology and biotechnology. The rise of venture capital to finance much of this new economy has been critical to this knowledge boom.

The major funding agencies such as the National Institutes of Health and the National Science Foundation have seen significant increases in their budgets for basic research over the past decade. The federal laboratory system responsible for defense and civilian research has experienced considerable re-structuring to address the new goals of a post-Cold War, post-Kyoto environment. The university research system and its production of highly-qualified personnel (often touted as the best in the world) has also undergone various reviews as efforts are underway to develop stronger partnerships with the private sector and foreign institutions. The US view of a more global setting within which to develop R&D partnerships is also changing. Debate over how better to use foreign policy to take advantage of the growing opportunities in such areas as environmental technology and health research is also active.

Finally, a great deal of effort is underway in the United States (at both the federal and state levels) to reform and accelerate the school system's use of information technology and to introduce science into the classrooms from K-12. Along with health care, this investment in future citizens and their skills is seen as the major avenue for continued prosperity and global leadership in innovation.

2.2 Canada and its Changing Image as a Knowledge-Based Society

Over the past decade, Canada has engaged in a fairly ambitious (some would say radical) series of ventures to improve its long-standing weaknesses in science and technology. While the country maintains one of the most generous industrial R&D tax credit systems in the world, Canadian private sector performance in industrial R&D still lags most of the G-8. Structurally, because of its highly decentralized nature, the federal and 10 provincial governments often find themselves squabbling over issues of regional innovation, skills and capacity-building (e.g., the country has no Federal Ministry responsible for education with each province responsible for its own educational system).

Despite this, Canada has embarked in a series of institutional initiatives designed to make the country a smart, connected society. The Canada Foundation for Innovation has a mandate to help renew research infrastructure in Canadian universities, colleges and hospitals. Virtual Networks of Centers of Excellence have been established on a competitive basis in 22 areas of research, especially in the life sciences and biomedical research where Canada has a well-established expertise. Cdn$900 million over five years has been set aside to establish 2000 Canada Research Chairs to help retain world-leading academic talent and attract researchers from around the world. In the area of biotechnology, a new organization at arm's length from the government called Genome Canada has been established to help Canada catch up in genomic research.

By 1998, the country's SchoolNet program has succeeded in connecting every public school and library in Canada. Over 140 000 computers and software packages have been delivered to schools through the Computers in the Schools program, with a goal to have every classroom (250 000) with a computer by early 2001. The world's first nation-wide Internet built directly around wave- division multiplexing technology (National Optical Network) is being developed and deployed. A Smart Communities program has led to financing of pilot projects in each province and territory with a goal to have information technology transform community and social development.

Continued efforts are underway as well to examine key areas of the country's innovation system, including the country's role in international S&T; skills in critical industry sectors; science advice for improved decision-making; and an enhanced role for federal government laboratories, including the premier public R&D institution in the country, the National Research Council of Canada.

Long-term issues will continue to be debated on the continued need for Canadian industry to access foreign technology from abroad especially its largest trading partner, the USA. How to grow clusters of research excellence and large technology-based multinational firms in the face of its high foreign ownership are issues that decision-makers from all sectors will face. Indeed, in 2001, Canada was about to engage in a large-scale national consultation on the future of innovation in the country, with the objective of outlining a strategy for new investments in 2002.

2.3 Establishing Mexico's Place in the Innovation Arena

Faced with a new government (the first non-PRI administration in many decades), it is difficult to predict where Mexico's focus in S&T and innovation will be placed. It is clear that efforts will likely be developed to improve the performance of the industrial sector's R&D capability. After Greece, Mexico has the largest share of publicly funded R&D as a percentage of domestic R&D in the OECD. Much of this focus will take in the small and medium-sized sector which account for 98% of the country's manufacturing sector. Networking and alliances between government and the academic sectors to help stimulate innovation is also a priority area. A Law introduced in 1999 on the promotion of scientific and technological research has been designed to establish guidelines for the government promotion of S&T; earmarking funds for research; enhancing the authority of public research centers; and improving linkages between research and teaching.

The Knowledge and Innovation Programme (with a budget of $US500 million) aims to increase private sector involvement in S&T with a view to increasing investment in innovation; through such mechanisms as joint industry-university projects; support for firms to upgrade their technical capacity; and a technology foresight exercise to identify critical technologies that require further development .

Mexico (through its Council for Science and Technology – CONACYT) continues to support international bilateral cooperation agreements in S&T and sponsors networks of researchers and scholars for studies abroad. In 1998, for example, CONACYT funded 18 000 graduate students pursuing training in Mexico and abroad. CONACYT supports 27 research centers integrated into a system known as SEP-CONACYT, with research distributed throughout the country in such areas as the natural sciences, humanities and social sciences and engineering and technology.

Although the Mexican capacity for S&T and innovation has grown considerably over the past decade, the key areas of weakness in Mexico remain in the production of highly-qualified personnel and the weak connection between the production of knowledge and its application. For this reason, Mexico's international partnerships (especially in the North American arena) and networks will be seen as indispensable inputs to the major national effort underway.

3. A North American Research Diaspora?

'Mexicans, Americans and Canadians already have a strong sense of their own identity. The challenge will be to develop a North American "footprint" that treads lightly enough that it does not crush the existing landscape formed by distinctive histories and cultures... My sense is that a North American community would be institutionally much lighter and more flexible than the European model... a community that serves North Americans but that is also open to the world.' (The Honourable Lloyd Axworthy, speech to the Canadian Institute of International Affairs, Toronto, October 1998)

There is nothing inherently 'natural' about scientific cooperation or technical exchange in the evolving North American region. This may seem a surprising statement given the strong economic integration brought about by trade flows and global activities of transnationals, and given the historical proclivity of science towards international cooperation. Nevertheless, unlike the European research area (launched as a concept by the European Commissioner responsible for science and research, Philippe Busquin) that is well developed and based on geography, culture, language and natural trading patterns, the North American 'innovation space' remains essentially an artificial one. Clearly, trade patterns are well established, with both Mexico and Canada having in the USA their largest trading partner. (Canada's exports to the USA represent almost 88 percent of that country's total exports worldwide, whereas trade with the USA represents almost 80 percent of Mexico's total exports). Much 'integration' debate has also surfaced and will continue about such issues as currency (a common monetary union), a social charter for North America, a new architecture for transboundary environmental challenges, and immigration patterns, including mobility of students and scholars among the three countries. But these are not issues that have yet surfaced as priority issues among the domestic agendas of each country. Rather, when the subject of cooperation does surface, border issues tend to dominate.

One of the major ironies of this potential for regional collaboration is that economic integration in the North American arena has preceded scientific and technical cooperation. In the past, scientific, cultural and educational relations had often formed the precursors for initial forays into bilateral trade or foreign relations with new partners. The NAFTA that came into force in January 1994 has little to say about the role of technology, or R&D in its agreement. To be sure, there are related issues such as intellectual property questions, subsidies regime, accreditation of professions and environmental protection issues; but these are grounded in 'rules of the game' that emerged in the 1970s, in an era when science and technology were rarely viewed as prime engines of economic and trade development. Today, the context for global cooperation and competition is essentially a knowledge-driven one with a currency based on scientific and technological skills and assets.

With the growing convergence of science and technology support within the North American region, especially in innovation policy, (or what is now more fashionably referred to as a national system of innovation) one can expect that opportunities for dealing with information gaps and improving efficiencies in research and genuine international cooperation will emerge. One should also expect that as the educational and labor systems of each country generate a skilled workforce familiar with new information technologies and learning software, a common platform or 'language' among the next generations from each country will emerge.

4. What are Some Key Elements of a Regional Research Space?

What are the conditions required for creating a regional research space that is fully functioning? In reviewing those that exist around the world, such as the EU's initiatives centered on a pan- European research area, or the Asia-Pacific ventures in cooperative science and technology, or even the Nordic and Arctic nations' regional experiments in research, certain prerequisites appear necessary for collaborative activity at the supranational level. What are some of the more important of these conditions? (borrowed in part from the European research area concept paper):

- ongoing networking (virtually as well as physically) of existing research centers of excellence in areas of national/strategic importance;
- a common approach to the needs and means of financing research facilities and information exchange on issues that are transboundary in nature;
- establishment of a common system of scientific and technical reference for the implementation of policies, including common indicators for S&T;
- mobility of researchers and introduction of pan-regional scientific studies programs and scholarships;
- promotion of common social and ethical values in scientific and technological matters, and in the dissemination of sound scientific advice through consultative bodies;
- regular exchanges or fora among senior officials in the design of innovation policy and research initiatives.

What follows is a discussion of some of these elements to outline where the North American region has progressed and what further steps need to be taken.

5. Emerging Signs of North American Collaboration in S&T

5.1 The Data Picture

Considerable activity among the three countries' scientific communities has already taken place over the past several decades, but with the emerging exchange of opportunities among Mexican, USA and Canadian researchers, trilateral cooperation will likely increase.

A review of some of the data on formal cooperation provides a preliminary picture (informal cooperation is extensive but difficult to measure). While the volume of scientific output from Mexico is far inferior to that of its two North American partners, one should not overlook that the US is both Mexico's and Canada's largest scientific partner. According to one source, since 1993, US government agencies have spent more than $100 million a year on projects involving cooperation with Canada and/or Mexico. Further, according to1996 data that measures the degree of scientific collaboration via research publications, the United States accounts for over 38% of Mexico's international scientific cooperation and 41% of Canada's. Mexican scientific cooperation with Canada represents about 6.4% of its international linkages while US cooperation with Canada accounts for about 10% of its total international cooperative activity.

This cooperation differs from field to field; for example, Canada-USA scientific co-authorships in clinical medicine are quite strong with close to 30% of the total scientific cooperation in this area (in 1996, over 4330 scientific publications were co-authored by researchers from Canada and the US). The scientific partnership collaboration between Canada and Mexico on the other hand is not extensive and uneven from year to year. (Between 1981 and 1996, the volume of scientific co-citations by both Canadian and

Mexican authors had climbed from only 18 in 1981 to 75 in 1996, a large portion of this in the chemistry and clinical medicine fields).

While still young, the trilateral relationship in scientific publication is growing. According to the data, between 1980 and 1991, only 56 articles were produced with authorships from the three countries. Again, here, collaboration in clinical medicine is the leader. Funding for trilateral activities has also increased. In 1993, only 3 trilateral projects were funded by US agencies and government organizations. In 1997, this number had increased to 20 for a total of $20 million. As the relationship takes hold, one would expect that this trilateral activity will increase.

A study by the Science and Technology Policy Institute of RAND in the US has attempted to paint a more detailed picture of cooperative R&D in North America. By reviewing data of expenditures of cooperative research activities between US and Canadian and Mexican government research organizations, the study finds that from 1993-1997, the US federal government spent on average $62M (US) per year with Canada and with Mexico, $26M per year. Such cooperation takes many forms: binational scientist-to-scientist collaboration; multinational collaboration, trinational collaboration and other cooperative activities such as standards development, technology transfer, database development, etc... For example, Canadian and Mexican researchers benefit from funding by both the National Institutes of Health and the National Science Foundation, in addition to the funds from their own existing national granting agencies.

With respect to Canada-US collaboration, biomedical research is the largest area of cooperative activity, in part because of the extensive shared databases and clinical research between the two countries. According to the RAND report, 37% of joint projects between these two countries cover this area of research, followed by earth sciences. Canada and the US also share common activities in multinational projects involving several other nations. In 1997, over $22 million of US funding for multinational projects involved Canada as a partner, in such areas as atmospheric sciences, biomedical sciences and physics. Among such multinational projects are the Inter-American Institute for Global Change Research and the Gemini Telescopes initiative involving Canada, the UK, the US along with Brazil, Argentina and Chile.

In the case of Mexico/US cooperative activity in research, binational projects during the 1993–97 period focussed heavily on environmental sciences which account for 31 % of the total number of projects. Another important part of this relationship is in materials sciences and engineering, and biomedical sciences. In the aggregate, environmental, earth, biomedical and agricultural sciences together accounted for an average of 70 % of total funding from the US over the period 1993–1997. Further, the US and Mexico have developed a common bilateral S&T framework through the US-Mexico Science and Technology Foundation established in 1992. Canada has no such formal agreements with either Mexico or the USA.

With these and other indicators of collaboration, the North American partners have moved to establish common measures of the level of collaborative activity. Each country has used the various protocols established through institutions like the OECD and the OAS to provide data and indicators that can be compared and assist each country to better understand the scale and scope of collaboration.

5.2 On-going Networking Among Research Centers

There is little evidence that a great deal of cooperation is taking place between the three North American countries in support of formal networking in the context of research centers. To be sure, there are small steps between taken to engender this type of cooperative activity. For example, a virtual materials center, NAFTA Net, has been established with funding from the National Science Foundation for North American researchers in materials sciences to share information and initiatives. This venture is the outcome of a 1995 Trilateral Materials Workshop that took place in Saltillo, Mexico and a 1997 follow-up workshop in Toronto, and is designed to establish a network that facilitates cooperation and interaction among materials researchers in the North American region. The NSF has since expanded this concept to include other countries in the hemisphere.

Early efforts in the establishment of environmental transboundary issues (including scientific research) have led to the formation of the Commission for Environmental Cooperation under the auspices of the North American Agreement on Environmental Cooperation. The CEC provides analysis and recommendations to the adhering members on how troublesome environmental issues can be addressed. Aqua NAFTA 2000 is another example. It is a consortium of Canadian, American and Mexican organizations from the private and public sectors assembled to promote aquaculture development in North America. In the area of standards and metrology, a partnership of three research organizations responsible for technical standards has formed under the umbrella NORAMET. Part of the Inter-American Metrology System, NORAMET works to ensure the establishment of equivalence among the national measurement standards and of calibration and measurement certificates issued by the national metrology laboratories. It also engages in the training of technical and scientific personnel and in the collection and distribution of technical and scientific documentation.

These are early steps. No examples exist yet of truly collaborative arrangements among research centers involving significant funding. For example, each country has developed its own centers of excellence program. While somewhat different in their respective structures, the terms and conditions for these research initiatives have yet to permit explicit linkages internationally, but this may yet emerge as it becomes apparent that research issues will increasingly involve global aspects.

5.3 Education and Skills Mobility

As the globalization of research and education has proceeded, the three countries have developed extensive and growing exchanges of students, researchers and scholars. Language instruction, especially in Spanish for Canadian and American universities, is on the rise. For example, after French and English, Spanish is the most widely offered course in Canadian universities with 80% of Canadian universities surveyed reporting that Spanish has the fastest rising enrolment. North American studies programs are increasingly being offered throughout the university and college systems of each country, and the exchange of students and professors is increasing.

An Alliance for Higher Education and Enterprise in North America has been created to unite private sector interests and universities and colleges to sponsor joint activity in education, mobility and training in the region. This Alliance is the result of a series of major roundtables among the three countries, initiated in Wisconsin in 1992 and known as the Wingspread conference (so-called because of the name of the conference center in which the participants met in Racine, Wisconsin). The original Wingspread statement laid out a series of recommendations that would have the higher education and research institutions of each country develop a series of joint actions, including:

(a) inventorying existing programs and relationships;
(b) increasing the capacity and enhancing the capabilities of institutions with special emphasis on faculty development in the three countries;
(c) eliminating obstacles and reducing barriers to enhanced trilateral collaboration in the field of higher education;
(d) developing collaborative pilot projects where there exists strong mutual interest, such as management of trade relations, sustainable development, public health, North American area studies, and training in languages;
(e) taking advantage of information technologies such as distance learning, interactive video conferences, etc...
(f) enhancing use of people-to-people exchange programs;
(g) disseminating successful collaborative experience throughout the North American community;
(h) increasing and expanding student access to international education opportunities.

After considerable activity over the past eight years, (including major conferences in Vancouver in 1993 and Guadalajara in 1996) many of these initiatives have been put in place. Today, Wingspread has given rise to the Alliance and has moved the agenda forward especially in the areas of mobility of students and researchers and use of information technology for educational purposes.

5.4. Promotion of Common Systems of Scientific Policy and Technical Issues

There have been various attempts to establish more institutionalized linkages among the three countries dealing with common policy issues for science and technology. An early attempt in 1994 was an inter-agency meeting in Mexico bringing together 35 S&T policy makers from Canada, Mexico and the USA to develop a more informed awareness of respective policy efforts to promote science, engineering, medicine and technology in each country. Out of this came specific projects to have workshops on advanced materials and plant biotechnology among the three countries. Other steps followed.

Along with Mexico's Science and Technology Council (CONACYT), both Canadian and United States science agencies actively promoted Mexico's first National Science and Technology Week (the week of September 26 1994) to raise the profile of science and knowledge among the youth in that country. In 1994, the USA had celebrated its

10th year (in April) and Canada its fifth year (in October) and their respective expertizes were pooled for the first time to develop a common transboundary science theme with Mexico (e.g. the migration of the monarch butterfly and the gray whale) to promote public awareness of science and increase science literacy among youth in the different countries.

In addition to these selected trilateral initiatives, other efforts have been under way with the Deans of Engineering among the three countries taking stock of issues and opportunities; the academies of engineering in each country discussing the environmental agenda and related technology transfer; and the corporate-higher education from Canada and the USA working with Mexican organizations to establish a similar partner organization in Mexico.

6. Reaching Beyond the North American Sphere

Trilateral efforts do not limit themselves to cooperation among the three member countries. Indeed, partly as a result of the North American partnership, Canada, Mexico and the US have found themselves working closely in other multilateral fora. The North American S&T cooperation is expanding the sphere of cooperation to the Asia-Pacific region and to other parts of the Western Hemisphere. The fact that the USA, Canada and Mexico are all Pacific Rim countries and participating in such fora as APEC (Asia Pacific Economic Cooperation; as well as the Pacific Economic Cooperation Conference) has broadened the horizons of the science and engineering communities of the respective nations.

The development of a Pan-American effort in S&T, triggered by the Organization of American States' Common Market for Scientific and Technological Knowledge, the Inter-American Institute for Global Change Research, and the Pan-American Advanced Studies Institute program will also strengthen the capacity of science and knowledge in the developing and newly developing economies in these respective regions. The first initiative is designed to promote projects of scientific and technological integration (such as exchanges and relocation of highly qualified human resources; development of networks of centers of excellence); encourage efforts for joint research in sustainable development issues; as well as projects to support innovation in the Western Hemisphere (such as support to develop incubators for technological enterprises). The second provides a network of researchers to undertake research and training in climate change affecting the Hemisphere. The last of the initiatives is modeled after the NATO Advanced Studies Institutes and is designed to disseminate advanced scientific knowledge and stimulate training and cooperation among researchers in the Americas.

The potential liberating force of science and knowledge (with an increased capacity of science to deal with important social problems and an informed public to animate the debate) will be felt as the pace of this North American cooperation increases opportunities for both developing countries and the developed world. The spread of science through new regional, cooperative trading and economic fora will likely engender the need to re-think how foreign policy, trade relations, international arrangements and development assistance is produced. In this context, the increased hemispheric cooperation in S&T received a shot in the arm with approval of the action plan from the Miami Summit of the Americas and subsequent Summits in Chile and Canada. One of the key elements of the plan approved by heads of government was the need to improve S&T cooperation and to review existing efforts in promoting S&T in the region, including the greater use and dissemination of information technology to enhance both economic and social goals. The latter was a major cross-cutting theme of the Summit of the Americas hosted by Canada in Québec City in April 2001 with the release of a Statement on Connectivity.

But none of this impetus will develop fruitfully unless national governments recognize the merits of knowledge spillovers in the region and the need to promote and enhance international linkages in education, science and technology. Preoccupation at the policy level over so-called 'national/ domestic security' to protect knowledge will only succeed in hampering this agenda, and possibly harm the development of a new model for economic and scientific integration. As each nation responds to the issue of growing knowledge-based economies, the spirit of a North American and, eventually, American partnership, should emerge as a central plank of respective policy agendas.

7. An Agenda for a North American Innovation Approach

As the discussion above has shown, cooperation among the research systems of the three North American countries is increasing. The indicators of formal cooperation show a rise in trilateral activity of various sorts, though these measure only national activities. There may be early signs as well of emerging cooperative activity at sub-national levels as provinces and states in the three countries look to enhanced cooperation to strengthen their respective clusters of technical and research excellence.

Economic integration is clearly on a path that has put forward a series of questions and debates as to the scale and scope of impact on national priorities and social and cultural issues. Despite some early institution-building of North American instruments for cooperation (in environment, and trade) little exists in other spheres of cooperative activity. Much of what does take place in S&T for example is bilateral in scope or multilateral involving other members of the Hemisphere.

Should there be a North American Foundation for S&T Cooperation? Should there exist a North American Academy of Sciences and Humanities? A North American University? A trilateral blue ribbon commission on common research infrastructure? North American prizes and awards for scholarships and post-graduate studies in S&T? Should the three countries get together to discuss key approaches to mobility of skilled personnel and accreditation of key professions?

What advantages can emerge from a trilateral approach to technology foresight or technology roadmaps among private sector firms and research organizations in each country? Are there benefits to sharing the design of new policy initiatives to advance innovation and R&D in North America?

These are questions that have been little debated in the region, in part because the cultural specificities of each country have led to little common language for this discussion. One thing we do know: with the increasing use of electronic communication among the research communities of each country, the enhanced mobility of researchers and scholars, and the growing trend to cooperate on transboundary issues requiring scientific and technical advice or input, we can only expect that such instruments of coherence and coordination will gradually arise. Maybe not in the form that will be highly structured (as in the European case), but at least from the perspective that will place S&T at the service of the North American community.

Glossary

APEC:	Asia-Pacific Economic Cooperation.
CEC:	Commission for Environmental Cooperation.
CONACYT:	National Council of Science and Technology of Mexico.
NAFTA:	North American Free Trade Agreement.
NATO:	North Atlantic Treaty Organization.
NORAMET:	North American Metrology system.
NRC:	National Research Council of Canada.
OAS:	Organization of American States.
OECD:	Organization for Economic Cooperation and Development.

Bibliography

Stann J., ed. (1993). *Science and Technology in the Americas,* Washington, DC: American Association for the Advancement of Science, pp144. [Reviews and provides a series of examples of potential for Pan-American research collaboration, this report is the result of a conference held in May 1992 in Washington D.C. organized by the American Association for the Advancement of Science, the National Science Foundation and the Organization of American States.]

Cimoli M., ed. (2001). *Developing Innovation Systems: Mexico in a Global Context,* London: Continuum, pp385. [An examination of Mexico's innovation system with an extensive description of key industry and technology sectors.]

CEC (2000). *Towards a European Research Area,* Brussels: Commission of the European Communities, pp18. [Consultation paper prepared by the European Commissioner responsible for research and has provided the context for enhanced cooperation among EU members States in this evolving area of research and innovation.]

CONACYT (1997). *Indicadores de Actividades Científicas y Tecnologicas.* México, pp262. [One of the annual reports produced by Mexico's national science and technology council on the statistics associated with expenditures and investments in S&T. A more recent compendium of statistics for S&T in the whole of the Spanish-speaking world can be found in *El Estado de la Ciencia: Principales indicadores de ciencia y tecnología Iberoamericanos/ Interamericanos 2000,* www.ricyt.edu.ar]

Dufour P. (1995). North American collaboration in science and technology: a developing model for the Americas, *Interciencia,* **20**(2), pp. 101–104. [Provides a series of examples of cooperative activity among the three North American countries in research, education and innovation.]

Dufour P. et al. (1998). Using science and technology as strategic instruments for Canada's foreign relations with Latin America, *Canadian Foreign Policy,* **5**(2), pp. 129–147. [Makes a case for why Canadian foreign relations and diplomacy should pay more attention to S&T cooperation with Latin America to enhance education, economic and trade development, and technological cooperation and other issues.]

Knight J. (2000). *Progress and Promise: The AUCC Report on Internationalization of Canadian Universities,* Ottawa, pp97. [Provides the results of a survey of Canadian universities and the scale and scope of their agreements with Latin American countries.]

North American Higher Education Cooperation: *Identifying the Agenda* (1992). Statement from the Racine, Wisconsin Wingspread Conference, pp4. [This is a statement of senior university officials, private sector representatives and research organizations on the need to enhance cooperation in education and research in North America.]

Wagner C. S.,and Berstein N. (1999). *US Government Funding of Cooperative Research and Development in North America,* Washington, DC: Science and Technology Policy Institute, RAND, pp47. [Provides an extensive description of spending by US departments and agencies in international collaborative R&D ventures with partners across the globe; a 2001 update to this inventory is also available at www.rand.org]

Welsh J. (2000). Is a North American generation emerging? in *Isuma,* **1**(1), pp. 86–92. [A speculative piece on policy issues affecting thinking about a North American region.]

Biographical Sketch

Paul Dufour is the Ministerial Assistant to Canada's Secretary of State for Science, Research and Development, where he is responsible for providing advice on matters affecting Canada's S &T policy as well as serving as an interface with the scientific and technological community. Prior to this position, he was senior analyst with the Science and Technology Strategy Directorate at Industry Canada where he was responsible for advising the Government on domestic and international S&T matters, especially with regards to implementation of Canada's S&T strategy. He is also international S&T relations' adviser with the Secretariat to the Prime Minister's Advisory Council on Science and Technology. Mr. Dufour is a Senior Fellow of the International Science Policy Foundation and Research Associate with the Programme of Research on International Management and the Economy, University of Ottawa. Born in Montreal, Mr. Dufour was educated at the Université de Montreal and Concordia University in the history of science and science policy, and has had practical S&T policy experience for over two decades. He lectures regularly on science policy, has authored numerous articles on international S&T relations and Canadian innovation policy. He is series co-editor of the Cartermill Guides to World Science (Japan, Germany, Southern Europe and the United Kingdom are the most recent books in this series) and North American editor for the revue Outlook on Science Policy.

SCIENCE AND TECHNOLOGY POLICY IN JAPAN

Yukiko Fukasaku
Organization for Economic Co-operation and Development (OECD), Paris, France

Sachiko Ishizaka
International Council for Science (ICSU), France

Keywords: Japan, science, technology, public policy, modernisation, industrialisation, development, R&D, innovation, technology transfer, education and training, innovation system.

Contents

1. Introduction
2. Features of the Early Development of Science and Technology Policy
3. Post-war Catch up and S&T Policy
4. Japanese Science and Technology in the Recent Decades
5. Framework for Science and Technology Policy Making and the Current Policy Trends
6. Issues for Science and Technology Policy in the New Century
Glossary
Bibliography
Biographical Sketches

Summary

Japan was the first country in the non-European cultural sphere that succeeded in industrial development based on modern science and technology. This was done through the process of catching up to the advanced industrial countries of the world, by building up a research and innovation system particularly suited to the goal of catching up, stressing learning processes such as reverse engineering and improvement engineering. Because resources were concentrated on catching up, R&D investments have been heavily concentrated in and by the industrial sector, in terms of both funds and human resources. Because of the need to learn quickly, researchers trained in the engineering disciplines predominate in the industrial sector.

1. Introduction

The development of science and technology in Japan is distinct in that it was the first country in the non-European cultural sphere to attempt to absorb the western scientific and technological tradition, and along with it, launch the process of modern socioeconomic development. The process started in the mid-nineteenth century, and as witnessed in the 1970s and 80s, after the devastation of the country during the Second World War, the nation achieved industrial success based upon superior technological advances, unprecedented for a country of non-European origin. Public policies toward science and technology played a key role in this industrial development.

2. Features of the Early Development of Science and Technology Policy

The modernization in the mid-nineteenth century, started with the sociopolitical developments that opened the country from the isolationist policy under the Tokugawa feudal regime in the 1850s to its overthrow and the launching of the full modernization process by the new Meiji government starting in 1868. The isolationist policy under the feudal *ancien regime*, meant that for more than two hundred years up to the mid-nineteenth century, the country was shut out from foreign trade as well as foreign intellectual influences, save for the very limited trade with China and Holland, by which some knowledge of scientific and industrial developments in Europe trickled through to the few Japanese scholars dedicated to western learning.

This line of inquiry, *rangaku*, provided the intellectual basis and resources once Japan launched the modernization process and the absorption of European science and technology. Beside this line of intellectual inquiry, the *ancien regime* had left the legacy of well-educated population and the continued aspiration for more education and training in the new Meiji era, which provided a fertile soil on which western science and technology could be transplanted.

The transplantation process even on a fertile soil was not easy. Here, public policies toward science and technology played a key role from the early days of the Meiji era. The modernization process in general was highly centralized, and the government took leadership in replacing traditional intellectual scholarship that was mainly based on schools of learning in the Chinese tradition by western scientific disciplines. The adopted policy was to import western science and technology in the restructured and/or newly founded universities and other educational institutions, government research institutions and to encourage industrial development using western technology in the private sector, as well as in the public enterprises. European and American scientists and engineers were employed in the universities and other educational/research institutions as well as in many industrial enterprises, both public and private. The government took initiatives in sending students abroad, and returning students rose up in the academia eventually to replace the foreign scientists and engineers. In this westernization process, the traditional sciences and professions, notably Chinese medicine was side-tracked as para-medicine, and the certified medical profession fell exclusively to those educated in the western medical sciences.

Thus, as a 'late-comer' to industrialization and modern scientific and technological development, the strategy adopted in the second half of the nineteenth century was catching up to the advanced countries of the west through transplantation of western scientific inquiry and technology transfer. Although a latecomer, the Japanese catching-up process in modern science and technology started not much later than in the United States or in Germany, and was facilitated by the institutionalization of science and technology which had started in Europe and America in the nineteenth century. Therefore, compared to the developing countries of the twentieth century, especially in the post second world war era, Japanese catch-up had started in a privileged epoch in the development of modern science and technology.

In the research sector, the universities were structured according to European universities, professional societies were founded in analogous disciplines, and the knowledge frontier was surveyed and brought back by researchers who normally spent a part of their research career in a foreign university or research institution. Up to the Second World War, Japan may not have produced world famous scientists, but some researchers did achieve first class research, in the physical sciences as well as biomedical sciences.

On the industrial technology side, public policies supported intense efforts to import technology and improve upon them to build up indigenous capacity to innovate. These efforts were facilitated by training of appropriate scientific and engineering human resources, in the universities and technical schools in the formal education sector, and on-the-job training of technicians and workers. Also, in some large manufacturing enterprises, as exemplified in the case of Mitsubishi Nagasaki Shipyard, in-house research became an institutionalized activity in the early years of the twentieth century, which facilitated the technological learning and adaptation of imported technologies. Therefore, training and research were identified as important elements of the technology transfer process, and the institutions that underpinned the technological learning process were laid in the decades since early Meiji and were in place well before the Second World War. The basic scientific and technological policy as well as the scientific and technological infrastructure was revived after the second war and facilitated the catch-up process after the Second World War.

3. Post-war Catch up and S&T Policy

In the post-war phase, intense efforts to catch up to the USA and the advanced European countries in the level of economic development and investments in science and technology were pursued. These efforts, as in the pre-war phase, were underpinned by a close co-operation between the government and the large enterprises sector. In public policy, this co-operation was the result of the articulation of S&T-centered policies adopted by the Ministry of International Trade and Industry. Rather than basing long-term development strategy on traditional theory of comparative advantage, but policy makers rather sought solution for Japan's post-war difficulties in enhanced technical efficiency and innovations in production and the promotion of the most advanced technologies through new investments. For this a mode of working which depended upon a continuing dialogue on questions of technological development with university and industry, in effect, a mode of continuous consultation with S&T community was systematized.

In practice, the catching up process, as in the pre-war phase, was based on even more intensified technology importing and improving. Technological learning and adapting was pursued through such processes as 'reverse engineering' and 'improvement engineering' that have been used to characterize the Japanese mode of building domestic technological capability through technology imports. These processes, in turn, enhanced the integration of production and innovation within large firms, which had already started before the war.

Increase in investments in science and technology was more rapid than in other advanced countries, and the increasing supply of university-trained graduates in scientific and especially, the engineering disciplines supported these developments in the industrial sector. The post-war catching up in Japan reveals a distinct institutional and social framework for innovation, a 'national system of innovation', characterized by long-term public industrial development policy, the integrated approach to production and innovation in large firms, and the public and private investments in education and training and R&D to support industrial development, which were different from other industrialized countries.

4. Japanese Science and Technology in the Recent Decades

The oil crises of the 1970s posed a real threat to the overwhelmingly petroleum-dependent Japanese economy. It did put an end to the so-called 'high growth' era, but compared to other advanced countries, it became clear by the 1980s, that Japan was able to absorb the adverse effects of the oil crisis and her industrial sector was showing strong competitive performance. This was often attributed to the innovative capacity of the industrial sector, which was enabling Japan to advance ahead of the other advanced countries by rapidly adopting information technologies in the industrial sector. However, the spectacular performance of the 1980s gave way to a serious downturn in the 1990s. Weakness of the public research base is pointed out to be a serious issue to be addressed in order to preserve innovativeness of the economy. This section reviews the trends in indicators of science and technology as well as new policy initiatives that Japan is launching to activate its research base. (Unless indicated otherwise, the data source in the section is *White Paper on Science and Technology, 2001*, compiled by Science and Technology Agency of Japan.)

4.1 R&D Expenditures

Figure 1 shows the trend in R&D expenditures of the major OECD countries since 1975. This does show that as a proportion of GDP, R&D expenditures in Japan increased most rapidly since the mid-1970s, and is at the highest level of 3.25% among the OECD countries. In the absolute level of funding it is second to the United States. In the 1990s, the total R&D expenditure decreased in the first half, but has made a remarkable increase since 1995, reflecting the increase in public R&D investments.

A comparison with other major R&D powers in the share of R&D financing and performance (Fig. 2) shows a larger contribution of industry in R&D activities in Japan, especially in financing of R&D. The government funding for R&D has been increasing since 1996, as a result of the current policy of promoting basic research; however, the share held by the government sector is still relatively low compared to other major R&D powers. One reason for this is the very low share of defense research in government R&D investments in Japan. For industrial sector R&D, the influence of a low tax rate and favorable tax treatment of R&D, which has stimulated R&D investment in the private sector, is by no means negligible.

SCIENCE AND TECHNOLOGY POLICY - Vol. II – *Science and Technology Policy in Japan* - Yukiko Fukasaku and Sachiko Ishizaka

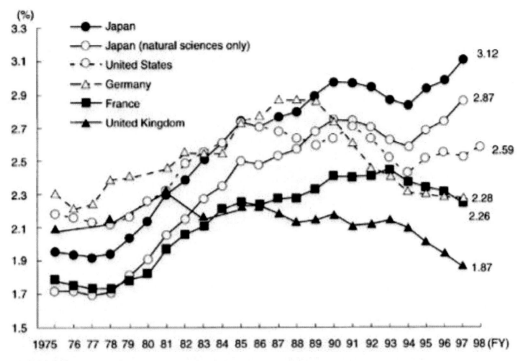

Note:
1. For comparison, statistics for all countries include research in social sciences and humanities.
    The figures for Japan show also the amount for natural sciences only.
2. In Japan, the software industry became a new survey target in FY 1996.
3. U.S. figures are for calendar years, while figures for FY 1997 and FY 1998 are provisional.
4. Germany in FY 1996 and FY1997, and France FY 1997, show provisional figures.
5. Fiscal years where there are no statistical figures for the U.K. and Germany are
    linked in a straight line between the fiscal years before and after.

Fig. 1. R&D expenditures as a percentage of GDP in selected countries.

In R&D performance, industry spends approximately two thirds of the total R&D expenditure in Japan which is similar to other R&D powers. However, the flows of R&D funds between sectors, in particular between the government and universities and between government and industry, are less dynamic in Japan. In Japan 48% of government R&D funding is distributed to the universities, 40% to government research institutes and only 0.12% to industry. The industrial sector spends 99% of its own R&D funds. Share of business in the funding for research performed by government and university is also very low in Japan. In countries such as the USA and France, a larger flow of funds between the public and private sectors is partly explained by defense-related contracts.

©Encyclopedia of Life Support Systems (EOLSS)

171

(1) Financing

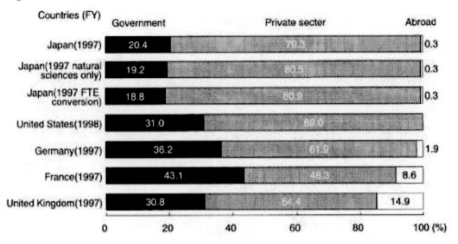

(2) Performance

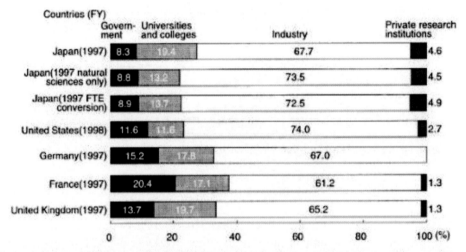

1. For comparison, statistics for all countries include research in social sciences and humanities. The figures for Japan show also the amount for natural sciences only.
2. In the (1) Financing column, the private sector includes any sector other than the government and abroad.
3. U.S. figures are for calendar years and provisional. German figures are provisional. In addition, Germany's research expenditures at "private research institutions" under (2) Performance, are included in "government" research institute.

Fig. 2. Share of R&D expenditures financing and performance sector in selected countries.

4.2 Human Resources

Total number of researchers in Japan is reported as 733 000 in 1999, which is the second largest as a single country following the USA. Japan had the highest number of researchers per 10 000 population as well as the number of researchers per 10 000 labor force among major R&D powers in 1999. These numbers are increasing steadily in

Japan in contrast to a slower trend in other countries such as USA, Germany and UK. Approximately 60% of those researchers are in the private sector, while 36% are in universities and less than 5% in the government sector. A greater percentage of researchers in industry also characterizes the distribution of scientific human resources in USA and UK (58.9% for Japan, 79.9% for US and 56.8% for UK). The number of researchers in the industrial sector is increasingly more rapidly (at an annual rate of 3.25%) than in the universities or in public research institutes.

In industry, 90% of the researchers are concentrated in manufacturing, out of which about 30% are in the area of telecommunications, electronics and electrical instruments in 1998. As for the academic background of researchers in industry, 65% have engineering background and only 22.5% in basic sciences. Particularly low is the percentage of 1.2% in the biological sciences. This contrasts with the pattern in public sector research whose researchers show more balanced academic background profiles (Fig. 3).

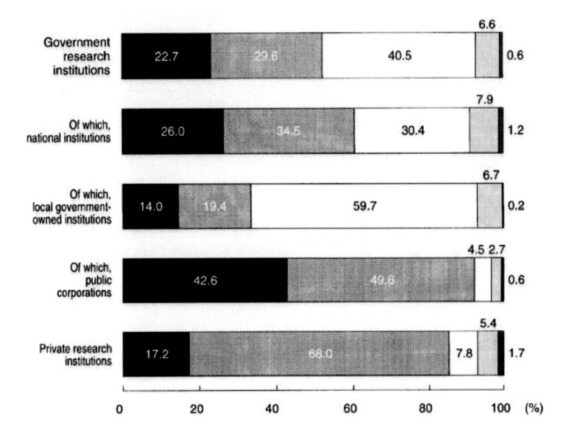

Fig. 3. Share of researchers in research institutions by organization and field (1998).

In universities, by field, health ranks 55.3%, followed by engineering at 28.2%, physical sciences at 10.1% and agricultural sciences at 6.4% (1999). The number of researchers in natural sciences and engineering is increasing rapidly, in particular in the fields of life sciences and electronics (Fig. 4). Japan is second in the number of university degrees awarded following the USA and has a higher ratio in engineering. In Japan, the number of postgraduate degrees awarded has been increasing in recent years.

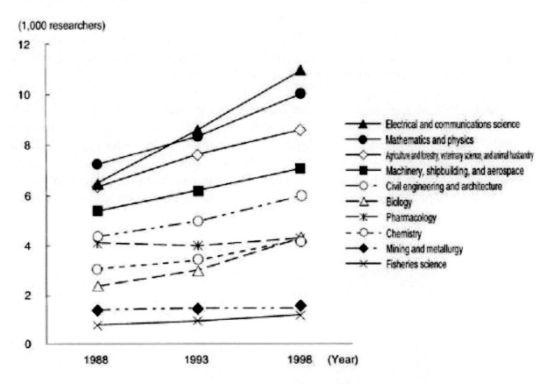

Fig. 4. Trends in the number of researchers at universities and colleges, by sector.

4.3 Research Outputs

Scientific publication is a major indicator for evaluating the performance of researchers in many countries. Scientific publications are increasing rapidly in Japan. Between 1988 and 1998, there was an increase of 2.5 times. In terms of the number of publications, Japan has ranked second in the world since 1989; however, the number of citations in scientific papers are slightly lower than the average of major R&D powers. By field, relative citation frequency is 0.38 for computer sciences, and 0.7–0.8 for life sciences, while for chemical and agricultural sciences, astrophysics and materials sciences, it is 0.9–1.0 (international average at 1.0).

In terms of relative comparative advantage (RCA) which is calculated by dividing the share of the number of papers in a field in a country by the corresponding share of the field in the world, Japan has a relative comparative advantage in chemistry and physics/materials sciences, but that of chemistry is decreasing rapidly. Clinical medicine is increasing rapidly, but its RCA for 1998 is less than the world average at 0.86.

As for the number of patents applied in major OECD countries (Fig. 5) which includes Patent Co-operation Treaty (PCT) applications and European Patent applications (EPC) in 1997, Japan has the second largest patent applications after the USA. Japan, in fact, was the leader until 1989, but since has been exceeded by the US and the EU. The ratio of patent applications by the Japanese in foreign countries in 1997 is 47% which is relatively small compared to 80–90% in other major R&D powers. Patent applications by foreign inventors are about 16% of the total applications in Japan, which is relatively low compared to more than 40% in US and more than 80% in France and the UK. On the other hand, patenting activities of Japanese are active abroad. In the USA, Japan is the country of origin of inventors with the highest number of patents awarded. According to the NSF Science and Engineering Indicators 2000, 1997 data show that in the US, Japanese inventors are prominent in technology classes associated with photography, office machines and consumer electronics industries as well as information technology.

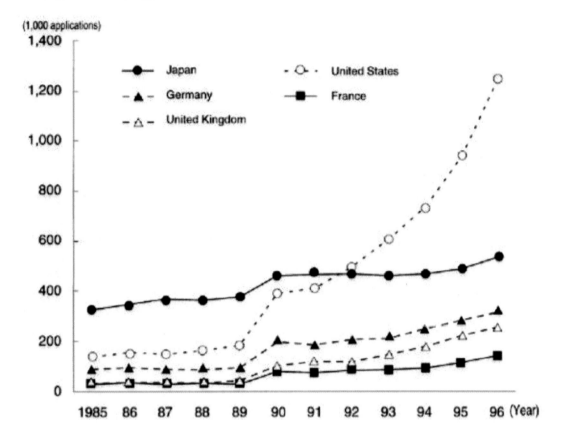

Fig. 5. Trends in the number of patent applications in selected countries.

Japan's technology trade balance with other advanced countries has been improving over the long run (Fig. 6). Technology imports from North America and Europe account for most of the Japanese technology imports. The US, in particular, has accounted for 71% of total technology imports. Technology exports exceed imports in most manufacturing sub-sectors including transport equipment, chemicals, iron and steel, textiles and electronic machinery. The surplus in the motor vehicle industry mainly accounts for the surplus in the transport equipment industry.

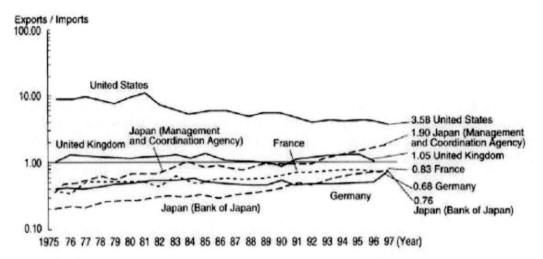

Fig. 6. Trends in technology trade balance of selected countries.

Notes:
1. The amounts are converted into US dollars, based on IMF exchange rate.
2. Bank of Japan refers to the "Balance of Payments Monthly," compiled by the Bank of Japan. Management and Co-ordination Agency refers to the "Report on the Survey of Research and Development," compiled by the Statistics Bureau, the Management and Co-ordination Agency.
3. The figures are totals for the calendar year; the fiscal year is used only for the fugues of Japan Management and Co-ordination Agency.
4. The major reasons for differences between the figures provided by the Bank of Japan and those provided by the Management and Co-ordination Agency are as follows.

The size of high-tech product exports is an indicator of a country's international competitiveness in science and technology, since high-tech industries require large investments in R&D as well as sophisticated technology during their manufacturing process. In 1996 Japan held the second largest export market share for high-tech products among the OECD countries at 17.6% while the share of US, Germany, UK and France were respectively 22.4%, 12.3%, 9.3% and 7.9%. By type of industry, Japan had the highest share in 1996 in communications equipment and electrical equipment. In pharmaceuticals and aerospace industry, the share is relatively low. As for balance of trade in high-tech industries, Japan's balance of trade ratio was larger than other countries, but now is coming closer to balance because of increase in imports.

5. Framework for Science and Technology Policy Making and the Current Policy Trends

5.1 Administrative Structure

The current framework for public policy making toward science and technology was defined in 1959 when the Council for Science and Technology was set up. This advisory body is chaired by the prime minister and is comprised of major cabinet ministers including ministers for science, education, industry, economy and finance in addition to several distinguished experts from academia and industry. The council responds to the Prime Minister's inquiries on basic science and technology policy including comprehensive long-term research goals and measures to promote essential research areas. The Council annually draws up the guideline on the priority for promoting science and technology, which is normally reflected in the research budget allocation for the following year. These guidelines also set the priority areas of funding of the Special Co-ordination Funds for Promoting Science and Technology, a budget item created in 1981. This accounts only for about 1% of the government research related budget, however, these funds are not allocated to a specific ministry or agency, but are allocated directly by the Council to projects conducted jointly by national research institutes and universities as well as private companies.

Policies related to science and technologies fall under the responsibility of several ministries including Ministry of Education (Monbusho), Science and Technology Agency (STA), Ministry of International Trade and Industry (MITI), Defense Agency, Ministry of Agriculture, Forestry and Fisheries, Ministry of Health and Welfare, Ministry of Post and Telecommunications, Ministry of Construction, Ministry of Transport and Environment Agency. Each ministry and agency has jurisdiction over the promotion of its own S&T policies and programs according to its missions. Largest percentages are claimed by the first three ministries. Of the total government S&T budget of 3,284 billion yen in fiscal year (FY) 2000, 42% was allocated to Monbusho, 23% to STA and 16% to MITI.

Monbusho's budget is mainly used to finance academic research in universities through grants-in-aid, which provide a core funding in areas of fundamental research. The ministry also provides special support to basic research in universities and research institutes in selected areas including astronomy, accelerator science, space science, nuclear fusion study, information-related research, global environmental science and life sciences. It also provides support for establishing Centres of Excellence and postdoctoral fellowships.

Science and Technology Agency, since its establishment in 1956, has been responsible for the formulation and implementation of government S&T policies and the co-ordination of policies and programs related to science and technology developed by other ministries and agencies. The STA is responsible for advancement of large-scale research projects in nuclear energy, space and ocean research, as well as for promoting R&D in pioneering fields such as earth sciences, disaster prevention, special materials, life sciences and aeronautical engineering. The agency also serves as the secretariat of the Prime Minister's advisory bodies which in addition to the Council for Science and Technology, includes the Atomic Energy Commission, the Space Activities Commission, and the Council for Ocean Development.

MITI is charged with the development of comprehensive industrial technology policy of Japan. Its Agency of Industrial Science and Technology assumes major responsibilities in the formulation of industrial science and technology policy and related R&D programs and conducting R&D on industrial science and technology in its 15 research institutes. Its R&D programs have included the so-called Large Scale Projects which included the VLSI, the very large-scale integrated circuit project, an internationally well-known projects that is acclaimed to have fostered the very competitive semiconductor industry in Japan. The agency is also responsible for the development of budget allocations, investments and loans, and tax incentives for the promotion of R&D activities in the private sector.

A major administrative reform of the Japanese Government was implemented in January 2001, with the objective of streamlining the various ministries and agencies and reducing the number of government employees by 25 percent by 2010. In particular as of January 2001, Monbusho and STA were merged and a new ministry, Ministry for Education, Culture, Sports, Science and Technology Agency (Monbu Kagakusho, or MEXT) was born. Several of the nuclear regulatory functions of the former STA were transferred to MITI, whose name was changed to METI (Ministry for Economy, Trade and Industry). More significantly, the Council for Science and Technology Policy was established within the Cabinet Office, replacing the former Council for Science and Technology. The purpose of the new council, which has the support of full-time staff members within its secretariat, is to provide strengthened policy-making and co-ordination capabilities, through the appointment of an independent state minister for Science and Technology Policy.

5.2 The Science and Technology Basic Plan

The past decade has witnessed a re-orientation of policies toward science and technology in Japan. As clear from the resource allocations for research discussed in the previous section, compared to other large advanced countries, the share of the government in total financing of research is relatively low. This implies the relative weakness of the basic research sector. It may be noted also that most researchers in the industrial sector are trained in the engineering disciplines.

Strengthening of the basic research base by increasing public investments in research and revitalising the university and the public research base is the aim of the Science and Technology Basic Law that took effect in 1995. There has been a recognition that the globalization of the economy, the rapid aging of the population, and the global scale problems that the humankind was faced with called for strengthening of the role of science and technology. The law laid down basic framework for future science and technology policy in Japan, aiming at achieving a higher standard of science and technology with an emphasis on creativity through the implementation of comprehensive and systematic science and technology policies. The law has recognized the increasing role of the government in the promotion of basic research that would meet the needs of economic development and social welfare.

In accordance to the provision of this law the Science and Technology Basic Plan was adopted by the Cabinet in 1996. The plan's primary aims are:

- To promote R&D that responds to social and economic needs, for example, R&D leading to the creation of new industries in response to the rapid progress in IT technologies, and meet the need to solve global-scale problems, improving human health and preventing natural disasters.
- To promoted basic research that contributes to enriching the knowledge base, i.e., discovery of new fundamental scientific laws and principles.

In order to achieve these goals, the plan identified several issues that were crucial to the Japanese research and innovation system, including 'flexible, competitive and open environment for science and technology', 'upgrading of overall industrial, academic and governmental R&D capabilities', and 'facilitating the use of R&D results by the public'. Key policy recommendations of the plan are found in the text box.

Key recommendations in the Basic Plan

- Securing and Training Researchers and Improving the R&D System:
  - securing young researchers by improving education and research in graduate schools, expanding scholarship opportunities, and increasing number of postdoctoral researchers;
  - increasing research supporting staff in national research institutes and national universities;
  - introduction of the fix-term employment system for researchers in national research institutes;
  - enhancing co-operation and exchange system among sectors, regions and nations, including review of regulations regarding the leave of absence, side job permission to promote co- operation with the private sectors;
  - developing the mechanism of appropriate review and execution of fair and strict evaluation.
- Development and improvement of R&D Infrastructure:
  - improving research infrastructure in national institutes and universities including research facilities and equipment, such as IT infrastructure;
  - expansion of various types of funds;
  - expansion of competitive research fund offered based on applications.
- Upgrading Research in Private Universities.
- Promotion of R&D by the Private Sector; Application of R&D Achievements by the Public Sector:
  - encouraging and facilitating transfer of research results to private sectors, allowing the priority license;
  - revising job-invention regulations to allow researchers of national institute to hold the patent right individually.
- Promotion of International Co-operation.
- Promotion of Science and Technology in Regions.
- Promotion of Learning and Understanding of, and Interests in Science and Technology:
  - improving science and engineering education to enhance public S&T understanding, in particular, the younger generation, and establishing the national consensus on promotion of S&T.

Text Box 1.

To implement these recommendations, the plan recommended doubling of government research investments, making its percentage to GDP comparable to that of other major industrialized countries by the beginning of the twenty first century, namely increasing the public S&T expenditure to 17 trillion yen in the first five years of the plan. The plan encourages researchers to seek research funds from different ministries and agencies as well as the industry for the purpose of exposing researchers to a more competitive environment, provide opportunities to use research funds which allow researchers to have more flexibility in terms of their use, and to give researchers more freedom to choose sources of funding. The plan also emphasizes training of researchers, researcher exchange and improving R&D infrastructure.

Following are some programs that have been implemented in accordance to the recommendations of the plan:

- *Fostering Young Researchers.* In view of the fact that the number of post doctoral researchers are much fewer than in the US and Europe, a 'Programme to Support 10 000 Postdoctorals' was launched in FY 1995 to foster young researchers at the postdoctoral level. Accordingly, the number of postdoctoral researchers supported by government sponsored fellowship programs has more than doubled since 1995.

- *Competitive Research Funding Programs.* One recommendation of the Basic Plan was to introduce more diversity and competition in funding basic research, in addition to increasing funding through the existing programs of Grants-in-Aid for Scientific Researchers by Monbusho and Special Co-ordination Funds for Science and Technology. Hence several new programs have been created by several ministries to support basic research in universities through competitive selection of publicly solicited research proposals. These programs amount to a total of nearly 60 billion yen in 1999 and includes, for example, the STA 'Core Research for Evolutionary Science and Technology'.

- *R&D Evaluation.* Following another recommendation of the Basic Law, a National Guideline on the Method of Evaluation for Research and Development was adopted in 1997. In accordance to this guideline, several ministries and agencies have conducted an evaluation of R&D activities falling under their responsibility. A review of these evaluations conducted by the Science Council of Japan revealed insufficiency of follow-ups to research evaluation activities. This leaves the adequate integration of evaluation results a future issue.

- *Promotion of Co-operation and Exchange between sectors and Transfer of Technology.* Some programs have been adopted to facilitate the transfer of the results of publicly funded research in universities and national research institutes to the private sector. Monbusho has launched a project to support the establishment of venture business centers in national universities to facilitate joint research projects between university researchers and entrepreneurs and initiate venture businesses. In addition, a new law (Law for Promotion of Transfer of Research Results at Universities to Private Sector Enterprises) was promulgated to provide support measures to institutions aiming at facilitating transfer of inventions at national universities to private industry for commercialization.

- *Promoting Public Understanding.* In order to promote understanding of the general public in promoting science and to respond to the diminishing interest in science and technology among the young people, programs have been designed to improve the science curriculum in schools and to familiarize public to science and technology through, for example, increased support to science museums. More funding is being allocated for activities to increase public awareness of science on such issues as cloning, organ transplants, genetic engineering and nuclear energy development.

The reforms undertaken under the Basic Plan through FY 1999 have been assessed by the Council for Science and Technology. The target increase in total government R&D investment of 36% in five years has been achieved. The number of post-doctoral researchers has hit the target of 10 000. The amount of competitive research grants has also increased considerably. However, the targets in the improvement of research infrastructure at national universities, including renovation and extension of facilities, replacement of equipment have not been met, since the budget allocated for these was crowded out by other items such as financing competitive research. Also, the introduction of fixed term employment for professors at national universities has remained at 0.1% of the total employment. Although share of Japan in the number of research papers and citations have increased, citation frequency per paper is still lower than US and the major European countries. These trends imply room for upgrading the quality of research.

Based on these assessment results, the government is preparing the second Basic Plan for 2001-2005. The key issues to be addressed in the new Plan include the following:

- Continued increase in total government R&D investments.
- Strategic investment of funds in focused areas such as IT, life sciences, material and environment sciences.
- Enhancement of research co-operation among sectors, including removing administrative obstacles for the flow of researchers, funds, and outcomes, by increasing flexibility in working hours and income of public researchers. Fixed term employment of young researchers needs to be expanded.
- Continued increase in the share of competitive research funds in the total government funding to the level of the major industrial countries.
- Enhance support measures for young researchers.
- Fostering research assistants to enable researchers to concentrate exclusively on research activities.
- Improving facilitates and equipment which are obsolete, and expanding space devoted to research activities in universities.

6. Issues for Science and Technology Policy in the New Century

Through the process of catching up to the advanced industrial countries of the world, Japan built up a research and innovation system particularly suited to the goal of catching up, which stressed learning processes such as reverse engineering and improvement engineering. Its scientific and technological resources were concentrated for this purpose. The imperative for catch up explains the relative concentration of R&D investments in and by the industrial sector, in terms of both funds and human resources. The learning imperative explains the dominance of researchers trained in the engineering disciplines in the industrial sector.

The excellent Japanese industrial performance in the 1980s demonstrates the efficacy of its research and innovation system for catching up. In contrast, its mediocre performance in the 1990s also demonstrates its inappropriateness of the current challenges facing the science and technology system in Japan. As the trends discussed above demonstrates, in the process of catching up, the research and innovative capacities became over concentrated in the industrial sector. Because innovation in the new century depends upon developments in new technology areas, such as information and communication technology and biotechnology, which are linked more directly to basic research, the Japanese research and innovation system clearly needs to be re-oriented.

One direction that science and technology policies for the new century need to take is the strengthening of the public research base and the promotion of basic science. As discussed above this has become the main thrust of science and technology policy in the 1990s and is a policy reform in the right direction. However, simple increase in funding and other financial support measures may not be adequate. The basic research system needs to be redesigned to enhance both competition and co-operation. Increasing share of competitively allocated research funds in the public sector is a good start, however ultimately, the entire research system including the universities needs to become more internationally competitive. Internationalization of the research system, or making the research system more open to international co-operation as well as competition should be an integral part of policy reforms. Introducing more flexibility in funding and execution of research, including the administrative reforms to remove barriers in flow of funds and researchers is also an issue to be addressed.

Another more difficult challenge is to how to deal with the R&D capacities that are 'locked-up' in the industrial sector, especially the large manufacturing firms. The Japanese industrial sector has been innovative, but its innovation system and the corporate structures are not necessarily geared toward developing and exploiting new technologies fully. In the 1980s, the Japanese industrial sector, especially the manufacturing subsector was competitive to the extent Japan seemed to open up a technological gap *vis-a-vis* other advanced countries by harnessing information technologies in manufacturing processes. In other words, Japan seemed to be moving quicker to a knowledge-based economy than other countries. However, the developments in 1990s demonstrated that this innovative trend was limited to the large manufacturing firms. The Japanese industrial sector clearly lacked the flexibility to exploit potentials of information and communication technologies and biotechnology in other sectors, also in fostering smaller firms that produced a range of new goods and services based on new technologies.

Therefore, a big challenge for science and technology policy in the new century is to facilitate the restructuring of the industrial sector so that the innovative capacities accumulated in the large manufacturing firms are liberated and reharnessed. This will imply restructuring the large enterprises to smaller specialized firms that are better adapted to develop and exploit the potential of new technologies. This means that the new innovative dynamism, may not come directly from the public research base which probably needs sometime to revitalize itself, but from tapping the innovative capacities of the industrial sector, the most innovative segment of the Japanese science and technology system.

This policy challenge is likely to call for new forms of public/private co-operation and partnerships. In addition to moving the public research base closer to the market by enhancing university entrepreneurship, for example, some policy measures will have to be taken to make the research base more responsive to the needs of the industrial sector. This would mean creating new forms of public/private research collaboration that would enable private sector researchers and entrepreneurs to tap public research capacities to develop new technologies jointly, which in some cases may lead to creation or restructuring of firms.

Glossary

R&D:	research and development
GDP:	gross domestic product
OECD:	Organisation for Economic Co-operation and Development
RCA:	relative comparative advantage
PCT:	Patent Co-operation Treaty
IT:	information technology
STA:	Science and Technology Agency
MITI:	Ministry of International Trade and Industry
FY:	fiscal year
VLSI:	Very Large Scale Integrated Circuit project
S&T:	science and technology

Bibliography

Freeman C. (1987). *Technology Policy and Economic Performance: Lessons from Japan,* London: Frances Pinter. [Attempts to explain the post-war Japanese industrial growth in terms of industrial research and development and technical innovation, and draws out the characteristics of the Japanese 'national system of innovation' which is also based on social and institutional innovations.]

Fukasaku Y. (1992). *Technology and Industrial Development in Pre-war Japan: Mitsubishi Nagasaki Shipyard 1884–1934.* London: Routledge. [Examines the development of shipbuilding industry in pre-war Japan, how the capabilities were built up to adopt and adapt advanced technologies through training and research, which generated a technological learning process.]

Goto A. and Odagiri H. (1997). *Technology and Industrial Development in Japan: Building Capabilities by Learning, Innovation and Public Policy,* Oxford: Clarendon Press. [Examines how Japanese industries developed innovative capabilities through technological learning process, and the role public policies played in the process.]

Hiroshige T. (1973). *Kagaku no Shakaishi* (social history of science), Tokyo: Chuo Koronsha. [Compares the process of the institutionalization of science in the nineteenth century in Europe, America and Japan. Argues that Japan, as a latecomer kept up well with developments in Europe and America in terms of institutionalizing science since the late nineteenth century.]

Nakayama S. (1995). *Kagaku Gijutsu no Sengoshi* (post-war history of science and technology), Tokyo: Iwanami Shoten. [Comprehensive history of the development of science and technology in post-war Japan, and the examination of the role played in industrial development.]

Science and Technology Agency (2000). *New Developments in Science and Technology Policy: Responding to National and Societal Needs,* White Paper on Science and Technology, Tokyo: Japan Science and Technology Corporation. [English translation of the White Paper on Science and Technology 1999. It analyses the trends in the related activities based on the comparison of various statistical data on S&T activities of Japan with those of other countries, as well as policy measures taken to promote S&T during the Fiscal Year 1999.]

Biographical Sketches

Yukiko Fukasaku received her first university education in chemistry and biology and worked in the Mitsubishi Kasei Institute of Life Sciences in Tokyo for a few years before turning to studies in the history and social studies of science and international relations. She received her Doctor of Philosophy degree from the University of Sussex in UK in science and technology policy studies. She has worked for international organizations and research institutes including UNESCO, ILO, JETRO, and Centre de Sociologie de l'Innovation of the École Nationale Supérieure des Mines either as staff or independent researcher. She currently works in the Science and Technology Policy Division, Directorate for Science, Technology and Industry of the OECD.

Sachiko Ishizaka graduated in biology from University of Tokyo in 1984, and further studied International Relations at Yale University obtaining an MA in 1991. From 1985, she has been working in the Science and Technology Agency of the government of Japan, engaged in policy-making in earth, nuclear and biological sciences; also, involved in establishing new initiatives for international S&T collaboration such as Human Frontier Science Program. Between 1996 and 2000, she was a project Coordinator of the OECD Global Science Forum. Currently she is the Science Program Officer with the International Council for Science (ICSU) in Paris, engaged in planning, coordinating and implementation of international scientific programs.

CHANGING INNOVATION SYSTEM OF ECONOMIES IN TRANSITION (CEE)

A. Inzelt
Innovation Research Center (IKU), Budapest, Hungary

T. Balogh
R&D Strategy Department, Ministry of Education, Budapest, Hungary

Keywords: National innovation system, transition crisis, CEE, macro economic indicators, GDP per capita, Foreign Direct Investment (FDI)

Contents

1. Introduction
2. Global Position of the Region
3. International R&D Consequences of Opening the Borders
4. Some Changes in the National Innovation Systems
5. Moving out from Transition Crises
Glossary
Bibliography
Biographical Sketches

Summary

The paper gives a general description on the macro economical situation and on the R&D sector of the Central and Eastern European countries in the last decade. There were radical changes and now there are several national ways to find the solution. The global position of the region, which was not very promising at the beginning of the 90s, is improving fast. Some statistical data and trends describe this process. The difficulties of transition are highlighted from the point of national innovation policy. The importance of EU co-operation is emphasized. A general view is given on the two typical phases of transition, pointing out that there are more and more signs that a positive scenario with a knowledge-driven catching-up process is a realistic goal for the CEE region. There are perhaps time lags but the chance is open for every CEE country to find a suitable role and place in the global economy.

1. Introduction

At the beginning of the new millennium, science and technology and a highly educated workforce are becoming the most important sources of national and regional competitiveness. Natural resources tend to be overcome by human factors like creativity, motivation and overall cultural level. This shift to knowledge-driven economy should give good chance to the Central and Eastern Europe countries where education has valuable traditions and achievements, especially in the field of mathematics and natural sciences. However, the heritage from the second half of the 20th Century history is still a burden in some respects and it is not easy to drop a heritage being present in the economy, infrastructure, and, in certain respect, also in the minds of people. This sophisticated situation and the present trends are analyzed in the study.

Before 1945, the frames of national economies and the national innovation systems differed among Central and Eastern European economies. Some countries from the region were closer to the Western European economies and societies than others, having better chances for institutional and personal contacts, while others had less chance and openness. However, all of them were more or less backward compared with the leading industrial countries.

Since 1917 the Soviet Bloc has been increased in several steps. The ideology-driven economy and the growing political gap between the two parts of Europe lead to a strong isolation, being seen at various levels in the individual countries of the region. There was a forced international co-operation between these countries, aiming at the unrealistic goal in the early 60s to reach and then exceed the technology level of the USA within two decades.

In the CEE economies there was a major contrast between the relatively well developed and successful science and the ineffective and rigid enterprises. Located in a 'socialist world system', this type of latecomer enterprise was dislocated from the main international sources of technology and from the real market messages. Its surrounding technological and industrial infrastructure was poorly developed.

The situation was different from other backward regions, because science was well developed and could offer a remarkable knowledge base even if the average of local universities and other educational and research institutions were poorly equipped. Of course, there were also high-level achievements in some technology fields, mainly related to the defense sector, but having little or no influence on the technology-based economic development. The remarkable reduction of the technological gap among military industries was not able to reduce the gap in core civilian technologies. Innovative spirit was pushed back by bureaucracy and equalist enviousness. The system was *per definition* very weak in diffusing knowledge and commercialization because it denied the market as a whole. There were some economic reforms in the 60s, for example in Poland and Hungary, using elements of market economy carefully within the narrow political limits determined by Moscow.

The collapse of the Berlin Wall opened up the Central and Eastern European economies and societies. Deep structural changes were necessary in practically all fields. Transformation of innovation systems is one of the crucial factors to improve their competitiveness and well being of their societies.

Central and Eastern Europe countries (CEECs) are developing fast with regard to economic and social transition to market economy and aiming at achieving economic stability and sustainable prosperity but the transformation has a high price: loss of jobs and decreasing social cohesion. Common pattern of all CEECs that has emerged in the first decade of transition is an inherent imbalance in reforms, which is characteristic even of more advanced transition economies. Finding a sustainable balance among development, transition and macro-economic stabilization is the most challenging task of the process. The relationship among them is strong but very different. It is obvious that stabilization and transition cannot be sustained without technological and industrial restructuring.

The transition of systems offers new possibilities for the economics of modernizing their system of innovation. Some aspects of a market economy can and have been created quickly, especially organizatory and legal processes. Competitive financing of R&D and innovation is a more difficult task and still a problem in the whole region. Transformation of innovation systems from a one-way linear model towards the feedback-loops model is an even longer process, depending not only on science and technological advance but on the overall business climate and the demand of the economy, assuming an established and well functioning micro economy.

The question was what to do with the scientific resources left behind by command economy and the Cold War era. The institutional system was not efficient enough to support economic development, and some of the research and development (R&D) organizations, and fields lost their previous purpose with the weakening of COCOM constraints and thus lost their reason for existence. A new mission was needed.

The CEE stock of knowledge is a common global value, but most of these countries are not able to use it effectively. More efficient usage is important for every country competing in the global economy. The future success or failure of these economies greatly depends on how their knowledge-creating institutions can get their regional industry and foreign industry to work in tandem, to form international research partnerships pooling their expertise to develop and commercialize new products and discover new ways to join global networks.

2. Global Position of the Region

2.1 General

This chapter briefly summarizes the most important macro economic indicators and technology indicators of the countries in the CEE region with special focus on the science and technology related issues. The different development levels of national economies and of the S&T sectors are analyzed and the main trends are highlighted briefly. Data of the following country groups are analyzed below:

- *Central Europe (*Croatia, Czech Republic, Hungary, Poland, Slovak Republic, Slovenia);
- *Baltic states (*Estonia, Latvia, Lithuania);
- *South-Eastern Europe* (Albania, Bosnia and Herzegovina, Bulgaria, FYR Macedonia, Romania).

The authors try to avoid lengthy tables and provide the reader with the most important and most interesting data only.

Unfortunately, many data are missing. This is a topic where the methodological difficulties can be traced due to the insufficient comparable data on the CEE region. It is hard to find reliable data on the transition period. Countries and international organizations (OECD, Eurostat) have made great efforts to create comparable statistics but it takes time, and current statistics provided by CEECs have different content, classification and methodology from comparable data available in market economies. Just like the economic systems, statistical systems in these countries are also in a transitional phase.

2.2 Macro Economic Indicators

2.2.1 GDP per Capita

Among the three country groups, the highest GDP per capita values can be observed in the Central European region, the 1997 list ranges from Slovenia (US$9779) to the Slovak Republic (US$3887). The differences are smaller among the Baltic countries: from Estonia (US$3593) to Latvia (US$2622). There are lower values in the South-Eastern European are: from Romania (US$1695) to Albania (US$930). The general price level in all of these countries is lower than in most OECD countries, consequently, the GDP/capita values calculated on a PPP (purchasing power parity) base are considerably higher. This applies also for the comparison among these countries: the differences are largely not proportional with the nominal numbers.

Regarding the trends of the last decade, due the disintegration of CMEA trade, the collapse of the Soviet Union being their largest trade partner and the ongoing economic recession, all CEE countries lost a considerable part of their GDP in 1990–1992. In 1990 all the mentioned countries had 'negative growth', from –6.2% to –27.7%, and still in 1991 Poland was the only one with real growth (+2.6%), all others had a decrease between –3.1% and –34.9%. In 1992 there were still more countries with a decrease than a increase but in 1993 the trend turned and, since 1993, the region as a whole, with local ups and downs, shows sustainable growth. According to the 1998 data, two countries, Romania (–4.0%) and Croatia (–0.5%) had a decrease, five countries, the Czech Republic, Estonia, Lithuania, Bulgaria and FYR Macedonia had no growth and the other seven countries had real growth from Bosnia and Herzegovina (+12%) to Latvia (+1.5%). In spite of this development, there are still countries in the region having lower GDP per capita value than 10 years ago. This is a clear evidence for the long-lasting existence of difficulties related to the transition.

The macro economic indicators at this and following sections are taken from EBRD Transition Report.

2.2.2 Private Sector Share of GDP

At the beginning of the transition, the private sector contributed typically 10–15% of the GDP in the region (extremes: 30% in Poland, 5% in Albania). During the 90s a tremendous privatization wave took place and by 1998, 50% was the lowest value (Slovenia) and 85% the highest (Hungary). The share of foreign owners differs country by country; the highest degree of internationalization can be observed in Hungary where companies with foreign ownership produce over 65% of the industrial export. Countries in South-Eastern Europe are somewhat behind in this respect. It has to be mentioned that not all privatization transactions were connected with real cash inflow; in some countries there were other forms (e.g. coupons) creating partially formal privatization without the change of the 'socialist' management and without the necessary restructuring at company level. This way seemed to be smoother and least radical but in turn the structural changes lasted and last longer: the relatively high growth rates achieved in the early 90s proved to be not sustainable in some cases.

2.2.3   Foreign Direct Investment (FDI)

The level of foreign direct investment in the CEE region was almost ignorable before 1990. After the political turn in 1989–1991, the FDI inflow started at different intensity and at different time. Between 1990–1995 Hungary was the first attracting considerable FDI and per capita she has still the highest stock in the region (about US$2000). Estonia has also good results per capita. Poland started to receive massive FDI inflow from 1995 on and reached a value of over US$6.5 bn in 1997 and 1998. After some years of moderate success, the Czech Republic attracted US$2.5 bn in 1997 and US$3.5 bn in 1998. The FDI inflow to Romania accelerated from 1996, reaching US$2.04 bn in 1997 and US$1.35 bn in 1998. In all countries not mentioned here, there was no single country reaching the US$1 bn level in any year up to 1998. It means, there is still a huge potential for FDI which is also necessary because FDI means not only capital but a complex system of technical knowledge, quality requirements, cultural change, advanced management methods and global market access, as it can be seen in the example of some successful CEE countries.

2.2.4   External Debt/Exports

The difficulties of transition lead almost inevitably to the increase of external debts. The lacunae in the infrastructure, the high unemployment levels and loss of traditional markets made debts necessary. Three countries had high external debt/exports level in 1990 in the CEE region: Albania (860%), Bulgaria (430%) and Hungary (245%). An interesting two-way equalization process took place during the 90s: these countries with a relatively high debt could successfully decrease this value by 1997: Albania (424%), Bulgaria (235%) and Hungary (129%) and the decreasing trend is going on. In the other countries of the region having practically no debt at all in 1990, an increasing trend can be observed, and in 1997, this indicator was over 100% for most of the countries. The two positive exemptions are Poland (30%) and Slovenia (54%) keeping external debts at a low level. The financing of external debt is a large load on the yearly budgets in most CEE countries, making the primary and secondary budget balances very different and decreasing the development resources.

2.2.5   Investment Rate

Investment rate is a good indicator showing the share of future in the expenditures given for past, present and future together. It is especially interesting after the external debt giving good indication on the expenditures in the past. In 1990, Czechoslovakia was the country with the highest investment rate (31.3% in Slovak and 26.3% in the Czech part). The Baltic countries had also high rates (Latvia 27.6%, Estonia 23.8%) during the late Soviet times. In 1997, the highest values belong to the same countries: Slovak

Republic (38.6%) and Czech Republic (30.7%), further increasing. The lowest value can be observed for FYR Macedonia (7.2%) and Bulgaria (11.6%), but there are several further countries in the CEE region with investment rates under 20%. There was a strong reduction in Hungary from 1992 to 1997: the investment rate decreased from 30.4% to 16.3%. Latest news show that investments are growing again after 1997. In general, fast economic development and successful catching-up is impossible without a reasonable investment rate, as we saw in the example of the small East Asian economies.

2.2.6   Unemployment Rate

Unemployment was an absolutely new phenomenon for societies that were used to full employment by definition for several decades. The main issue is that it does not raise merely economic questions but also a question of morale, optimism in the future and social cohesion. Long-term unemployment can ruin the chance of people to return to work any time and unemployment concentrated with poverty in some regions and social groups can distort values and increase criminality. In 1990, four countries in the CEE region reported unemployment rates over 10%: FYR Macedonia (19.2%), Croatia (13.2%), Poland (11.8%) and Bulgaria (11.1%). In 1997, most countries in the region reached 10%. The highest rates were seen at FYR Macedonia (34.5%), Bosnia and Herzegovina (20.0%), Albania (17.7%) and Croatia (17.2%). The lowest rates were observed in Lithuania (6.4%), the Czech Republic (7.5%) and in Slovenia (7.9%). Radical reforms with high immediate unemployment seem to give the chance for later decrease, as the examples of Poland and Hungary show. Slow restructuring implies hidden problems, leading to the increase of unemployment several years later, when the reform countries have already improving indicators.

2.2.7   Change of Labor Productivity in Industry

In 1990–1991 all CEE countries had a loss in productivity due to the general economic recession. Poland began the impressive development in 1992 (+12.5%), Hungary joined the trend in 1993 (+16.3%) and Slovenia in 1994 (+11.4%). From 1994, practically all CEE countries show improving numbers. By the end of the decade, the so-called 'Visegrád countries' (Hungary, Poland, Czech Republic, Slovak Republic) reached the productivity level of the 'cohesion' EU countries and were improving at a good rate. Other CEE countries like Croatia and the Baltic Countries showed also significant improvement in the late 90s. The productivity growth is especially spectacular in the new production facilities established in the region by foreign capital and expertise.

2.2.8   Share of Exports to Non-transition Countries

After the shock-like collapse of the Soviet Union being the most important export market for almost all CEE countries, the region was able to find a new foreign trade orientation quickly. Many data are missing for the years 1990–1992, but most of the existing ones show low percentages at about 11–12%. By 1993, most countries exceeded 70% in this respect and the ration grows further. Surprisingly, Lithuania is the only country where the share of exports to non-transition countries decreased year by year. In 1997, Romania had a share of 88.2%, Hungary 80.7%, Poland 77.5%, Slovenia

76.7%. Lithuania (49.9%) and the Slovak Republic (49.3%) showed the lowest figures. In the case of many countries in the foreign trade to the EU countries exceeded a critical level and their economy is *de facto* integrated in the EU through the complex micro-economic relations between and within thousands of companies.

2.3 Technology Indicators

2.3.1 Gross Domestic Expenditure on R&D (GERD)

Financing R&D has a strong impact on the activities of the national system of innovation. The allocation of R&D resources can encourage or discourage industrial innovation. In the 80s, CEE economies mobilized relatively large proportions of national resources to invest in R&D and education.

Since the late 1980s, a sharp decline can be observed in gross domestic expenditure on R&D and its relation to GDP in every country. It is a common feature that the decrease in R&D expenditures (GERD) was larger in all transition countries than the decrease in the GDP. In the CEE countries, the typical GERD was 2.0–2.5% of the GDP and in the years of crisis it was only 0.4–0.7% of the (lower) GDP. CEE countries spend a decreasing share of their GDP on R&D. However, there appears to have been stabilization in some of countries. This may be the sign of an end to the drastic reductions in R&D during the transition process. Apart from the enormous methodological problems, the shrinking of funding sources is a common factor among all these countries. Business R&D expenditures declined faster than governmental expenditures. In terms of industry-financed R&D, CEE countries are considered the lowest funded among OECD countries. Businesses maintain a modest funding as both financiers and performers of R&D. While it is true that part of the R&D financed by the business sector has dropped sharply because of the transition, the economic freedom of businesses exists and has its own value.

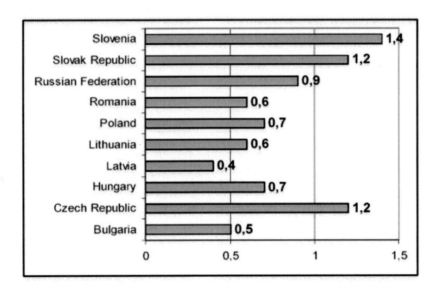

Figure 1. Gross domestic expenditures on R&D in some CEE countries as a percentage of GDP (1997). Source: EUROSTAT.

Gross domestic expenditure on R&D (GERD) as a percent of GDP is a widely used indicator, showing that CEE countries do not invest enough in R&D. The 1997 data are shown in Fig. 1.

Taking into account that the average GERD/GDP level of the EU is 1.8% and, according to a recent document (*Towards a European Research Area*) it is unacceptably low compared to the global competitors, these values are extremely low. This level of R&D financing is not enough even for keeping the relative position, and absolutely insufficient for catching up.

The low percentages do refer to the current low GDP values; therefore, it is also interesting to compare them at real value in ECU per capita (Fig. 2). In reality, just a part of R&D costs can be covered by cheap R&D labor, but materials, measuring equipment and infrastructure cost at least as much in the CEE countries as in the USA or in the EU.

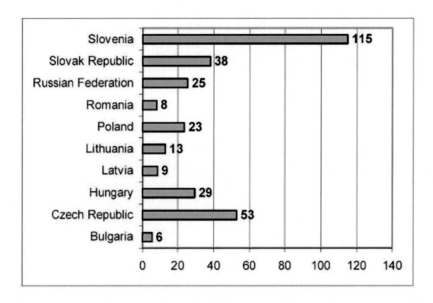

Figure 2. Gross domestic expenditure on R&D in some CEE countries in ECU per capita (1997). Source: EUROSTAT.

The problem is not only the lack of R&D financing but that there are no proper financial systems for innovation, development activities. Interest rates are high, few funds are available, and venture capital companies and business angles are rare or missing. There is one small source among funding, foreign state support, which has a special role in the transition process. Foreign aid as 'seed money' can provide many opportunities. Seed money can encourage the restructuring of S&T to adapt itself to the global environment. This money can also help to preserve the accumulated resources (scientific 'schools', R&D skills, laboratories, knowledgeable people, etc.). The development of innovation-friendly business environments, which is an inevitable precondition of a well-functioning knowledge-driven market economy, needs several more years or decades in the CEE region.

2.3.2   Total R&D Personnel

Independent experts' evaluations usually emphasize that the intellectual assets of CEE countries still have a significant value. There are many reasons: institutional, economic and historical ones. The number of research scientists, engineers and supporting staff has steadily decreased in every CEE country during recent years.

Concerning the CEE countries, few reliable data are available from the early 1990s. According to the Second European Report on S&T Indicators (1997) European Commission (Eurostat, 2000) the full-time equivalent (FTE) numbers of R&D personnel are as follows for 1997:

Central Europe	
Czech Republic:	23230
Hungary:	20758
Poland:	83803
Slovak Republic:	16365

South-Eastern Europe	
Bulgaria	18625
Romania:	54436

Russian Federation:	1053013

As a general trend, the number of R&D personnel decreased in the 90s together with decreasing financing. The loss was moderate in the higher education sector where positions had to be kept at least for maintaining education also in the years of crisis. State research institutes had higher loss rates due to the decreasing financing. It is a general problem that talented young people do not apply for researchers' jobs because the emerging market economy provides sometimes 200%–500% more salaries in the business sector for highly trained young people with good foreign language command and co-operation ability, which is typical for young researchers.

Most lossage has taken place in the field of experimental development at the companies. The typical crisis management in the late 80s and early 90s began with closing the R&D branch of the companies; the rest of this R&D activity disappeared during privatization. In the late 90s growing numbers of multinational companies decided to take over or re-establish R&D facilities in Hungary, based on the local brains receiving traditionally a high level of education in mathematics, physics and chemistry. Similar trends can be foreseen in other CEE countries as soon as foreign investors discover the value of local universities and researchers.

2.3.3   Higher Education R&D Personnel

The university sector seems to be somewhat more transparent than the economy as a whole, because there are more data on the higher education personnel in the CEE. Based on the data of EUROSTAT, the full-time equivalent (FTE) numbers of higher education R&D personnel are the following for 1997:

Central Europe	
Czech Republic:	3981
Hungary:	7210
Poland:	40977
Slovak Republic:	5041
Slovenia:	1443

Baltic states	
Estonia:	3230
Latvia:	2143
Lithuania:	6020

South-Eastern Europe	
Bulgaria	3308
Romania:	4318

Russian Federation:	102627

The ratio of higher education personnel within the total R&D personnel is very different in different CEE countries, for example, this ratio is about 50% in Poland while only 10% in the Russian Federation reflecting an inherited S&T organizational structure.

2.3.4 US Patents

Patents registered in the United States are good and comparable innovation indicators. It is well known that several innovators do not publish at all. Table 1 contains the data on the CEE countries in the above order, based on the data of the US Patent and Trademark Office, between 1995–1999, including the full transition period. By 1995 several former states were split. Some countries have signed the Trade Related Intellectual Property Rights (TRIPs) Agreement by that time and new (modified) laws on IPRS came into force.

Country	1995	1996	1997	1998	1999	Total
Central Europe						
Croatia	0	0	4	5	3	12
Czech Republic	1	5	14	17	25	62
Hungary	51	43	25	52	39	210
Poland	8	16	11	19	20	74
Slovak Republic	0	1	3	2	6	12
Slovenia	5	13	7	20	13	58
Baltic countries						
Estonia	2	1	0	0	1	4

Latvia	0	0	0	1	3	4
Lithuania	1	0	2	3	2	8
South-Eastern Europe						
Albania	0	0	0	0	1	1
Bulgaria	1	1	5	4	3	14
Romania	3	4	1	3	4	15
Other						
Russian Federation	99	118	112	194	185	708

Source: United States Patent and Trademark Office

Table 1. Central and Eastern European patents granted in the USA.

Table 1 shows interesting trends and ratios. It is clear that in spite of all economic and financial difficulties, splitting of the USSR and the mass brain-drain from the region, Russia remained number one in the CEE region in the field of R&D output. The number of Russian patents granted in the USA is about 50% higher than the sum of all other countries in the region. It shows the tremendous power of human capital, backed by the high level mathematics and science education.

The other successful country to be mentioned in the CEE region is Hungary, showing continuously good performance per capita (10 million inhabitants). Slovenia performs also well per capita (1.9 million inhabitants). Czech Republic and Slovak Republic together seem to exceed the performance of former Czechoslovakia (not listed in the Table), with considerably more success in the Bohemian part of the former federal state.

The Baltic countries just began this activity in the late 90s. Albania achieved her first granted USA patent ever in 1999. The performance of Yugoslavia and the former Yugoslav republics is hardly affected by the wars, civil wars, instability and partial international isolation. Two former Yugoslav republics, Bosnia and Herzegovina and FYR Macedonia are not included in the data of the USA Patent and Trademark Office at all, obviously due to the lack of granted patents.

This patent performance of the CEE region is not so excellent in a wider international comparison. For example, Austria had 505 USA patents granted in one year (1999). The patent indicators of the CEE region are, however, not bad if the per capita R&D expenditures are also taken into account (typically 10–50 Euros per year per capita, see Fig. 2). Strengthening of the CEE economies, especially business R&D activities is the necessary precondition for a considerable improvement of the region in the field of foreign patenting.

3. International R&D Consequences of Opening the Borders

3.1 Importance of International R&D Co-operation and Problems in the Past

Science and technology are *per se* international activities, with an inevitable need for knowledge flow and competition. In the CEE countries there are several examples how isolated national S&T systems can distort the local priorities and can raise values being significantly different from the internationally accepted ones. The first factor of isolation was the lack of mobility. In some countries there were hard administrative burdens to block or at least filter the foreign trips of researchers. In more open countries where travelling was theoretically free, strict financial regulations or simply the lack of 'hard' currencies blocked mobility.

The isolation was also earmarked by the COCOM list, excluding this region from a large number of technologies, therefore several reproductive and parallel R&D activities were carried out, with few scientific novelties in a global sense. But the so-called 'reverse engineering' activities needed at least as much creativity to reach the same result a different way: it was excellent brain training, and skill accumulation. Despite, or because of, lack of financial resources and the difficulties with available tools several top results were achieved in the CEE region.

3.2 Exceptions: 'Rare Flowers of Isolation'

In the following, there are two examples how the lack of financial resources and the difficulties with available tools could lead to top results in the CEE region.

Our first example is in software development. Having been disclosed from the latest hardware technologies and even, for financial reasons, from the commercially available technologies, but facing similar scientific and technological problems, CEE researchers and engineers had to solve the problems by using the existing hardware. Sophisticated modeling problems, laboratory data acquisition systems and finite-element method calculations were managed successfully by using cheap game computers being available for any schoolboy in the West. In this case the lack of technical background was compensated with a surplus of creativity. This is, over and above the high level of mathematical education in the CEE region, a reason why several new software development companies became so successful in the region.

Our second example is also about computers, poverty and creativity. It is not often when hard discs are damaged that it is worth to invest several expert hours to fix them. A new one should be bought. Yes, it is true in an established market economy, but what to do of there is no import permission, no hard currency but the hard disc is needed badly? CEE researchers invested tremendous efforts to work out the ways to fix damaged hard discs and they used this knowledge for years for their own laboratory only. After the political turn they recognized that this knowledge was unique; no one dealt with this problem successfully. As a spin-off, they established a company in Central Europe that is the only one in the world repairing damaged hard discs (e.g. fire or mechanical damage) and able to restore the majority of all the stored data. They receive discs by DHL or other worldwide services and send back the fixed discs in a matter of days.

This is another example how poverty-driven creativity can lead to global success. For better or worse, this source of creativity has been exhausted after 1990, but in turn, thousands of researchers were allowed to join international bilateral and multilateral R&D projects. Of course, the great results of this open co-operation cannot be compared to the above-mentioned 'rare flowers of isolation', but these examples can prove interesting to obtain a more detailed view on CEE research and development.

3.3 1989–1991: The Iron Curtain Disappears

After the global political turn in 1989–1991, the iron curtain disappeared and most of these countries become members of European R&D co-operations such as EUREKA, COST, CERN, and EMBO. CEE countries, being many of them newly independent, having new national identity, placed considerable emphasis in building up the international network of R&D. COST and EUREKA offer especially good chances to join international R&D activities and develop their own professional network. CEE representatives are also active in the organization committees, umbrella organizations, etc.

Several bilateral intergovernmental R&D agreements were born after 1990, reinforcing the contacts of CEE countries with dozens of other countries, including each other. There are many agreements where both sides are CEE countries but, in spite of that, there is a growing concern that the Central and Eastern European R&D system shows clear disintegration. Every country turns to the same Western partners and forgets about its direct neighborhood. Foreign languages are good examples: it is very rare that people study the language of the neighbor country, most concentrate on English, German or French.

Four countries (Czech Republic, Poland, Hungary and Slovak Republic) joined OECD, increasing the number of member countries to 30 in this way. In the field of R&D the work in the Committee of Science and Technology Policy (CSTP) and its subcommittees is of central importance.

Other countries participate in the OECD Partners of Transition (PiT) Program, maintaining a continuous exchange of information and consulting, preparing country studies and distributing the achievements embodied in further OECD studies recommendations.

Three countries (Czech Republic, Poland, Hungary) became NATO members in 1999 and there are good chances for Romania, the Slovak Republic as well. The chances of Slovenia and Croatia are developing also. Countries that are still not NATO members participate in the Partnership for Peace (PiP) Program and in the NATO Science Program. This program also enables mobility and further intensification of international science and technology co-operation. It is worth while to give more details concerning the EU RTD Framework Programme being the most important tool for integrating CEE research and development into the European mainstream.

3.4 CEE Participation in the EU 5th RTD Framework Programme

The EU Fifth Research and Technological Development (RTD) Framework Programme, being open to all legal entities established in the Member States of the European Union and in any of the countries formally associated with the program, is perhaps the most important tool of international R&D co-operation for the CEE countries. The CEE countries associated with the framework program comprise Bulgaria, Czech Republic, Estonia, Hungary, Latvia, Lithuania, Poland, Romania, Slovakia and Slovenia. Other Central and Eastern European countries (including those from the former Soviet Union, and Albania, Bosnia-Herzegovina, and the Former Yugoslav Republic of Macedonia) can also participate as can international organizations and any country that has concluded a bilateral S&T agreement. Finally any entity from another country can be part of a consortium if its participation is of substantial added value for the program. A total budget of 14.96 billion Euro has been allocated to implement the Program.

CEE research teams have already been taking part in the 4th RTD Framework Program between 1994–1998 on a project-by-project basis with fairly good results. The rights and obligations concerning the CEE participation are the same as those of the EU member states. Concerning the financing mechanism research teams of the associated countries may also participate with the same conditions in any of the specific research and mobility programs. Associated countries may send researchers to the Joint Research Centers (JRC) of the EU, can delegate members with consultative rights to the meetings of the Program Committees, in the External Advisory Groups of the specific programs and their experts may participate in evaluating proposals.

In order to own these rights, the associated countries contribute to the budget of the FP, the participants of a project prepare the financial management according to the rules concerning the members of the EU and the rules of Intellectual Property Rights must be kept. Concerning the GDP proportional contribution to the budget of the Framework Program, a temporary allowance was offered to the newly Associated countries.

In order to ensure the successful participation in the 5th FP, CEE governments built up a special supporting network and implemented special financial measures to enable the best possible results. One of those is the network of the National Contact Points (NCPs). They are the main links in their professional fields with the EU. In some countries there is a network of liaison offices to promote participation in the FP5. Liaison offices are located in the different regions within the related CEE countries Their main task is to give advice to possible participants on the research program and its specific key-actions, on the administrative rules and instructions of the EU related to proposal writing, reporting, negotiating or on the evaluation criteria.

Associated countries established consortia that formed the FEMIRC (Fellow Member to Innovation Relay Centres) to disseminate information throughout their home country on the European research possibilities. The Innovation Relay Centres were formed in each Member State of the EU. Fellow Members were appointed after an evaluation procedure in the Central and Eastern European Associated countries, to give advise to organizations – especially small and medium-sized enterprises and RTD institutions – on technology and innovation. They help institutes and companies to identify their technology needs, the suitable technologies to match these needs, give assistance on exploitation and advice on research and technology programs.

The CEE participation was rather successful in the first calls, as – taking into consideration the number of the successful projects, the CEE participants, the planned amount of Community funding, and the CEE contribution to the budget of FP5 in 1999. Concerning the total amount of the handed in proposals – at the end of December, 1999 after the first call – the number of CEEC proposals was about 5.6 percent of the total participation.

Three countries performed at an acceptable level: Poland with 855 submitted and 167 accepted proposals, Hungary with 692 submitted and 129 accepted, and the Czech Republic with 571 submitted and 127 accepted proposals. Small Slovenia is good here again with 319 submitted and 78 accepted proposals, showing the results of the highest per capita R&D expenditure in the CEE region. All other participant countries (Bulgaria, Estonia, Lithuania, Latvia, Romania, and Slovakia) should take efforts to promote project participation. However, all countries in the region have to do so in this matter. On one hand, it is a hard financial question, because the normative country contribution is always a goal to bid back, especially in countries with very low R&D expenditures. On the other hand, the successful participation in the Framework Program is also an indicator for the international acceptance of the researchers and of the whole national innovation system.

In the case of newly associated countries there were always difficulties first: rules of the program are new for them, their partner connections are limited. However, this is the only way to join the European mainstream in the strategic R&D fields. It is often said that science and technology is a front runner and engine of the EU accession process. It is true, because the CEE countries having been associated to the EU 5th RTD Framework Programme are de facto members of the EU in this sense, having similar rights and obligations as the full members. Therefore, participation in the Framework Programme has large political importance and therefore enjoys full support in the CEE countries.

4. Some Changes in the National Innovation Systems

The transition of the Central and Eastern European countries was not so much a planned process, but it occurred due to the worldwide crisis of command economies and the collapse of the URSS. This situation gave chances to all countries in the region to manage in parallel the transition, stabilization and development for the well-being of their societies.

In the following sections we try to characterize the nature of the two phases, giving thus an inside view of the recent years in the Central and Eastern European R&D sector. A certain abstraction is aimed: it is not so important to point out the differences between the individual countries as to find some common and generalizable features.

4.1 Changing Institutional System

At first glance, we can observe a lot of similarities in the problem of the transformation of institutional and legal systems. Tremendous difficulties in redeployment are also common. Structurally, there were other similar trends in the CEE countries.

The national systems of innovation can be divided to four typical parts:

- Higher education institutions (universities and technical colleges, 'Fachhochschulen').
- R&D institutes.
- Sectoral R&D institutes.
- Company R&D units.

Higher education institutions had to suffer budget cuts but higher education as an essential task had to be continued everywhere. Consequently, higher education institutions survived, furthermore, developed in many sense, because the increase of students numbers was an important political goal to provide the human resources for the future knowledge-driven society, or, as not published so often, a useful tool against mass youth unemployment. Higher education cannot be maintained without laboratories and without research, therefore, some higher education research remained in all related countries. The quality of the research infrastructure declined in the 90s because there was only a few chances to purchase new equipment. It is typical that the infrastructure is still suitable for gradual education but not advanced enough for research. There were some positive changes as well; some international aid programs and loans (e.g. EBRD) helped the development and restructuring of several higher education institutions.

The national R&D institutes, represented typically by the institutional network of the national Academy of Sciences, suffered more loss than higher education institutions. In the Soviet-type science structure it was typical that the national academies of science had their own large research institute network. These networks represented high quality and were sometimes shelters where talented researchers being removed from universities for political reasons, but working in the academy institution successfully. Many R&D institutes were merged or ceased, but many survived and represented high scientific value in international comparison. It is interesting that in the 2000 Centers of Excellence bid of the European Union for the CEE region, what a high percentage of the winners were institutes of the Academy of Sciences (in Hungary, for example, 5 out of 6). Institutes of the Academy lost many researchers leaving for the economy or for foreign R&D institutions. The brain-drain loss ratio was the highest in the academy research institutions, exceeding 50% or even more of the researchers in some institutes, especially from the young and the middle generation, being flexible and having good foreign language command. Of course, these were potentially the best research fellows letting a large vacuum in the domestic R&D sector in many countries.

Sectoral R&D institutes were important part of the national innovation system in the CEE countries, supervised by the relevant ministries. The task of sectoral institutes was to conduct applied research, design, experimental development, and testing for manufacturing industry, agriculture, health care, environmental protection, energy, and telecommunications, primarily on a contract basis. The state was an important, and in many ways the major, contractor, and the founders and supervisors of these institutes were usually the relevant ministries. Their number decreased significantly. Several went bankrupt and were dissolved, while others were transformed into economic associations or small and medium-sized private enterprises. Some began to transform themselves into private sector engineering companies, R&D service firms, or knowledge-based business organizations that ceased their R&D activities. Changes in demand (caused by shifts in markets, easier technology transfer, changes in size of firms, growing FDI, the presence of multinationals, etc.) have a strong impact on reorienting the activities of these institutes from product and technology-oriented research toward R&D services.

The company R&D units represent the fourth sector. At the beginning of economic difficulties, still in the command economy era, closing the R&D unit was the typical first cost reduction step of the management. Many engineers became 'forced' entrepreneurs because their job had been deleted suddenly. Some of these new enterprises became successful SMEs, some even successful large companies, but their majority fights continuously for survival. Many people took jobs at the local branch of multinational companies, working at lower level than their education. It was typical that engineers with Master degree work in technician's jobs in two or three shifts, for still several time higher salaries than in a researcher job at an academy institution. In the first period of transition the company R&D activities show continuous erosion, which is amplified by the privatization process. New domestic owners rarely had the strategic will to build on R&D, rather the quick utilization of the property that had been bought at very low price. New foreign owners had typically more strategic thinking but they also had their own technologies and researchers from their home country. This is perhaps the main reason why company R&D units show the most loss in Phase I of the transition.

At the beginning of transition, universities and state research institution networks were reorganized; several new institutions were created. The political changes in 1989–1991 in the CEE region launched basic structural changes in the science and technology institutional system in all countries. Universities were reorganized and several new universities were established. CEE countries obtained huge amount of advice and aid through bilateral and multilateral channels from abroad in modernizing higher education. Many of the CEE countries were so successful in the transition that they can offer courses in foreign languages based on local professors, attracting a large number of paying students from abroad.

There were also efforts to establish new research facilities according to Western examples, like the example of the Fraunhofer Gesellschaft in Germany (e.g. Bay Zoltán Foundation in Hungary). Of course, it is extremely hard to clone successful structures working in such significantly different conditions. These new institutions have a large number of unforeseen difficulties because there is no advisor who could see what happens with a proven structure in an unreliable and fast changing environment.

Despite new liberal economy theories winning governing rules in the government policies, which are easy to understand in the recession period, new structures like technology transfer offices, incubator houses, technology centers and entrepreneurs' advisory institutions were established, strengthening the innovation infrastructure in most CEE countries. Every country established data banks for helping the technology transfer. Their presence shows that countries are likely to adopt a market economy model, in which these organizations have a growing role in the dynamic process of innovation. They are working on filling the gaps of interaction between firms and universities, and firms and contract research organizations. It appears that they can bring industry closer to the work going on at universities, and give them a window on new ideas and access to emerging scientific and technological developments. They can help the post-socialist countries to use their intellectual assets much more efficiently.

Institute/university/industry partnerships were initiated to pool their expertise to develop and commercialize new products and discover new and better ways of co-operation.

International co-operation is also of crucial importance for the CEE countries from the beginning and ongoing. It is important not only for the individual topics treated but also the way of working, the way of thinking and the style of a tender-oriented researcher behavior.

4.2 Mass Brain Drain from the CEE Academic Sector – Typical in Phase I

Under the command economy, a scientific career was one of the very few possibilities for unfettered intellectual activity, openness and international recognition. The developing market economy offered many more new possibilities with better prospects. Democratization has opened the door to international mobility and economic factors also affect emigration of scientific personnel. This phenomenon has been observed for several years in countries with a similar level of economic development. Besides higher salaries, better research laboratories, better working conditions, and better jobs – all of which contribute to a better professional situation – attract scientists from this part of Europe to the West. Most young researchers went to the USA but many to the EU countries, especially to Germany and the United Kingdom. Some people return after three or five years and they represent high value for the local R&D. The problem is that many people willing to return home cannot find a proper job providing salary and infrastructure that is comparable to the Western one. These people stay typically abroad but they are still not lost from the point of national R&D because they initiate many international activities and joint projects. Final solution to the brain-drain problem can only be found in the long run with the general recovery of the CEE economies and with improving financial and infrastructure conditions.

5. Moving out from Transition Crises

Due to the economic recovery and growing role of strategic thinking, government R&D expenditures begins to grow again along new priorities based mainly on the lessons from OECD and EU. Governments begin to spend more R&D. Universities and surviving research institutes ask for more development sources to overcome the development deficit that has been collected in the first years of transition, and it is necessary to remain competitive. Imported technology help build regional infrastructures, developing networks and clusters. Regional co-operation activities over borders tend to widen, using combined financial sources from the local and the national governments, the business sector and more and more from the EU. Daily working contacts with the EU countries lead to a secondary cultural development that are not easy to trace but are inevitably important.

Company R&D has started to develop in several countries as well, mainly powered by the foreign direct investments, creating local R&D facilities as part of global business networks. Foreign direct investment (FDI) is important for all transition economies, to overcome several burdening problems associated with introducing innovation. FDI is influenced by a complex set of economic, sociological, legal, etc., factors. With closer relationship with advanced markets, leading firms can prevent CEE countries remaining on the periphery, outside of commercial networks. Fast technological development and close integration into the global networks are the basic ingredients of future economic growth. An increasing number of foreign investors recognize the unique chance they have for a limited period: they can get skilled European labor available at a discount price. This situation cannot last too long, just a matter of years, because the wages have to increase gradually (and they do); they approach the wages of the EU countries sooner or later. Quality factors, training, education, quick response and overall cultural levels become key issues to sustain competitiveness.

The development of human resources in the CEE region becomes a critical issue to establish international competitiveness. The importance of graduate and postgraduate training and international mobility is already recognized in most CEE countries. New university curricula are developed, taking the needs of the global industrial and service partners into account, addressing new priorities, for example, the growing needs towards IT expertise. New institutional forms, such as company-financed university chairs, company universities, and university-based co-operative research centers emerge. The economy tends to reach a development level where knowledge-oriented networking with the R&D institutions and other companies is a factor of survival. The complex thinking of new network economy can gain a central role right after the transition phase, skipping the difficulties of the industrial and early postindustrial economy as a chance for latecomers. The CEE region can be a part of a knowledge driven Europe on an equal basis, overcoming the handicaps inherited from the 40/50/70 years of an unfortunate political and economic macro 'experiment'.

The above-described phenomena is already an established fact in some CEE areas, especially in the OECD member countries, and the trends show that this is a realistic scenario. We do hope that the actual differences can be attributed to time lags because there is no real reason why the development path of these countries should be significantly different. The region is scattered by many languages, religions and traditions, and some time is still necessary to recognize that there are more common features than differences.

Glossary

Activities excluded from the scope of research and development (R&D):	Research and development should be differentiated from a wide range of scientific and technical activities related to R&D, although these latter are often closely related to R&D in terms of the flow of information, economic operations and personnel. A few examples of activities not regarded as R&D are : training and education, general purpose data collection, routine testing and calibration, feasibility studies, medical care based on the usual application of medical knowledge, routine software development etc.
Applied research:	An investigation made to learn new things for some practical purpose. Its findings are only valid with respect to a single or a restricted number of products, processes, methods or systems.
Basic research:	Experimental and theoretical work, where the primarily goal is the accumulation of knowledge on the essence of phenomena and

	on observable facts without the objective of any concrete application or use. The findings are usually not marketed, but can be used for developing theories and laws.
Demonstration:	The initial but still experimental application of a new technology in practice on an industrial scale (e.g. in test operation, pilot plant).
Development:	A regular activity which relies on existing knowledge gained from research and practical experience, where the objective is to introduce new materials, products, new processes, systems, services or to substantially improve the ones already introduced.
Innovation:	This covers the introduction /application of all new or modified product or technology including new releases which are only new in a given sector or a given organisation. Innovation furthermore covers all corporate, organisational and marketing solutions (e.g. buying, marketing) which improve productivity and profitability.
Research and development (R&D):	A creative activity pursued on a regular basis, the goal of which is to expand knowledge, including the knowledge of man, culture and society, and the use of this knowledge to develop new ways of application.

Abbreviations

CEE:	Central and Eastern Europe
CEECs:	Central and Eastern European post-socialist countries
CMEA (or COMECON):	Council for Mutual Economic Aid
COCOM:	Coordinating Committee for Multilateral Export Controls
CERN:	European Laboratory for Particle Physics
COST:	Cooperation on Science and Technology in Europe
CSO:	Hungarian Central Statistical Office
CSTP:	Committee of Science and Technology Policy at OECD
EBRD:	European Bank for Reconstruction and Development
EMBO:	European Molecular Biology Organisation
EC/EU:	European Communities/Union
EU RTD:	European Union Research and Technology Development Programs
EUREKA:	European Strategic Research Initiative
EUROSTAT:	Statistical Office of European Community
FDI:	Foreign Direct Investment
FTE:	Full-time equivalent
FYR:	Former Yugoslav Republic
GDP:	Gross Domestic Product
IPRs:	Intellectual Property Rights
ITD:	Hungarian Investment Development Agency, Hungary Ltd
NATO:	North Atlantic Treaty Organisation
OECD:	Organisation for Economic Co-operation and Development
PIT Program:	OECD Program for Partners in Transition
PPP:	Purchasing Power Parity

R&D:	Research and Development
S&T:	Science and Technology
TRIPs:	Trade Related Intellectual Property Rights

Bibliography

The most important data sources on S&T and related matters are OECD regular publications for OECD members of transition economies and the following publications for non-members:

OECD (1999a). *Science and Technology Main Indicators and Basic Statistics in Slovenia, 1991–1997,* OECD CCNM/DSTI/EAS (99) 71, Paris. 60 pp. [This volume covers the main indicators and basic statistics for science and technology on Slovenia. This publication contains 40 tables of which the majority concern resources devoted to research and experimental development and some other indicators of output of scientific and technological activities.]

OECD (1999b). *Science and Technology Main Indicators and Basic Statistics in the Slovak Republic, 1992–1997,* OECD CCNM/DSTI/EAS (99) 22, Paris. 67 pp. [The volume on the Slovak Republic covers the main indicators and basic statistics for science and technology. This publication contains 40 tables of which the majority concern resources devoted to research and experimental development and some other indicators of output of scientific and technological activities.]

OECD (1999c). *Science and Technology Main Indicators and Basic Statistics in the Russian Federation 1992–1997,* OECD CCNM/DSTI/EAS (99) 23, Paris. 62 pp. [This volume on Russian Federation covers the main indicators and basic statistics for science and technology. This publication is based on data provided by the Russian Centre for Science Research and Statistics in reply to the OECD international survey on resources devoted to R&D. This publication contains 40 tables of which the majority concern resources devoted to research and experimental development and some other indicators of output of scientific and technological activities.]

OECD (2000). *Science and Technology Main Indicators and Basic Statistics in the Russian Federation 1992–1998,* OECD CCNM/DSTI/EAS (2000) 69, Paris. 63 pp. [This volume on Russian Federation covers the main indicators and basic statistics for science and technology. This publication is based on data provided by the Russian Centre for Science Research and Statistics in reply to the OECD international survey on resources devoted to R&D. This publication contains 40 up-dated tables of which the majority concern resources devoted to research and experimental development and some other indicators of output of scientific and technological activities.]

Some publications of European Union offer moderately comparative data on R&D and innovation statistics on transition economies:

European Commission (2000). *R&D and Innovation Statistics in Candidate Countries and the Russian Federation Data 1996–97,* Luxembourg: Eurostat. 138 pp. [This publication contains a statistical data base on R&D and innovation activities in the candidate countries (Bulgaria, Cyprus, Czech Republic, Estonia, Hungary, Latvia, Lithuania, Poland, Romania, Slovak Republic and Slovenia Republic)]

Transition report 1999, Ten years of Transition: Economic Transition in Central and Eastern EBRD (1999). *Europe, the Baltic States and the CIS,* European Bank for Reconstruction and Development. 288 pp. [This special issue within the annual series, takes stock of developments in transition over past decade. The analyses have pointed to broad patterns across the region. The report gives an overview on all transition economies.]

Commission of the EC, *Towards a European Research Area.* Communication from the Commission to the Council, the European Parliament, the Economic and Social Committee and the Committee of the Regions, EN 99109-C, COM (2000) 6 final, Brussels: 18.1.2000. 86 pp. [This publication contains a set of methodologies and indicators for benchmarking of national research policies in Europe in the frame of the creation of European Research Area.]

Biographical Sketches

Annamária Inzelt is an economist specialized on S&T and innovation analysis. She is director of *IKU, Innovation Research Centre* (Budapest University of Economics). She has formerly been head of a research team at the Institute of Economics (Hungarian Academy of Sciences) and research fellow at the *Economic Research Institute* (Statistical Office). Dr. Inzelt has been the Hungarian representative at OECD NESTI group, a member of the Committee for Industrial Economics at Hungarian Academy of Sciences and member of the OMFB (National Committee for Technological Development) Council, Honorable Citizen of Nebraska USA, and since 1997 holds the position of Széchenyi Professorship. She has undertaken many research projects and activities related the analysis of innovation (Translation/review of OECD Frascati Manuals, assessment and contribution to employ OECD measurement – R&D, innovation, S&T mobility – in Hungary). Professor Inzelt has been a guest professor and consultant at University of Campinas, Brazil, Visiting Research Fellow at SPRU. She has held the chair of the Scientific Committee of IATAFI (International Association of Technology Assessment and Forecasting Institutions); advisor at 'Research Policy'; advisory Board Technological Forecasting and Social Change (an International Journal); Member of the Standing Committee on East European Affairs and Hungarian Representative at E.E.A. European Economic Association She is member of the several international organisations and she has been organizer of some international conferences. She has participated in many international research groups in industrial economics and technological development analysis on subjects such as the Emergence of New Firms in Czechoslovakia and Hungary, the Institutional Support for Technological Improvement and Industrial Policy Priorities' (World Bank), research projects (INCO-COPERNICUS) on 'Technical Innovation and Industrial Restructuring in Central and Eastern Europe' and 'Information Society and Industrial Development in Central and Eastern Europe', 'Innovation in Eastern Europe and Russia' (CERNA), 'Analyses of Strengths and Weakness of Biotechnology in Selected Sectors' (FhGISI-IKU), 'The Relationship between technological strategies of MNCs and national systems of innovation' (European Union STRATA programme), an EU project on 'The Brain Drain' with (MERIT and CNR).

Tamás Balogh is a civil engineer and economist of the Budapest Technical University. His work focuses on technological development. He was member employee of the National Committee for Technological Development (OMFB); after a merger, it became the R&D division of the Ministry of Education. He works on innovation policy, technology audit, R&D evaluation, regional development, and international co-operation. Since 2000, he is head of the R&D strategy department and invited lecturer at three Hungarian universities. His engineering activities have focused on applied civil engineering (structural designer at an engineering company) and he has experience in research and higher education in civil engineering at the Hungarian institute for building science, at the Budapest Technical University, at the University of Illinois at Urbana/Champaign, USA, and at the University of Stuttgart, Germany. He is member of the Hungarian engineering chamber, of the architectural science committee at the Hungarian academy of sciences and of the Hungarian association of economists. Dr. Balogh has published over 50 papers in civil engineering and over 30 papers in science and technology policy in periodicals, conference proceedings and three books.

SCIENCE AND TECHNOLOGY POLICY IN UNESCO: A HISTORICAL OVERVIEW

Vladislav P. Kotchetkov
UNESCO consultant, Russia

Keywords: science and technology policy, science ethics, international organizations, United Nations system, UNCSTD, ACAST, UNESCO, CASTASIA, CASTALAC, CASTARAB, CASTAFRICA, MINESPOL, regional networks, training

Contents

Summary

Since its creation in 1945 UNESCO dealt with the development of scientific knowledge. At that time a strong belief existed that through application of scientific knowledge and technology transfer from the industrialized to the developing countries it is possible quickly rise welfare of society and spur economic transformation of the newly independent countries in need. Development of science and technology potential and devising strong policy became the important priorities of each government. Since early 1950's experts have been sent to the UNESCO's Member states to assist in establishing science and technology infrastructure, improve research performance and planning in the developing countries. The full fledged programme on Science and Technology Policy was adopted in 1964. The programme existed nearly 30 years and contributed in paving the ways for the science's orientation towards national socially oriented objectives. The article analyses the UNESCO's acievements in this field till the termination of the STP programme in 1994.

1. Genesis of science and technology policy in UNESCO.

When the founding fathers of the United Nations Educational, Scientific and Cultural Organizations met in 1945 in London, the old world laid in ruins, with many of its educational, scientific, economic and political institutions shattered or in disarray. The founding fathers who met in London had a vision of the dawn: they believed that conflicts could be prevented through the educational efforts, spread of knowledge and through international co-operation.

"Since wars begin in the mind of men", they wrote in the UNESCO constitution, "it is in the minds of men that the defenses of peace must be constructed". The Organization had been set up, they went on to say, "for the purpose of advancing, through the educational and scientific and cultural relations of the people of the world, the objectives of international peace and common welfare of mankind".

It is a historical fact that the acronym UNESCO narrowly missed not having the "S" for science. The new Organization supposed to deal exclusively with education and culture. The letter "S" was added in November 1945 by the preparatory commission under the pressure from scientists' groups. Joseph Needham, who became the first head of the Natural Science Section of the newly created organization, must be praised for this addition of "S" in the name of UNESCO. The appointment of Sir Julian Huxley, who was himself a distinguished scientist and accomplished popularizer of science, as the first Director General, assured that activities in the field of science would play an important role in UNESCO.

Sir Huxley recognized the potential of science and technology for development in his first publication about UNESCO in 1946. "The application of scientific knowledge", he wrote, "provides our chief means for raising the level of human welfare". These words remain even more valid in our times.

The task of promotion literacy, in general, and science and technology literacy, in particular, has challenged UNESCO since its inception. At the third session in 1948 UNESCO General Conference instructed the Director General of UNESCO to complete the preparation of reports on the popularization of science(which is   one of important components of science and technology policy), disseminate them to Member States and to draw the attention of governments to the methods of popularizing science. Said one of the earliest UNESCO publications: "It is impossible to become an effective protagonist in the microelectronic age with large segments of populations remaining illiterate and excluded from political and cultural life". In pursuing this task a famous scientific review Impact of Science on Society, which discussed the problems of contemporary sciences and their impact on life of ordinary people, and understandable to a layman manner was launched in 1950. It was published in eight languages and contributed to raising level of scientific literacy over the world. After more than forty years of its existence it was replaced by a review "The World of Science", the first issue of which appeared in 1994, and became a good educational and informative instrument for policy-makers and researchers. The prestigious UNESCO award for scientific popularization, the Kalinga Prize, named after old Indian emperor, was initiated in 1952. The first Kalinga prize-winner was the Nobel Laureate Louis de Broglie. Among the Kalinga prize-winners were such prominent figures as Julian Huxley, Bertrand Russel, Arthur C. Clarke, Pierre Augier, Alexander Oparin, Sergei Kapitza, Abdullah Al Muti Sharafuddin and many others.

"It is of great importance", wrote Albert Einstein, "to give the great public the opportunity to experience, consciously and intelligently, the efforts and results of scientific research. It is not enough that each result is obtained, prepared and applied by a few specialists. Restricting knowledge to a small group diminishes the philosophical spirit of the people and furthers their spiritual poverty". To me, these words of a genial scientist and philosopher provide a key to UNESCO efforts in the field of informal science popularization. To raise scientific literacy of society means not only to increase the level of human welfare, but also to boost spiritual level of society and its members, to improve their ability to make multivarious political, environmental and ethical choices with which scientific discoveries and its consequences are confronting, to resist antidemocratic attempts to impose on them wrong decisions, and to promote, through this, the spread of peace and democracy through society.

Profound geo-political changes have taken place in the world after the World War II when the decolonization process began and the emergence of newly independent states ensued. The course of the World War II, that was determined by use of scientific and technological achievements (radar, antibiotics, nuclear fission, operational research), demonstrated the possibilities which science and technology engendered in the reconstruction of society at the end of the war. Hence, an importance of science and technology was widely recognized as a crucial element of national development and a lack of science and technology capacities in the newly independent nations was considered as a major obstacle to the successful economic development. The theoretical works, in particular that of J.D. Bernal about social function of science, has crystallized opinions towards that end. Some economists call this period as a "science push" phase, the main objective of which was the establishment of scientific and technological infrastructure consisting of laboratories, research institutes, universities, and science and technology councils. Governments were ready to spend large expenditures for these purposes. The creation of a famous European Centre for Nuclear Research (CERN) that followed several meetings and an intergovernmental conference organized by UNESCO may serve as a good illustration to those early preoccupations of national governments and UNESCO. On 24 September 1954 CERN's Convention came into force, thus establishing the world's leading fundamental physics research institution. Such "science push" model triggered optimism among the governments of developing countries. They were of a view: once the infrastructure of research and development was created, scientific personnel trained and policy-making institutions put in place much of the development problems could be solved.

2. Science policy consultancy services

UNESCO as well as the whole newly created UN System (United Nations was officially chartered in 1945 at conference in San Francisco, CA, USA) has tried to respond to the pressing needs and demands of their member-countries. Since early 1950's experts have been sent to the UNESCO's Member states in order to submit recommendations for the improvement of science and technology infrastructure, research performance and planning in the developing countries.

A standard procedure for a mission was the following. A consultant proposed by UNESCO's Secretariat and approved by the corresponding government visited a country in question, worked in the country in close relationship with local authorities during a certain period (a medium stay for those consultancy missions was 22 days, as recorded in one evaluation report) and thereafter submitted a report with concrete recommendations. After being considered by a corresponding government and the UNESCO's secretariat a report was eventually approved. According to the existing rules, the reports from such consultancy missions were not generally available to the public – only from the 1960's have reports been available after they were agreed by concerned governments. The first UNESCO consultancy missions were mainly focused on the establishing of scientific and technological infrastructure in the newly independent developing countries.

The 1960's witnessed an emerging interest on the part of governments to science policy. It was the work of scholars (S. Dediger, K. Olszewski, J.D. Bernal, D.de Solla Price, H. Brooks) and of different research groups, which gave an academic and institutional legitimacy to science policy. As F. Sagasti correctly noted, the use of the term "science policy" did not prevent those researchers from undertaking also technology policy issues. Such trend was reflected in the UNESCO's working programmes. The evolution of UNESCO's approaches, operations and formal mandate in science policy can be traced by reference to the science policy resolutions adopted by UNESCO General Conferences. Thus, the twelfth session of the UNESCO General Conference held in Paris in November - December 1962 for the first time included "information on the science policy of Member States" among the priorities of its science programmes. Education and science sectors were established within the UNESCO secretariat. It should be noted that in 1962 the UNESCO memberships counted as much as 114 Member States in comparison with 37 founding Members in November 1945.

UNESCO actively participated in the preparation of the important international event in 1963 , the United Nations Conference on the Application of Science and Technology for the Benefit of the Less Developed Areas (UNCAST), organized by the UN secretariat in collaboration with UNESCO. This conference was regarded as an initial step in focusing the UN system on science and technology policy as an important instrument in attaining the development goals. According to Juma, it was based on the view that the developing countries could "leapfrog" across various development stages by adopting technologies developed in the industrialized countries. (more details about UNCAST can be seen in the article about STP within the UN system).

A very important milestone at the thirteenth session of UNESCO General Conference (12 October-20 November 1964) was a decision to accord the UNESCO's programmes on natural sciences and technology an importance similar to that given to education. Member States were invited to formulate and implement national science policies with a view to increasing their scientific and technological potential, to improve science teaching at all levels. So, to be able to give effect to new orientation for science programmes, the divisions dealing with science received increased budgetary allocations and their reorganization was embarked upon. The full fledged programme for science and technology policy was worked out, the corresponding Division was set up. Yvan de Hemptinne, a chemical engineer and a functionary from Belgium was appointed as the first Director of Science and Technology Policy Division (STP) of UNESCO.

Lack of political constituency for the representation of science and technology in government affairs of newly independent countries was the major reason why the emphasis of the UNESCO's science and technology policy programme in the sixties and early seventies was mainly in the establishment or strengthening the governmental machineries for science and technology policy –making.

According to the UNESCO reports released early in 1980s, after 20 years of offering advisory services to more than 80 Member States , the situation in this field has changed and by the mid- seventies a majority of the developing countries was equipped with governmental structures for the formulation of their national science and technology policies. The examples are many: the « Consejo Nacional de Ivestigaciones cientificas y tecnologicas» (CONICIT) was established in Venezuela by the law of 26 June 1967, the Tanzania National Scientific Research Council was established by the Government Act in October 1968, the Nigeria Council for Science and Technology (Federal Military Government Decree of 3 February 1970) which has been replaced in 1977 by the National Science and Technology Development Agency (Federal Military Government Decree of 20 January 1977) and later transformed into a Ministry for Science and Technology (National Science and Technology Act, 1980), the National Council for Research (NCR) of Sudan was set up by the Governmental Bill of 15 April 1970, the "Centre National de Planification de la Recherch Scientifique et Technique" of Morocco (Dahir No 1-76-503 of 5 August 1976).

These activities have been substantially increased after the 1979 UNCSTD. In 1985 UNESCO reported that the number of developing countries and intergovernmental organizations which benefited from UNESCO's advisory services in science and technology policy after UNCSTD were: in 1979 -1980- 26 Member States and five regional organizations, in 1981-1983 -51 Member States and four regional intergovernmental organizations and in 1984 - 20 Member States and two regional intergovernmental organizations. These services mainly related to the preparation of scientific and technological development plans and budgets consonant with the overall development plans of the countries concerned.

UNESCO secretariat has always emphasized that UNESCO did not apply a "blue-print" in its consultancy work. The needs and wishes of a government in question guided the work of an expert but not preconceived models. Among the UNESCO's consultants were the representatives of different school of thought and partisans of different economic models. Nevertheless, in spite of remarkable achievements in the efforts to strengthen national science and technology policy potential, some criticism about UNESCO's high reliance on the role of central planning bodies and perhaps a kind of simplistic, institutional planning philosophy could be heard. The critics said that the established national system had frequently become too bureaucratized and incapable in providing leadership in the field of science and technology policy.

UNESCO not only assisted in establishing national science policy-making machinery but also monitored changes that occurred in this field with a view to meet better the changing needs of developing world. One of the notable world trends in the mid 1960s and 1970s was the growing regionalization of scientific and technological co-operation including the setting up of regional instruments or mechanisms for the programming and financing of scientific and technological development. Regional groupings of Member States were increasingly formed in various parts of the world. This trend towards close association of countries belonging to the same (sub-) region raised the problem of harmonization or even integration of their national policies in many fields which are normally the sovereign province of the partner governments. UNESCO assisted these associations of Member States, sometimes with the financial support of UNDP or other financial institutions, with a view to elaborating and implementing science and technology policies for such communities of States. The fruitful work was carried out with the West African Economic Community (CEAO), the Convenio Andres Bello (SECAB), and Association of South-East Asian Nations. The setting up of the Caribbean Council for Science and Technology (CCST) is a sound example of the UNESCO's efforts in the regions. In 1979, a group of Caribbean countries approached UNESCO and UNDP for help in planning their scientific and technological developments. UNESCO arranged for a six-month work of a consultant whose advises were implemented by the countries. Furthermore, the UNESCO Secretariat proposed draft statutes for the Caribbean Council for Science and Technology (CCST), an independent intergovernmental organization. The CCST held its inaugural meeting in Barbados from 29 June to 3 July 1981. Its inauguration was a culmination of a number of UNESCO sponsored (sub-regional) consultations and advisory missions. At the time of inaugural meeting eight countries- Cuba, Dominica, Grenada, Guyana, Jamaica, St. Lucia, Suriname and Trinidad and Tobago- had accepted to membership. Currently the membership in the Caribbean Council for Science and Technology has grown to encompass 16 member countries of the wider Caribbean. The overall objective of the CCST, as it stands now, the development of cooperation in the field of science and technology. Among key areas of activity of CCST are the promotion of science and technology programmes in member countries, science and technology popularization , renewable energy and others. CCST is unique among intergovernmental science and technology bodies. Its members are nominated by the governments. Each member country is entitled to designate two Council members, one of whom must be a scientist chosen by the respective government from amongst senior officials of national bodies responsible for science and technology policies, or of national science and technology research councils, or, where such bodies do not exist, from amongst leading scientists or engineers, the universities, appropriate professional associations or the productive sector. One member sits in his/her personal capacity while the second Council member may be a governmental representative. A small CCST Secretariat updates researchers, planners and policy and decision makers from the region on its activities and topical science and technology-related issues via its quarterly newsletter - the Caribbean Council for Science and Technology Newsletter.

3. Science policy publications

To satisfy the needs of the Member States in latest information and exchange of experience UNESCO has launched in 1963 a famous series "Science policy studies and documents" (SPSD) as part of the Organization's "programme of information on science policy of Member States". The first issue "Science policy and organization of scientific research in Belgium" was published in 1965. The series existed nearly 30 years. The last issue No 74 "The management of science and technology in transition economies" was published in 1994. The series comprised four broad types of documents: country studies, regional studies, normative documents and reports of international conferences. The country studies, which were carried out by national authorities responsible for science and technology policy-making, contained policy statements and descriptive information on national priorities, organization and resources for science and technology . The selection of the countries in which studies on the national science policy were undertaken reflected the originality of methods used in the planning and execution of national science and technology policy, the extent of practical experience acquired in this field, and the geographical distribution. The regional studies were devoted to science and technology policy of the group of countries; it contains as well the allied documents covering regional analyses and reviews of scientific and technological development and UNESCO's activities in the world regions.

The normative documents covered a range of topics, from the methods of policy analysis to manuals on planning techniques and directories of institutions. The reports of international meetings included the proceedings and decisions of policy forums convened by UNESCO on priorities and cooperation in the field of science and technology. For the detailed contents of SPSD series see the

Type of document	SPSD No	Date of publication	Title of document
Country Studies	1	1965	La politique scientifique et l'organisation de la recherche scientifique en Belgique
	2	1965	Science policy and organization of scientific research in Czechoslovak Socialist Republic
	3	1965	National science policies in countries of South and South-East Asia
	4	1966	Science policy and organization of research in Norway
	7	1967	Science policy and organization of research in USSR
	8	1968	Science policy and organization research in Japan
	9	1968	Science policy and organization of scientific research in the Socialist Federal Republic of Yugoslavia
	10	1968	National science policy of the U.S.A. Origins, development and present status
	12	1969	Science policy and organization of research in the Federal Republic of Germany
	19	1970	National science policy and organization of research in Izrael
	20	1970	Politica cientifica y organizacionde la investigacion cientifica en la Argentina
	21	1970	National science policy and organization of research in Poland
	22	1970	National science policy and organization of research in the Philippines
	23	1971	La politique scientifique et l'organisation de la recherche scientifique en Hongrie
	24	1971	La politique scientifique et l'organisation de la recherche scientifique en France
	27	1972	National science policy and organization of scientific research in India
	32	1974	La politique scientifique et l'organisation de la recherche scientifique dans la Republique populaire de Bulgarie

	34	1974	Science policy and organization of research in Sweden
	36	1976	La politique de la science et de la technologie en Roumanie
	56	1985	Science policy and organization of research in the Republic of Korea
	57	1986	Science and technology policy and organization of research in the German Democratic Republic
Regional Studies	11	1969	La politica cientifica en America Latina
	14	1969	The promotion of scientific activity in tropical Africa (Engl., Fr)
	17	1970	National science policy in Europe (E., F.)
	25	1971	Science Policy and the European States (E.,F.)
	29	1972	La politica cientifica en America Latina
	30	1972	European Scientific Co-operation: priorities and perspectives (E.,F.)
	31	1974	National science policy in Africa (E., F.)
	35	1974	Science and technology in African development (E., F.)
	37	1975	La politica cientifica en America Latina
	38	1976	National science and technology policies in the Arab States (E., F.)
	41	1977	Science and technology in the development of the Arab States (E.,F.Ar.)
	42	1979	La politica cientifica en America Latina
	42 Add.	1980	Presupuestacion nacional de activities cientificas y tecnologicas
	43	1979	National Science and Technology Policies in Europe and North America (E., F.)
	45	1979	UNESCO's Activities in Science and Technology in the European and North American Region (E., F.)
	52	1985	Science and technology in countries of Asia and the Pacific- Policies, organization and resources (E., F.)
	54	1983	Informe nacionales y subregionales de politica cientifica y tecnologica en America Latina y el Caribe
	58	1986	Comparative study on the national science and technology policy-making bodies in the countries of West Africa (E., F.)
	62	1985	National and sub-regional reports on science and technology policies in Latin America and the Caribbean
	63	1985	UNESCO science and technology activities in Latin America and the Caribbean (Sp., E.)
	64	1987	Etude comparative sur les organismes directeurs de la politique scientifique et technologique nationale dans les pays d'Afrique centrale

	65	1986	UNESCO activities in the field of science and technology in the Arab States region (E., F.)
	66	1987	Comparative study on the national science and technology policy-making bodies in the countries of Eastern and Southern Africa
	74	1994	The management of science and technology in transition economies
Normative Documents	5	1967	Principles and problems of national science policies (Engl. , Fr.)
	6	1967	Structural and operational schemes of national science policy (E., F. , Arabic)
	13	1969	Bilateral institutional links in science and technology
	15	1969-1970	Manual for surveying national scientific and technological potential (E., F., Spanish., Russian)
	18	1970	The role of science and technology in economic development (E., F.)
	26	1971	International aspects of technological innovation (E.,F.)
	28	1971	Science policy research and teaching units (E.,F.)
	33(1)	1974	Science and technology policies information exchange system (SPINES). Feasibility study
	33(2)	1974	Provisional world list of periodicals dealing with science and technology policies
	39	1976	SPINES Thesaurus. A controlled and structured vocabulary of science and technology for policy-making, management and development (in 3 volumes)
	40	1977-1979	Method for priority determination in science and technology (F., E., Sp.)
	46	1979	An introduction to Policy Analysis in Science and Technology (E., F., Sp.)
	47	1981	Societal Utilization of Scientific and Technological Research
	48	1981, 1984	Manual on the national budgeting of scientific and technological activities (F., E.)
	49	1981	World directory of research projects, studies and courses in science and technology policy (E., F., Sp)
	50	1984,1988	SPINES Thesaurus- A controlled and structured vocabulary for information processing in the field of science and technology for development (in 2 volumes in F., E.)
	59	1984	World directory of national science and technology policy- making bodies (E., F., Sp.)
	60	1984	Manuel pour le developpement d'unites de documentation et de bases de donnees bibliographiques nationales pour la politique scientifique et technologique (F., Sp.)
	61	1984	Technology assessment: review and

			implications for developing countries
	67	1986,1990	Manual for surveying national scientific and technological potential. Full revised second edition of the Manual first published by UNESCO in 1969 (F., E.)
	68	1990	Methodes de programmation applicables a l'orientation et a la gestion de la R&D nationale
	70	1989	World directory of reserach projects, studies and courses in science and technology policy (E., F.)
	71	1990	World directory of national science and technology policy-making bodies (Second edition) (E., F.)
	72	1991	Practical aspects of scientific and technological research programming. Case studies from the USSR
	73	1993	World directory of academic research groups in science ethics
Reports	16	1969	Proceeding of the symposium on science policy and biomedical research (E., F.)
	44	1979	Science, Technology and Governmental Policy. A Ministerial Conference for Europe and North America – MINESPOL II (E., F.)
	53	1983	La Sexta reunion de la Conferencia permanente de organismos nacionales de politica cientifica y tecnologica en America Latina y el Caribe
	55	1983	Science, Technology and Development in Asia and the Pacific- CASTASIA II (E., F.)
	69	1988	Science , technology and endogenous development in Africa –CASTAFRICA II (E., F.)

Table 1.   Science Policy Studies and Documents series of UNESCO.

Analysing the UNESCO's SPSD series we have plausible speculations that UNESCO had successfully initiated and fulfilled its role of clearing house in the field of science and technology policy. It was the only organization within the UN system that dealt in a systematic manner with the comprehensive science and technology policy issues. The most notable allusion to the UNESCO's work in the UN official documents was the "Chapter VI. Science and Technology Policies and Institutions" of the World Plan of Action for the Application of Science and Technology to Development prepared by ACAST in 1971. In order to further the contribution of the World Plan of Action in national development ACAST set up in 1972 a working group for analyzing UNESCO's science and technology programme. As the Group observed, until the countries could develop effective science policies it would be difficult for them to make sound decisions between the multitude of choices provided by the Plan of Action. The Group, which was constituted of six ACAST members and chaired by M.G.K. Menon from India, visited UNESCO in December 1972. The UNESCO's work in this field came under scrutiny. The abandon documentation including the consultancy reports, SPSD series documents, reports of regional meetings and like, testified to the UNESCO's activities, much of that

had not been known to the ACAST members, led the Working Group to express its
satisfaction and favorable impression regarding the orientation and scope of the
UNESCO's work in the field of science and technology policy. The Group noticed an
intellectual drive, qualities of leadership and team effort implied by the work so far
accomplished by the UNESCO Division of science and technology policy. The ACAST
Group insisted that more should be done to display world-widely the available
UNESCO's science policy documents. The Group concluded: "The standards which
UNESCO sets itself in its science policy activities- and hence the standing of its science
policy division within the organization, in the United Nations system, and in the world
at large- must be maintained, since science policy is a field of critical importance to the
less developed countries." The most popular publication of this series was N°46 "An
introduction to Policy Analysis in Science and Technology" well known to national
policy-makers.

The second observation, derived from the analysis of SPSD series, is an increasing
comprehensiveness and complexity of issues covered by the series. Having started from
the relatively simple stock-taking and description of the scientific and technological
potential in a given country, the series gradually addressed itself to the complex
methodological issues of planning and programming of science and technology
activities. Thus, for example, UNESCO developed a "priority exercise", a method
addressed to planners and scientists for the determination of national research priorities.
The main purpose of that method was to collect national views on how to develop
research system in order to fulfill national priorities. The method has been successfully
tested in such countries as Indonesia, Jordan, Ghana and Australia. Account of it is
given in N°40 of a SPSD series(see Table 1). Nevertheless, such methodological work
provoked a rather hostile attitude on the part of some UNESCO opponents, which were
of opinion that the UNESCO's work in the field of science and technology policy had
been of questionable value and too theoretical. We shall notice here that such criticism
was heard mainly from the developed countries, which have had themselves in
abundance of their own intergovernmental organizations, professional societies and
scientific groups dealing with theory and practice of science and technology policy. For
example, an assessment made in 1984 by the USA scientific organizations on the impact
of U.S. withdrawal from UNESCO stipulated that though the science and technology
policy programme of UNESCO had been a subject of a general interest (S&T planning
and impact of S&T on society) however, it was not particularly productive. In my
view, this was a cheap shot on the part of critics who have no burden of responsibilities
for the situation in developing countries. The majority of governments from the
developing countries were partisans of this UNESCO programme and their demands for
science policy consultancy services could not be satisfied by the UNESCO secretariat
because of their abundance and the shortage of funds available. Other serious critical
remarks were that UNESCO ignored and therefore had not been able to integrate
science and technology policy in development planning process, i.e. a cooperation
between the educational, industrial, economic and other sectors , on the one hand, and
scientific and technology policy planning machineries, on the other, were insufficient in
developing countries. We could address those critics to the UNESCO publication
"Science for Development. An essay on the origin and organization of national science
policies" published in French in 1969 and in English in 1971. In this essay, written by a
team of specialists under the direction of Jacques Spaey, the authors, while analysing

the currents of thoughts and action which characterized national science policies of leading European countries, clearly outlined those problems. One of the chapters of the essay entitled "Science policy as part of the general policy of the government" referred to relationship between science policy and education policy, science policy and economic policy, science policy and foreign policy. The authors concluded that science policy was at the cross-roads of economic, social and foreign policies and was, therefore, the responsibility of the whole government and even of the head of the government. Is there any reason to believe that UNESCO has ignored those links and that the failures in the integration of science into development process at the national level could be attributed to a wrong UNESCO's approach? In reality, some UNESCO advises have ended up making it difficult to certain governments to implement them, partly because of incompetence, or lack of means, both financial and human available to those governments. On the other hand, we can agree with those critics who noticed that intended integration of science with technology did not prove trustworthy: UNESCO did not possess a capacity to deal in detail with technology component of policy when the time came in the mid-1970s and 1980s, identified by some experts (Sagasty, Salomon, Krishna) as the "innovation and technology policy implementation" phase. Certainly, a number of the UN organizations like UNCTAD, UNIDO, ILO and others have all carried out substantial work in the area pertaining to their specialization responsibilities. UNCTAD, for example, established in 1970 an Intergovernmental Group on Transfer of Technology and became a leading organization within the UN system dealing with this problem. However, only UNESCO has been dealing with all aspects of science and technology policy from the strengthening of managerial and planning capacities of developing countries, assisting in development of national policies and in building up the corresponding institutional machinery, the facilitating of regional cooperation in this field, the developing, as and when required, of analytical methods and techniques applied by policy-makers, as well as the popularizing of science and technology and dealing with acute ethical problems of science and technology application.

4. Regional Ministerial Conferences

Probably the most known UNESCO's activities in the field of science and technology policy were the regional ministerial conferences (CAST/MINESPOL). These conferences were regarded as important forums for information exchange between the countries of one region and as vehicles for general political recognition of importance of science and technology to national development.

The first in the series of CAST conferences was the Conference on the Application of Science and Technology to Development in Latin America (CASTALA) held in September 1965 in Santiago de Chile. It was the first of its kind ministerial conference in Latin America. 18 of 24 countries of the region participated in CASTALA. The Conference did bring about awareness amongst the national leaders, planners, administrators and scientific circles in Latin America of the need for strengthening science policy at national and regional levels. Among the 87 recommendations of the CASTALA were those proposing the creation of bodies for the formulation of national science policy, coordination of activities in the field of science and technology at the national level and arrangements for the financing of research. Pursuant to the recommendations of CASTALA, UNESCO organized and convened six sessions (1966, 1969, 1971, 1974, 1978 and 1981) of the Standing Conference of national science and technology policy-making bodies in the Latin America and the Caribbean. These meetings greatly contributed to the training of experts and administrators of science and technology policy, thus promoting the creation of teaching and /or research units within governmental and semi-governmental organizations concerned, and in a number of universities or foundations in the region.

The Conference on the Application of Science and Technology to the Development of Asia (**CASTASIA I**) was held in New Delhi, India from 9 to 20 August 1968. It was organized with the cooperation with the UN Commission for Asia and Far East (ESCAFE). The delegations of 14 of the 25 countries represented were headed by ministers concerned with scientific development, ministers of education and ministers of planning. The conference broadened its scope in comparison with CASTALA. The 153 recommendations adopted by the conference established a framework of a concept in the field of application of science and technology to development. This concept was further elaborated in the following UNESCO's ministerial conferences and even in the World Plan of Action published by the United Nations three years later. Thus, CASTASIA I considered the improvement of science education and the popularization of science among the common people as the most important aspects in the creation of a social climate favourable to the application of science and technology to development. It recommended to formulate and implement a purposeful national science policy as a high priority measure in any government programme and to ensure the integration of science policy decisions in the overall national development plans. The financial targets for the governments that had been set forth at CASTASIA I, e.g. the target of 1% of GDP devoted to research and development by the year 1980 were difficult to implement in the region in the existing then economic conditions. However, as a result of CASTASIA I, science and technology policy received a high priority by the top political authorities of the countries of this region.

In the European region the first Conference of Ministers of European Member States Responsible for Science Policy (**MINESPOL I**) took place at the UNESCO Headquarters in Paris in June 1970. Pursuant to a recommendation adopted by the this conference and in accordance with resolutions on European cooperation adopted by the UNESCO General Conference in its fifteenth Session, it was decided to establish within UNESCO an appropriate mechanism to act as a regional European bureau for scientific co-operation. Such bureau was created and has been operating since 1971. It was responsible for coordinating the organization's activities in the region in the fields of science and technology. It also encouraged contacts between scientific organizations from different European countries.

MINESPOL I served also as a basis for the implementation of the UNESCO's long term science statistics programme. In 1968, while preparing MINESPOL I, UNESCO circulated a questionnaire to collect research and development data in both Eastern and Western European countries. Although the results of a survey prepared on the basis of those statistical data were far from satisfactory in terms of coverage, definitions and classifications, it was the first time the same instrument had been used for countries with different socio-economic systems. This survey thereafter served as the model for the UNESCO's statistical biennial surveys.

The first Conference of Ministers Responsible for the Application of Science and Technology in Africa (**CASTAFRICA I**) was held in Dakar, Senegal in January 1974. It was the time when many African countries had been involved in setting up national policy mechanisms for science and technology and in developing explicit national science and technology policies. We already mentioned examples of Tanzania and Nigeria. Côte d'Ivoire set up a Ministry of Scientific Research in 1970, Senegal had a Délégation Générale pour la Recherche Scientifique et Technologique (DGRST) by 1974. Ethiopia's Science and Technology Commission was established in 1975. Ghana, Mali, Niger had all set up their national research councils by the end of 1970s. Later these were transformed into ministries for higher education and scientific research, for instance, in Senegal, Burkina Faso, Cameroon, and Benin. Somalia inaugurated an Academy of Sciences and Arts in 1979. Since CASTAFRICA I many universities and similar institutions of higher education and research have also been created in the African countries. However, the efforts to develop African scientific and technological potential have fallen short of expectations. The latency of governmental efforts may be ascribed mainly to unfavorable economic conditions, lack of finances, absence of long-term commitments on the part of political forces and, sometimes, rather superficial understanding of the role of science and technology in the socio-economic transformation of society. One of 86 adopted by CASTAFRICA I recommendations , for example, urged the African States to take measures needed to attain, if possible, before 1980, the target of between 1,000 to 2,000 scientists and engineers per million inhabitants of whom 10 percent (100-200 persons) should be engaged in R&D activity. Yet, statistical evidence indicated that this target had never been reached. The average number of African specialists engaged in research and development activities in 1980 was of order 45 persons per million inhabitants. Another CASTAFRICA I recommendation which called upon the African States to devote by 1980 one percent of their Gross National Product to R&D and scientific and technological public services has also proved difficult to achieve. In fact, it was only 0,36 percent in 1980. A study undertaken by UNESCO in 1981-1982 in West Africa revealed that many of national science policy-making bodies, though in existence for many years, had not been able to perform the functions for which they were created for lack of qualified personnel, equipment and financial resources. All African countries suffered from the absence of reliable social, cultural and economic data indispensable for realistic science policy that made it difficult to undertake efficient planning of scientific and technological activity, since planning is a strategy of options, and therefore depends upon correct information. CASTAFRICA I strongly criticized the external assistance provided by the UN system, which was considered to be insufficient in terms of its volume, and capacity for delivery to sustain development programmes of many African countries. The conference brought about the importance of transfer of technology, scientific and technological services and technological forecasting (aspects that were almost absent in the previous ministerial conferences). As well the role of overall planning and links between science and technology policy on the one hand, and various national policies, such as educational policy, manpower development policy, industrial policy etc., on the other, had been clearly spelled out, thus reflecting a new trend in evolution of science and technology policy. One of the achievements after CASTAFRICA I was the establishment in

January 1980 of the African Network of Scientific and Technological Institutions (ANSTI). It was a network conceived by UNESCO in order to facilitate cooperation among African institutions for the purpose of national capacity building in science and technology. The purpose was to help African universities and research organizations engaged in training and research in S&T to establish such linkages among themselves to enable them to pool together their human and material resources and thereby contribute more effectively to the application of science and technology to development in Africa. The network disseminated information on S&T activities to ninety one   faculties of science and engineering in thirty three African countries.   Among the important activities of ANSTI was the identification of strategic issues of scientific and technological education in Africa. Some observers considered that it was probably the biggest scientific regional network in the world in terms of the number of countries involved and the intensity of the activities carried out.

The first Conference of Arab Ministers Responsible for the Application of Science and Technology to Development (**CASTARAB I**) that was held in Rabat, Morocco in August 1976 concluded the first round of the UNESCO regional ministerial conferences. CASTARAB I was organized by UNESCO with the co-operation of the Arab League Educational, Cultural and Scientific Organization (ALECSO). The Conference, that brought together ministers (though they all were from the educational field) from 18 Arab States, called upon the strengthening of regional cooperation, outlined some scientific programmes and adopted a resolution in which it proposed to establish an Arab Fund for financing scientific research. A target of $ 500 million dollars was set for this purpose. A working group was entrusted to carry out a feasibility study for this fund. The conference has also established the Castarab Continuing Committee (CCC) to follow up the recommendations of CASTARAB I and to prepare CASTARAB II, that was scheduled to take place in three years. UNESCO provided the secretariat services to CCC. Although the well-grounded feasibility study came up with positive conclusions, the Fund had never been implemented not because of a lack of funds, but rather owing to a lack of political will. Furthermore, in spite of intensive preparations made by UNESCO, the Castarab Continuing Committee and the Arab countries themselves the CASTARAB II never took place for similar reasons. This so-called fiasco could not be the only incentive to attach fault for what did not happen, however the facts witness that by the end of 1980s, very few Arab countries had national science and technology policies. Only Egypt and Iraq had formulated five-year plans of science and technology strategies. Most other countries in the Arab States region did not have national research plans for science and technology, and they therefore were lacking comprehensive plans and policies in this field.

The internal evaluation of the results of the first round of the regional ministerial conferences carried out by the UNESCO secretariat has clearly demonstrated that these conferences promptly aroused a general awareness of the role of science and technology policy in national development. The large number of sub-regional and regional meetings which have taken place in Latin America, Asia and Africa regions following the CAST conferences may be regarded as essential follow-up activities. Multitude of standing conferences, expert meetings, meetings of directors of national councils on science policy and research or similar organizations, seminars and workshops which took place in these regions served as a prove for such conclusion.

Amongst those the following might be mentioned : in Buenos Aires (July 1966), Algiers (September 1966), Colombo (December 1966), Caracas (December 1968), Addis Ababa (October 1970), Villa del Mar(August 1971), Cairo (October 1972), Nairobi (December 1972), Abidjan (January 1973), Lagos (February 1973), Yaoundé (February 1973), Algiers (March 1973), New Delhi (1973), San José (July 1974), Lima (August 1974), Cairo (October 1974), Jakarta (October 1974), Mexico City (December 1974), Kuwait (March 1975), Kuala Lumpur (November 1975), Bogotá (June 1976). It was also concluded that the impact on the national development t scene had not been as great as it might had been expected.

The second round of the regional ministerial science and technology policy meetings organized by UNESCO started in September 1978 when the Conference of Ministers Responsible for Science and Technology Policy in the European and North American Region (**MINESPOL II**) was convened in Belgrade, Yugoslavia. It was concentrated in the exchange of information and views on three major themes: recent developments in national science and technology policies; emerging issues in national science and technology policies; and the status of international co-operation in science and technology in the region and its prospects for the future. We should not forget that MINESPOL II was held a year ahead of the UN Conference on Science and Technology for Development (UNCSTD) when the preparatory process for UNCSTD reached its apogee. Many countries confessed later on that the preparation to MINESPOL II helped them in crystallizing their position for UNCSTD in 1979.

MINESPOL II adopted the recommendations on the strengthening of science and technology information systems and the future development of scientific co-operation in the region. Two years later, in 1980, each of the participating governments was invited to make assessment of the implementation of the MINESPOL II recommendations from the national standpoint. Many countries participated in such assessment. Following this exercise the Director-General submitted to the twenty-first session of the UNESCO General Conference a report on progress in implementing the recommendations which set out the main action taken by UNESCO since 1978 pursuant to the recommendations of the Conference. For example, the definitive text of the document "National science and technology policies in Europe and North America, 1978: present situation and future prospects" appeared as No. 43 in UNESCO's series "Science Policy Studies and Documents". In the spirit of another recommendation concerning the effectiveness of research and development, a group of five European countries took part in the International Comparative Study on the Organization and Productivity of Research Units (ICSOPRU) which had been conducted worldwide under the coordination of UNESCO. The multidimensionality of research performance has been one of the most consistent results of this study. A series of software programmes or packages for the management and analysis of survey data was developed and distributed free of charge to the participating countries.

It seems interesting to compare the assessments of MINESPOL II made by the Soviet and American governments. The USSR highly appreciated the results of the conference and proposed to convene MINESPOL III in 1985. The comments from the USA made on 30 May 1980 were rather opposite. They found that MINESPOL II had had virtually little effect on United States decisions involving science and technology policy.

Therefore, the American government anticipated that it would not be appropriate to recommend the holding of future MINESPOL conference.

CASTASIA II – the Second Conference of Ministers Responsible for the Application of Science and Technology to Development and Those Responsible for Economic Planning in Asia and the Pacific was held in Manila, Philippines in March 22-30, 1982. One hundred and thirty-six representatives from 22 countries in the region took part in this meeting. The demand for training of professionals (high-level officials from science ministries, administrators from policy planning groups etc.) was clearly spelled out. The conference called for strengthening of existing national institutions in the Asian region that were engaged in science and technology policy training and research, to connect them in an interacting network. In response to this recommendation a cooperative program ASTINFO, the Regional Network for the Exchange of Information and Experience in Science and Technology in Asia and the Pacific, was established in 1983. But the most known of the regional networks is STEPAN- Science and Technology Policy Asian Network. STEPAN was inaugurated in May 1988 at the Centre for Technology and Social Change of the University of Wollongong, Australia, following two planning meetings in Bangalore (1983) and Beijing (1985), organized by UNESCO.

The most significant impact of STEPAN activities on national science and technology policy activities has been in the area of training and human-resource development. STEPAN provided a vehicle for raising issues on other network agenda to the senior policy levels of government. STEPAN has also helped the Asian countries to review their S&T policies. For example, STEPAN-UNESCO mission reviewed the Mongolian science and technology policy. The consultancy team, which consisted of Australian, Indian, Korean, Chinese, and UNESCO experts, prepared a draft of the national science and technology policy plan for Mongolia, which was submitted to the Ministry of Science, Technology, Education and Culture of Mongolia.

The Conference of Ministers Responsible for the Application of Science and Technology to Development in Latin America and the Caribbean - **CASTALAC II** was held in Brasilia, Brazil in 1985. The conference promoted the creation in the same year of the Regional Programme for strengthening cooperation among national information systems and networks for Latin America and the Caribbean (INFOLAC). With the financial and substantive support of UNESCO, INFOLAC has grown since its inception into a broad and successful network involving national focal points in 25 countries.

The last one in the series of the UNESCO CAST conferences **CASTAFRICA II** was held in Arusha, Tanzania from 6 till 15 July 1987, thirteen years after CASTAFRICA I.

By the time of the second conference, 18 African nations had already established national science and technology policy bodies at the ministerial level. However, such increase in numbers did not necessarily implied efficiency. Some countries established science policy bodies without any scientific tradition or even infrastructure. The UNESCO World survey on teaching and research units in the field of science and technology policies revealed less than twenty such units in inter-tropical Africa while there were some fifty member states in this region. CASTAFRICA II established the following quantative targets: African countries should devote 1% of their GNP to R&D by 1995, and each African country should consecrate at least 0.4 - 0.5% of its GDP to research by 2000.

Following the request of the UNESCO General Conference in March 1987 the Committee of experts under the chairmanship of A. Shihab-Eldin from Kuwait carried out an external evaluation of the results of the CAST/MINESPOL conferences. The committee concluded that the major importance and value of the regional ministerial conferences had been as fora for information exchange and as vehicles for general political recognition of science and technology for development. They also helped in many cases to foster the development of national science and technology policy and infrastructure as well as regional cooperation. The Committee, however, did not find clear and indisputable examples of concrete examples of science and technology agreements among member states resulting from such conferences. It indicated that important improvements have to be made in all stages: preparation, holding and follow-up in order to make the future conferences more effective and useful. They should be made shorter in duration, with a preparatory phase attended by designated scientists, experts and civil servants; they should concentrate on decisions and recommendations leading to action; mechanisms should be set up for facilitating and monitoring the implementation of the decisions. The Committee felt that it was not meaningful to impose new cycles of regular regional ministerial conferences on all regions. The Committee was of opinion that the levels of conferences and meetings should be adopted, in a more rational, flexible and responsive way, to the needs in different regions at different times. Since that evaluation UNESCO has never been involved in the organization of the conferences of ministers responsible for science and technology policy.

5. Information Exchange and Normative–making Activities

Since 1972 UNESCO has started to work on a science and technology policy information exchange programme. A feasibility study (1974) proposing an international exchange system resulted in the resolutions adopted by the nineteenth and twentieth sessions of the General Conference of UNESCO in 1976 and 1978, which authorized the STP division to undertake a pilot programme on Science and Technology Policy Information Exchange System (SPINES). The main aim of the programme was to facilitate the management and exchange by member states of documents and factual data related to national science and technology policies not only at the governmental level but also at the level of the institutions executing scientific and technological activities or involved in technology transfer. Among the practical results of this programme was a SPINES Thesaurus, a controlled and structured vocabulary of science and technology for policy-making, management and development in 3 volumes was firstly published in 1976. It contained over 10,000 terms and 74,000 semantic relations covering all fields of science and technology, socio-economic development, policy-making and management.

The SPINES Thesaurus enabled the national libraries in indexing documents dealing with scientific and technological or socio-economic development; it also enabled the description of ongoing research projects and expert profiles for the use of decision-makers. This Thesaurus was later on updated and translated from English in Arabic, French, Portuguese, Russian and Spanish languages; it was very much appreciated by the practical librarians. Among other documents produced within the framework of the SPINES programme were: the world directory of periodicals dealing with science and technology policy, the world directory of research projects, studies and courses in science and technology policy, the manual for the development of the documentation units and data base of national bibliographical data in the field of science and technology policy.

Among the normative acts which influenced national science and technology policy was the Recommendation on the Status of Scientific Researcher adopted 20 November 1974 at the eighteenth session of the General Conference of UNESCO. Drawn up during the multistage process of international consultations this Recommendation became the first documented evidence of worldwide governmental recognition of mutual well-defined and interwoven responsibilities and rights of individual researcher and society. The Recommendation stressed that it is the researcher's responsibilities which constitute the foundation of his/her rights. The Recommendation frequently mentioned of intellectual freedom and academic freedom as the indispensable norms for scientific activity. It stipulated that in all parts of the world policy-making was coming to assume increasing importance for the Member States, therefore the Member States should develop machinery for the formulation and execution of adequate science and technology policies, thereby demonstrating recognition of the growing value of science and technology for tackling various world problems, strengthening co-operation among nations as well as promoting the development of individual nations. This document proved to be useful not only to governments and scientists from developing countries but also from developed ones as the situation of scientific researchers at that time felt far short of what the Recommendation proposed. The fact, mentioned by J. Dickinson in the UNESCO publication " Science and scientific researchers in modern society", that the Japan Scientists' Association had found it necessary to cite the Recommendation in several legal actions undertaken by it to protect the rights of individual scientific researchers speaks itself.

6. Training in science and technology policy.

According to the recommendations of all CAST conferences one of the major UNESCO's task had been in the training of experts at all levels of a decision-making process. The UNESCO's activities in this field consisted mainly in supplementing the training of policy analysts in science and technology and in the management of scientific research by giving special attention to the problems of the developing countries. Training activities, even in restricted field of policy-making in science and technology for development called for flexible approaches to accommodate the particular target groups. The purpose of such exercises was the development of a critical mass of competent analysts, planners and managers for the formulation of science and technology policies and plans in developing countries as well as the strengthening of institutional capacities for training and research in this area at the national, sub-regional and regional levels. UNESCO grants, study fellowships, a number of national symposiums, regional training seminars, and summer courses were regularly organized for science policy makers and research managers. In 1981 World Directory of research projects, studies and courses in science and technology policy was published as N° 49 in the UNESCO SPSD series. It listed relevant activities of more than 1100 units in 85 Member states of UNESCO. We already mentioned of the establishment in 1988 of STEPAN- Science and Technology Policy Asian Network. The most significant impact of STEPAN activities on national governments has been in the area of training and human-resource development. STEPAN has initiated a number of training courses for personnel from government, university and private sector institutions in wide ranging research policy, technology and innovation management topics. For example, STEPAN was involved in development of major training programs for Indonesia, Lao PDR and Thailand. The important event in Africa was the International Workshop on Science and Technology Management in Africa held in Ile-Ife, Nigeria in February 1990, which analysed the training needs of African countries and recommended the creation of an African regional network as a part of the UNESCO international decentralized scheme for training and research in science and technology management. Similar function were carried out by the Science and Technology Management Arab Regional Network (STEMARN). An interesting experience was observed in Latin America where the regional networks of industrial and scientific leaders as well as of junior scientists for training in science and technology policy-making and research management have been working with success.

7. Termination of UNESCO Science and Technology Policy Programme.

The International Scientific Council for Science and Technology Policy Development was established by the decision of the 24th session of General Conference of UNESCO in October 1987 for advising on the science and technology policy programmes of UNESCO. At its first session held in Paris in November 1988 the Council recommended UNESCO to allocate more funds to STP programme bearing in mind the resurgence of interest in science and technology policy all over the world.

Unfortunately with the withdrawal of USA, UK and Singapore from UNESCO's memberships of 1 January 1985 the budget available to science and technology policy programme had been gradually curtailed and the number of UNESCO's major programmes diminished. Starting from 1990 the Science and Technology Policy programme of UNESCO has became a part of a new Science, Technology and Society programme. The new programme consisted of three major components: popularization of science, strategies for science and technology development and ethical implications of science and technology. The most important event in the field of science popularization was the First World Conference of Science Journalists held in 1992 in Tokio, Japan and organized jointly by UNESCO, The European Union of Science Journalists' Association, the International Science Writers Association and the National Federation of UNESCO Associations in Japan. 165 participants from 31 countries took part of this conference. Based on the theme of "Science and Communication: The Pursuit of Science in the Service of Humanity", the conference not only assembled a cross-section of acting practitioners, but also offered tangible recognition of the role professional communicators played in creating public understanding of science and technology in the life of society. The famous marine researcher Jacques-Yves Cousteau from France opened the conference by presenting a keynote lecture" The Future of the Earth and the Role of Science Journalism". The closing of the conference featured the adoption of a "Tokyo Appeal", a broad statement of support for the advancement of science communication worldwide. UNESCO was called to expand its activities towards promoting science journalism, establish a global network of science journalists, organize their training and set up a special UNESCO prize in the field of science journalism. The Conference requested UNESCO to continue its efforts and organize the next conference in a 3-4 years period.

In the field of S& T strategies the high-level UNESCO colloquiums on Science and Technology for the Future of Europe held in September 1990 in Berlin, Germany, and for the Future of Latin America, held in December 1990 in Acapulco, Mexico, and International seminar on "Organizational Structures of Science in Europe", held in April 1992 in Venice, should be mentioned. They spelled an increasingly clear need for greater regional cooperation between industry, Universities and research centres, private enterprises and government organizations. The new role of governments in knowledge –based economies was discussed. Scientific community was called to take part in devising national science and technology policies and the international organizations- to assist in this. The programme of training in science and technology management was considered as one of the important action of UNESCO. Another important task assigned to UNESCO was the assistance in the adaptation of the legal systems dealing with science and technology in the countries in transition from central planning system to market economy. An important pioneering activity in this field was the UNESCO symposium "New Science and Technology Legislation for Europe in Transition: Role of Governments and Parliaments held in May 1993 in Siena, Italy. At the same time UNESCO started to compile comprehensive materials for the comparative study on the existing legislative frameworks in the field of science and technology in developed countries.

In the field of ethics the UNESCO Science and Culture Forum "Towards Eco-Ethics: Alternative Visions of Culture, Science, Technology and Nature was held in April 1992 in Belem, Brasil. The meeting of the international eminent philosophers and scientists generated an intense flow of ideas. These culminated with the Declaration of Belem and a Synthesis Document as the resource for reflections about the future of civilization on Earth. The Forum called to make a shift from "ego-ethics" towards "eco-ethics". It concluded that in spite of crisis that faces humanity we still have the capacity to create a sustainable future. To achieve this new morality drawn from many complimentary sources is required. These sources included the objective finding of science as well as the deepest feelings towards nature expressed in many cultures. While considering the importance of advanced technologies, the meeting felt that the invaluable aspects of the traditional cultures and their traditions provided important elements of the needed eco-ethics. The knowledge stored in the non-Western civilizations could be a source of inputs into modern science. Two essential resources for nature and humankind were identified as bio-diversity and cultural diversity. Eco-ethics calls for the preservation of both.

These events created a solid basis for further development of UNESCO's action towards new understanding of the role of science and technology in the development of society and direction of science and technology policies in new conditions. However, the UNESCO's highest authorities decided and implemented the termination of this UNESCO Programme in 1994. This was done in spite of the recommendations of the international forum of experts, created by the decision of the UNESCO General Conference, on reflection of the place of science management and science-society issues in the UNESCO programmes, held in June 1994 in Monsoraz, Portugal and the Genoa Declaration on Science and Society adopted by the European Conference on Science for XXI century, which both recommended the re-establishing of the UNESCO Programme on Science, Technology and Society. The positions of many national delegates (Pakistan, Russia, Malaysia, Turkey, Côte d'Ivoire, Kuwait, Oman, Republic of Korea, Uganda, Belarus, Ghana, Islamic Republic of Iran, Italy, Philippines, Portugal, Switzerland, Ukraine, Venezuela and others), who unequivocally spoke in favour of this programme during the 28th session of the UNESCO General Conference in 1995, have not also been taken into account. The UNESCO programme dealing with science and technology policy became a history.

Glossary

UNCAST:	1963 United Nations Conference on the Application of Science and Technology for the Benefit of the Less Developed Areas (the Geneva Conference)
UNCSTD:	1979 United Nations Conference on Science and Technology for Development (the Vienna Conference)
STP:	Science and Technology Policy
SPSD:	Science Policy Studies and Documents series of UNESCO
UNDP:	United Nations Development Programme.
UNCTAD:	United Nations Conference on Trade and Development
UNIDO:	United Nations Industrial and Development Organization
ILO:	International Labour Organization
ACAST:	UN Advisory Committee on the Application of Science and Technology for Development
CASTALA:	UNESCO Conference on the Application of Science and Technology to Development in Latin America
CASTALAC:	UNESCO Conference of Ministers Responsible for the Application of Science and Technology to Development in Latin America and the Caribbean
CASTASIA I:	UNESCO Conference on the Application of Science and Technology to the Development of Asia
CASTASIA II:	UNESCO Conference of Ministers Responsible for the Application of Science and Technology to Development and Those Responsible for Economic Planning in Asia and the Pacific
CASTAFRICA:	UNESCO Conference of Ministers Responsible for the Application of : Science and Technology in Africa
CASTARAB:	UNESCO Conference of Arab Ministers Responsible for the : Application of Science and Technology to Development
MINESPOL I:	UNESCO Conference of Ministers of European Member States :

	Responsible for Science Policy
MINESPOL II:	Conference of Ministers Responsible for Science and Technology : Policy in the European and North American Region
CCC:	Castarab Continuing Committee
SPINES:	Science and Technology Policy Information Exchange System
STEPAN:-	Science and Technology Policy Asian Network
STEMARN:	Science and Technology Management Arab Regional Network

Bibliography

Pati S.C. (1986). *Experiences with the Vienna Programme of Action*, 87 pp, Munchen ,Forschungsstelle Gottstein in der Max-Planck-Gesellschaft [This is study on the follow-up activities of developing countries on the recommendations of UNCSTD]

Cvallin J. (1982). *Science Policy and Planning. Some observations on UNESCO's Advisory Work on Developing Countries*, 63 pp. Stockholm, Svensca Unescoradets, scriftseri e: Nr 4/1982 [This is a critical report prepared by the Secretary of the Swedish National Commission for UNESCO , in which an analysis of the UNESCO's STP programme was made, I particular the advisory services carries out under this programme]

UNESCO Science *Programmes/ Impact of U.S. withdrawal and suggestions for alternative interim arrangements. A preliminary assessment.*, (1984), 64 pp.,Washington, D.C., National Academy Press [The report prepared by the Office of International Affairs of the National Research Council for the U.S. State Department contains the detailed analysis of three UNESCO science programmes and their budgets for 1984-1985 and discusses possible intrim arrangements in case the U.S. were no longer a member of UNESCO on January 1, 1985]

Sharafuddin A.M. et al., eds. (1988). *Role of UNESCO in Scientific and Technological Development*, 300 pp. Dhaka, United Nations Association of Bangladesh (UNAB). {This book has grown out of the International seminar and reflects the views of both experienced and younger scientists and academics from South and South-East Asia on the role of UNESCO in strengthening scientific and technological capacities of developing countries.]

D'Ambrosio U. and Kotchetkov V.,eds. (1993). *Towards eco-ethics: alternative visions of culture, science, technology and nature.* 144 pp, [The proceeding of the third UNESCO Science and Culture forum held in Belem, Brazil. The possible measures of bringing the gap between the needy and wealthy, the poor and the rich, the North and the South both in the concert of nations and internally even in the most prosperous societies are discussed].

Science and Social Priorities-Perspectives of Science Policy for the 1990's. (1990). 246 pp., Prague, Institute of Theory and History of Science, Czechoslovak Academy of Sciences. [This is a proceeding of the conference organized by UNESCO. It evaluates the changes in the nature of science and technology policy over a decade].

U.S. Background Preparations and Policy Formulation for the United Nations Conference on Science and technology for Development, (1979), Report prepared by the Science Policy Research Division, Congressional Research Service of Library of Congress for the Committee on Science and Technology, U.S. House of Representatives, ninety-six Congress, First session, 395 pp. Washington U.S. Government Printing Office [It contains background information dealing with U.S. preparation for UNCSTD, summarizes the history of preparatory activity, outlines policy position of the U.S. government and the other countries participating in the Conference]

Biographical Sketch

Vladislav P. Kotchetkov is a consultant to the UNESCO. He graduated from the Moscow Higher Technological University in 1959 and worked on satellite and missile technology. In 1969 he was awarded by the USSR Laureate State Prize.   He   received a diploma *cum lauda* in international economics from the USSR Foreign Trade Academy in 1977 and worked as a Director of a Division of International Scientific and Technological Co-operation for the USSR State Committee for Science and Technology. He was involved in the preparation of the 1979 United Nations Conference on Science and Technology for Development (UNCSTD), was a Secretary of the USSR Preparatory Committee for UNCSTD, represented the country at the sessions of the UNCSTD Preparatory Committee, was elected as a Rapporteur at the European Regional Conference for UNCSTD in Bucharest in 1978 and participated in the Vienna Conference in 1979. From 1980 he has been working for the UNESCO Science and Technology Policy Division, was a Chief of the UNESCO Science, Technology and Society Programme and a UNESCO representative at the UN inter–Agency Task Force on Science and Technology for Development. He also served as the Executive Secretary of the Moscow International Energy Club, of the UNESCO International Council for Scientific Communication, the UNESCO representative at the IUCN Commission on Education and Communication. He is a member of the UNESCO-EOLSS Joint Committee.

SCIENCE AND TECHNOLOGY POLICY IN THE UNITED NATIONS SYSTEM: A HISTORICAL OVERVIEW

Vladislav P. Kotchetkov
UNESCO consultant, Russia

Keywords: science and technology policy, United Nations system, 1963 Geneva Conference, World Plan of Action for the Application of Science and Technology to Development, UN Conference on Science and Technology for Development, Vienna Programme of Action, IGCSTD, ACAST, UN Financing System for Science and Technology for Development UNESCO, UNCTAD, CSTD, UN restructuring

Contents

1. The 1963 Geneva Conference
2. The UNCSTD preparation
3. The Vienna Programme of Action
4. UNCSTD Results and Post-Vienna Activity
5. Further restructuring of UN in the economic and social fields
6. Commission on Science and Technology for Development
7. Science and technology policy in the work of UN bodies
8. Conclusion
Glossary
Bibliography
Biographical sketch

Summary

The main thesis is that the restructuring of the UN system in the economic and social fields that happened in the 1990s lead to the practical disappearance of science and technology policy programmes from the UN scenery for a long period of time. It was not a mere coincidence that in 1992-1994 UN terminated the work of IGCSTD, ACSTD, abolished the CSTD, UN ECE SAST, UNESCO Science, Technology and Society programme, etc in spite of a support and needs of developing countries. To understand the reason for such evolution of science and technology policy programmes the author tried to construct a chronology of key milestones in the development of those programmes and demonstrate the various components of how the programmes were working and main results obtained. The major event in the UN science and technology policy scenery was the 1979 Vienna Conference on Science and Technology for Development. The conference preparatory process and its results were, in fact, a consciousness-raising experience for the majority of participating countries, which realized a need in bringing science and technology to the forefront of their economic strategies. However, the results of the post-Vienna restructuring and follow-up action demonstrated that modelling science and technology policy programmes of various UN bodies after the UNCTAD's technology policy imposed model might be in conflict with the real needs of developing countries. In the view of the author, this restructuring side-tracked the Vienna Programme of Action's aspirations. To day, more than a quarter of century after UNCSTD, the most notable allusion to the requirements first formulated by the Vienna Conference in 1979 can be found in the newly established programmes of such bodies as CSTD or UNESCO.

1. The 1963 Geneva Conference

Though the first scientific conference under the auspices of the United Nations was held as early as in 1949 and called the United Nations Scientific Conference on the Conservation and Utilization of Resources (UNSCCUR), it did not deal explicitly with science policy issues. The Conference was the outcome of a proposal made by the US representative to the UN Economic and Social Council (ECOSOC) in September 1946. It was envisaged as an exchange of views and experience among experts who would not necessarily represent the views of their governments, but would cover scientific topics within their competence on the basis of their individual research and experiences. Nevertheless, this Conference represented an important land mark of the beginning of science history in the UN.

The first attempt to treat science and technology policy for development was made at the UN Conference on Science and Technology for the Benefit of Less-Developed Areas. It was held in Geneva from 4 to 20 February 1963 and brought together 1,665 participants from 96 countries, as well as a multitude of international organizations. The national participants were mainly scientists and engineers. The purpose was to draw attention of policy makers to the advances of science and technology and their application to the solution of problems faced by developing countries in such sectors as agriculture, health, and transportation. As the UN Secretary-General U Thant indicated in his foreword to the Geneva conference proceedings, science could be a powerful force for raising living standards if the governments find the means and the political will. However, such acute problems for developing countries as acquisition, transfer and social impacts of technology were not discussed at length. This conference elucidated a dilemma of a right balance between profound discussion of existing links between science, technology and development and adoption of politically accepted practical action to be followed, as well as of a right balance in representation of scientists and politicians at such big world events. Joshua Lederberg, Nobel Laureate and a Chairman of the USA Carnegie Commission on Science, Technology, and Government, has rightly noted that science and politics are a hard match; while scientists can bring much to the political process, they, lacking the policy skills needed to relate their expertise to social action, often feel uncomfortable dealing with the political machinery. This was the case with the Geneva conference. Moreover, according to Raziuddin Siddiqui's evidence, who was then a member of the Pakistani delegation at the Geneva conference, there existed a strong desire on the part of some groups to take out the letter "S" from UNESCO and make it "UNECO". These groups wanted to make a new separate organization for science within the UN system. After a long debate extended over several days this idea was rejected. In spite of all these complications the far reaching consequences of this conference were not that bad. First of all, the Geneva conference, according to F. R. Sagasti, visualized the accumulated stock of science and technology knowledge which the developing countries could rely upon in order to solve their development problems. Secondly, it provoked the important institutional changes within the UN system and created mechanisms for discussing science and technology policy issues. As a follow-up to the Conference, ECOSOC established on 1 August 1963 an Advisory Committee on the Application of Science and Technology for Development (ACAST). This Committee was constituted of 24 members from both developed and developing countries, which were nominated by the UN Secretary General after consultation with corresponding governments. One of the major tasks of ACAST was to find an effective follow-up to the Geneva Conference. The Committee met on a regular basis for 16 years.

ACAST released in 1971 the World Plan of Action for the Application of Science and Technology to Development. The World Plan was prepared on the basis of material submitted by the United Nations system, by intergovernmental and non-governmental organizations and by individual experts. It consisted of two parts: part one dealing mainly with those priorities areas in which science and technology could make a resounding impact and the finances needed for the implementation of the Plan. The part two of the Plan referred to science policy, institutional and educational matters. Among the policy recommendations, the Plan of Action insisted that countries should make effective arrangements for formulating and executing national science and technology policy. The Plan made explicit references to science policy documents published by UNESCO. Following the recommendations of the UNESCO Regional Ministerial Conferences the Plan stipulated that developing countries should allocate one percent of their GNP to science and technology for development. However, as rightly noticed a former director of OST K.H. Standke, the World Plan of Action had never been transformed into a Plan for World Action. Nevertheless, all principal ideas of this Plan have been later expanded and included in the Vienna Programme of Action adopted by the UN Conference on Science and Technology for Development (UNCSTD) in 1979.

Since ACAST had an advisory function and in reality was not involved in policy making process ECOSOC decided to overcome this deficiency and established in 1971 an ECOSOC Committee on Science and Technology for Development (CSTD) where the politicians from 52 Member states discussed policy issues including those suggested by ACAST. The Office of Science and Technology (OST) was created as a part of the UN Secretariat to support both of these committees and assist in the implementation of the advice.

2. The UNCSTD preparation

2.1 Chronology

The second UN Conference on Science and Technology was proposed by the Romanian delegation during the 25th session of the UN General Assembly in 1970. In its resolution 2658(25) the General Assembly requested the Secretary General to apprise the results attained by the UN system in the field of science and technology and their application to development since the first UN Conference on Application of Science and Technology for the Benefit of the Less Developed Areas held in 1963.

The UN Secretary-General, in his report of 26 January1973, recommended to the ECOSOC Committee on Science and Technology for Development (CSTD) to consider a possibility of convening the second international conference on science and technology. Following this report ECOSOC, by its resolution of 16 August 1973, authorised CSTD to investigate a possibility of convening such conference. CSTD, at its second session in March 1994 proposed that the conference would be held in 1978 or 1979. ECOSOC at its 57 session in a resolution 1897 of 1 August1974 supported the idea of holding the Conference and set up an Intergovernmental working group for the examining of "the specific objectives, topics and agenda of such a conference". Frank Joao da Costa, a diplomat from Brazil, was elected as the chairman of the International working group. In October 1974 during the meeting of a Panel of experts jointly with ACAST it was proposed that the forthcoming conference should be: a) held at the level of the governmental representatives responsible for national science and technology policy,   b) devoted to the problems of developing countries, and c) preceded by the regional preparations. One of the experts- Alexander King, a founder and the future President of the Club of Rome, proposed that the UN regional conferences would be held before the major conference in order to discus a situation with science and technology at the regional level. Basing on these proposals the Intergovernmental working group, in its meeting held in Geneva from 21 April till 02 May1975, suggested the provisional agenda of the future conference. The General Assembly at its seventh special session adopted a Resolution 3362 (S-VII) of 16 September 1975, in which it agreed to convene the Conference in 1978 or in 1979 with the main objective of strengthening scientific and technological capacities of developing countries to enable them in applying science and technology to development and in utilizing their own scientific and technological potential to the solution of development problems of regional and global significance.

31 Session of the General Assembly in the resolution 31/184 of 21 December1976 decided to convene the Conference in 1979, approved the provisional agenda of the Conference and requested the UN Secretary-General to appoint a Secretary-General of the Conference. It also established the Preparatory Committee for UNCSTD (Prepcom) and invited all governments and organizations of the United Nations system to co-operate fully in the preparation for the Conference. 28 January 1977 Mr. Joao Frank da Costa was appointed as the Secretary-General of UNCSTD.   The Preparatory Committee hold its first session in 1977 and elected a Prepcom Bureau with Mr. A. Ramachandran (India) as its Chairman (later replaced by Mr. M.G.K. Menon from India), Nikolau (Romania), M.B. El-din Fayez (Egypt), Jankovich (Austria) as the Vice-Presidents.   At its thirty-second session the General Assembly in the resolution of 15 December 1977 decided to hold the Conference "at an appropriate time in 1979", affirmed the objectives of the Conference and requested the UN Secretary-General and the Executive Heads of the United Nations system to give priority to the preparation of the Conference. Although the representatives from the developed countries were first reluctant to the idea of convening second UN conference on science and technology they all voted for the resolution. It is interesting to recall that ten countries from the group of socialist countries abstained, one developing country (Ethiopia) voted against the resolution. The major reason for such position of the East European countries (at that time they were called "socialist") was a fear of possible proliferation of the United Nations systems and the creation of a new organization responsible for science and technology. Their position was that science policy issues should remain within the UNESCO's domain. By the resolution 32/184 of 19 December 1977 the General Assembly accepted an invitation of the Government of Austria to act as host to the Conference and decided that the Conference would be held in Vienna for two weeks in 1979. According to all assessments, the preparatory process for UNCSTD was not less

significant than the conference itself since it triggered the whole range of activities at the national, regional and international levels. Each participating country has been required to prepare "a national paper", that was to analyse the social and economic problems of the country that might be solved with the help of science and technology. ACAST drafted the guidelines for the preparation of national reports and urged governments to ensure broad participation of scientists in their preparation. The preparation of such reports, in its turn, in many cases stimulated extensive national debates among government officials, development specialists, scientists and technologists and industrialists, in other words – all groups of actors responsible for national science and technology policy. From 142 States participated in the Conference 130 presented their national papers that were attached as the "background documents" to the Conference proceedings. Some countries, while preparing the national papers, launched an internal process of formulating a national science and technology policy and incorporating this policy in development plans. According to many post-conference assessments, such internal processes might be considered among the major achievements of UNCSTD. Prepcom hold five sessions devoted largely to the preparation of the programme of action to be presented to the Conference.

2.2. UN Regional conferences

Apart from the Preparatory Committee meetings a large number of international, regional, sub-regional and national meetings were organized. In particular, the regional conferences on science and technology convened by the UN regional economic commissions in 1978 should be mentioned. They were held under the chairmanship of the Secretary- General of UNCSTD: for Europe in Bucharest, Romania from 26 to 30 June 1978, for Asia-in Bangkok, Thailand from 17 to 22 July 1978, for Latin America- in Panama from 16 to 21 August 1978 and in Montevideo, Uruguay from 29 November to 01 December 1978, for Africa-in Cairo, Egypt from 24 to 29 August 1978, and for the Arab States- in Amman, Jordan from 12 to 14 September 1978. The major task of these regional conferences was a selection of substantial subject areas for scientific and technological cooperation in each region. Reflecting on a relative failure of the 1963 Geneva conference for science and technology, which tried to cover too many scientific issues, it was envisaged that the common five subject areas of cooperation could be selected on the basis of regional choices and   discussed at the Vienna Conference. The results of the selection can be seen in Table 1.

SUBJECT AREA REGION	EUROPE	AFRICA	LATIN AMERICA	ASIA	ARAB STATES
Agriculture		x			
Construction		x			
Health	x	x	x	x	x
Transportation and communication		x	x		x
Energy	x	x		x	x
Natural	x		x	x	x

resources					
Food	x		x	x	x
Industrialization				x	x
Environment	x				

Table 1.   Priorities for regional co-operation in the field of science and technology in 1978

Unfortunately, the extensive negotiations in the course of selecting five subject areas in regions did not find its further utilization during the Conference itself. Therefore, these results have a historical interest to realise which areas were considered by the countries in 1978 among the major priorities for the regional co-operation in the field of science and technology.

2.3. Surrounding scientific meetings

The science-oriented meetings held in 1979 in affiliation with ACAST were organized in response to the request of the Secretary-General of the UNCSTD, addressed to the world scientific community, to participate in the preparation of the forthcoming conference.   The most significant among these meetings were:

"Trends and perspectives in Development of Science and Technology and Their Impact on the Solution of Contemporary Global Problems"-an international symposium organized by the USSR Academy of Sciences, held in Tallinn, USSR from 8 to 12 January 1979. The participants numbered nearly a hundred prominent scientists, public figures and representative of international organizations with substantial theoretical and practical knowledge of science, technology and global problems. The   Secretary-General of UNCSTD participated in the meeting. The symposium expressed its deep concern that the world-wide concerted action in respect of the contribution of science and technology to solution of global problems was insufficient. The meeting called for the establishment of interdisciplinary task-force arrangements within the UN system to allow for adequate treatment of the complex global problems.

An International symposium on Science and Technology for Development organized by the International Council of Scientific Unions (ICSU) in cooperation with 19 other international professional organizations was hold in Singapore from 22-26 January 1979. The meeting where 140 participants from 40 countries expressed their views endorsed the proposal of the developing countries for the creation of a science and technology development Fund. The symposium also stressed that more emphasis be given in development planning to the role of social sciences.

"Technology for Development" meeting organized by the UN Department of Technical Co-operation and Development took place in Abidjan, Ivory Cost from 9 to 13 April 1979. The Pugwash Executive Committee on Military Technology, Disarmament, and Development made a statement on science, technology and development based on the results of a workshop held in March 1979 in Baden, Austria. In order to achieve the science and technology goals of a New International Economic Order, it recommended examination of how military uses of science and technology may distort development. It called for setting up a special tax on military expenditures of all countries which would be pooled centrally and used for development purposes. The deliberations of these meetings were addressed to UNCSTD and utilized as a basis for the position platforms of various groups of countries at the Vienna Conference. In total more than seventy international scientific meetings and seminars devoted to the Conference preparation were held all over the world. These meetings constituted the scientific surrounding of the Conference.

The Conference itself and its Prepcom meetings have not been a scientific exercise where the substance of science and technology was discussed. They provided a place for political and governmental debate for identifying and removal of obstacles to the application of science and technology to development, be they political, social, economic, institutional or cultural. The debate underscored dramatic differences between the developed and developing countries in their proposals of solutions for overcoming existing obstacles. That explains why despite a long gestation period in preparation (according to some reports no other UN conference hitherto has had a preparatory period larger than that of UNCSTD) and overall expenditures estimated at some $50 million the Conference did not managed to attain the tangible results.

3. The Vienna Programme of Action

The United Nations Conference on Science and Technology for Development (UNCSTD) was held in Vienna, Austria from 20 to 31 August 1979. 1865 delegates and advisers from one hundred and forty-two countries as well as from the majority of the United Nations specialized agencies and bodies, intergovernmental and non-governmental organizations participated in the Conference. It was preceded by the International Colloquium on Science, Technology and Society: Needs, Challenges and Limitations organized by ACAST from 13 to 17 August in Vienna. The colloquium was an official UN event though not a part of the official Conference meeting. It was chaired by W.K.Chagula, distinguished African scientist and the Chairman of ACAST. 403 scientists and technologists from 95 countries participated in this colloquium. The meeting provided an opportunity to scientists to speak on science policy and to reach international consensus on scientific and technological issues affecting development. The colloquium worked out recommendations which suggested the directions for the UNCSTD deliberations and follow-up.

It should be noted that the United Nations Conference on Science and Technology for Development (UNCSTD) was a part of the efforts of developing countries to establish a New International Economic Order. UNCSTD tried to reach some compromise agreement on the following main points: 1) governing principles for the choice and transfer of technologies for development; 2) institutional arrangements within the United Nations system with a view of its strengthening and new forms of international cooperation in the application of science and technology and 3) financial system to implement the Programme of Action. These points were discussed in plenary meetings , chaired by Dr Hertha Firnberg, Federal Minister for Science and Research from Austria and in two committees chaired by Jacques Diouf from Senegal and M.G.K. Menon from India and at meetings of the Working Group on Science, Technology and Future chaired by M. Malitza from Romania. The outcome of negotiations resulted in the Vienna Programme of Action on Science and Technology for Development adopted by the Conference at its 16th plenary session on 31 August 1979. The principal thrusts of the Programme were strengthening the endogenous scientific and technological capacities of developing countries; restructuring international relations; increasing the role of the United Nations system; and providing an adequate financing for these purposes. The Vienna Programme of Action concentrated on policy, structural dimensions of science and technology in a rather general manner. It was a set of non-binding principles and guidelines. The Programme stipulated, for example, that governments of each developing country should formulate an effective national policy for science and technology. International organizations were called to provide meaningful consultations and exchange of experience at international forums on science and technology policies and planning and organize the training of experts in science and technology policy. However, the implementation of these general guidelines depended on the capability, initiative and political will of each government to do so. The advice incorporated in the Programme advocated by the leadership of federal ministries for science and the expertise of the groups like national academies of sciences in the advanced countries could not replace the real action of a developing country. There was a serious disagreement between developed and developing countries on issues of facilitating access to industrial information, and to patent rights and transfer of technology, on establishing of a new financing system. Because of sharp incompatibility of the positions of the developed and developing countries these disagreements remained unsolved during the conference in spite of around the clock negotiations and they were referred to the new Intergovernmental Committee on Science and Technology for Development (IGCSTD) established by UNCSTD. According to some assessments, the universality of science during the conference was submerged by the universality of politics: as a result, science lost and politics did not win.

4. UNCSTD Results and Post-Vienna Activity

In spite of widely spread opinion about the frustrating results or, even, a "classical failure", the Vienna Conference might be considered as a qualified success. First of all, the Vienna Programme of Action helped to crystallize the understanding of the multiplicity and complexity of issues involved in science and technology policy for development. The conference preparatory process was a consciousness-raising experience for the majority of developing countries, which realized a need in bringing science and technology to the forefront of their economic strategies. Many provisions of the Vienna Programme of Action remain valid till now. The programme has also outlined important changes in the structure of the UN governing bodies. These changes were made in an attempt to adjust the UN mechanisms towards the successful fulfilling the magnitude and complexity of their tasks in the application of science and technology to development. Among these important structural changes were the establishment of the Intergovernmental Committee on Science and Technology for Development (IGCSTD), a new Advisory Committee on Science and Technology for Development (ACSTD), an inter-agency ACC Task Force on Science and Technology for Development (ACCTF), the Centre for Science and Technology for Development (CSTD) and the UN Interim Fund for Science, Technology and Development (IFSTD).

4.1. Intergovernmental Committee on Science and Technology for Development

The Intergovernmental Committee on Science and Technology for Development (IGCSTD) was established as a subsidiary body of the UN General Assembly to provide policy guidelines and oversee the development and implementation of the strategies spelled out by the Vienna Programme of Action. It was conceived as a powerful political body in the area of international science and technology policy reporting to the General Assembly via ECOSOC. The IGCSTD included all states as full members, in contrast to its predecessor the Committee on Science and Technology for Development of the ECOSOC. The IGCSTD adopted in 1983 the operational guidelines for United Nations system support in eight subject areas and the first of them was "Science and technology policies and plans for development". However, in spite of more than ten sessions hold, the Committee never fulfilled its functions, mainly due to a lack of financing as well as a lack of political will. Thus, the 1987 cross-organizational review of the medium-term plans of the United Nations system noted that most of these plans specifically referred to neither the Vienna Programme of Action, nor to any of the eight programme areas intended by the Intergovernmental Committee as a framework for the system's science and technology efforts. While the review registered an ever-growing volume of United Nations system science and technology activities, it stressed the lack of common understanding within the United Nations system of what constitutes science and technology for development activity and, more specifically, the necessary characteristics it should entail for building endogenous scientific and technological capacities of the developing countries.

4.2. Advisory Committee on Science and Technology for Development

The new Advisory Committee on Science and Technology for Development (ACSTD) had as its task to advise on substantial questions related to the implementation of the Vienna Programme of Actions and to serve as the subsidiary body reporting to IGCSTD, but not the UN Secretary- General, as it was in case of the former ACAST. This new advisory committee comprised not only scientists, but also experts from government and business sectors as well as specialists in human affairs and policy establishments. They served in their personal capacities and did not represent their governments or organizations. The Advisory Committee has analysed such issues related to science and technology policy, as: human resource development for planning and management of science and technology; role of regional organizations in promoting research and development and popularizing science and technology; indicators of impact of science and technology for development , long-term perspectives on science and technology for development. At its seventh session held in Petropolis, Brazil in February 1987 ACSTD outlined as its priorities: focussing on more specific problems and potential remedies to the problems of specific regional, sub-regional and national contexts; continuity to the topics under study in order to build up a sufficient body of knowledge enabling to provide strong support for policy advise to the developing countries; enhancing the quality of its own advices in order to make them more useful to policy makers, etc. However, the real role and influence of this Committee was undoubtedly less that those of the former ACAST

4.3. ACC Task Force on Science and Technology for Development

The ACC Task Force on Science and Technology for Development (ACCTF) was constituted as a subsidiary body of the UN Administrative Committee on Co-ordination (ACC) with the purpose to increase co-ordination and avoid possible duplication within the UN System in the field of science and technology. The work of this group was particularly successful. For example, the eight session of the ACC Task Force was held in New York in December 1986. 31 representatives of the UN system participated in the work. The group considered the coordination of the UN activities in Colombia, Jordan, Kenya and Thailand. It selected eight joint projects for the implementation. It also issued recommendations on establishing a data base of the UN science and technology information system for wide dissemination in all countries to assist policy makers in those countries. From the early existence of the ACC Task Force UNESCO actively participated in its work and was recognized as the lead-agency for the UN joint activities in the field of science and technology policy. In 1990 UNESCO organized in Paris an Ad-hoc high-level inter-agency meeting to discuss further coordination and cooperation in the field of science and technology. The Assistant Director-General of UNESCO for science was the last Chairman of the ACC Task Force before this ACC Task force was abolished in 1993 as a result of the vast reorganizations within the UN system.

4.4. Centre for Science and Technology for Development

The Centre for Science and Technology for Development (CSTD) was established in 1979 to provide secretariat to the above three committees, to carry out policy analysis and secure necessary interactions with members states, UN bodies and intergovernmental and non-governmental organizations with the purpose of promoting policy dialogues in the field of science and technology. It replaced the Office of Science and Technology established in the wake of the Geneva Conference. During thirteen years of its existence the Centre established close working relations with the UN organizations and UN Regional Commissions dealing with different aspects of science and technology policy. It tried to coordinate joint UN activities in the field of science and technology in selected developing countries, like, for example, Colombia, Thailand, Jordan. It organized a dozen of international meetings which analysed different aspects of science and technology policy. The Centre initiated the Advance Technology Alert System (ATAS), a network designed to become a consulting instrument to developing countries on issues of technology choice and science and technology policy planning. The ATAS bulletins assessed a wide range of technologies to be used by the developing countries. The Centre was also involved in activities in the area of conversion of technology from military to civil use.

4.5. UN Financing System for Science and Technology

As an agreement on the UN Financing System for Science and Technology for Development (UNFSSTD) with a two-four billion US dollars capital had never been reached during the Vienna Conference, the UN Interim Fund for Science, Technology and Development (IFSTTD) was established, as a provisional system based on voluntary contributions to assist developing countries in strengthening their scientific and technological capacities. A sum of $250 million was named as the target for the first two years of operation to provide loans and grants to developing counties in order to fund their activities. An executive board was set up to review the project proposals and grant financial assistance. In reality, because of the debt crisis in many developing countries in the 1980s and negative attitude on the part of major donor countries, a voluntary fund collected only $40 million. After extensive consultations, the IFSTD was converted in 1982 to the Financing System for Science and Technology for Development (FSSTD). However, the expected contributions had never arrived and in 1987 the General Assembly terminated FSSTD and transferred its functions and resources to a new trust fund within the UN Development Programme, entitled the Fund for Science and Technology for Development (FSTD).

5. Further restructuring of UN in the economic and social fields.

In 1989 IGCSTD decided to undertake a review of the implementation of the Vienna Programme of Action. A review confirmed the validity of the Programme and outlined the increasing disparities in scientific and technological capabilities between industrialized and developing countries. The review noticed some progress made since the Vienna Conference in 1979. For example, there has been a substantial growth in the establishment of policy-making bodies at the ministerial level in Africa. By 1987 about two-thirds of African countries have created science policy-making bodies. However, an effectiveness of these bodies, due to lack of capital to engage in systematic policy research and of human capacities to undertake these research in order to remedy the economic situation of the country in question, had not been proven. Moreover, at the time when the review was finished and soon after that, an international environment has changed considerably. The increased internationalization of production, accelerated privatization, strengthening the private sector in developing countries, etc. began to spread over the continents. Technologies started to play a role of new driving force in economic development of developing countries. The expanded use of information and communication technologies influenced international competitiveness and services. Science and technology policies in developing countries, which were formulated in the pre-globalization age, were not able to meet the new conditions. Efforts to reform these policies by focusing on the role of specific sectors, such as information technology, have not been very successful. Very often an important economic dimension was excluded from STP, particularly in the UN programmes and in the science and technology policy of many developing countries. Few international institutions possessed the capacities to provide valuable advices as policy was concerned. The challenges associated with globalization and technological development demanded new approaches to national science and technology policy formulation. All this provoked restructuring the UN system in the economic and social fields. It was considered as the most extensive restructuring in the UN history.

6. Commission on Science and Technology for Development

In April 1992 ECOSOC abolished IGSTD, ACSTD and replaced them by the Commission on Science and Technology for Development (CSTD). The CSTD was conceived as an intergovernmental body, a functional commission of ECOSOC, with member countries elected by the ECOSOC, and with governments nominating experts to the Commission meetings. Among the major functions of the new commission was the focussing on technology transfer and capacity building, studying science and technology policies, particularly in respect of developing countries and providing technology assessment and related policy analysis. However, the result, in particular at the beginning of the CSTD work, was not that encouraging. According to its mandate, the CSTD shall report to the ECOSOC, and thus to the General Assembly. But in practice, the ECOSOC only took note of such reports, and there was no overall policy debate and conclusions on these important issues.

The Commission consisted of the representatives of 53 Members states met for the first time in April 1993 in New York, USA and subsequently met in Geneva, Switzerland, at its second, third, fourth, fifth and sixth regular sessions, in 1995,1997, 1999 and 2001, 2003 respectively. By its resolution 2002/37 the Economic and Social Council (ECOSOC) authorized the Commission to meet annually. Beginning with its seventh session in May 2004, the Commission is to hold its regular sessions on an annual instead of a biennial basis.

In 1998 CSTD was restructured once more time. The membership of the Commission was reduced from fifty-three to thirty-three with the following geographical distribution: eight members from African States; seven members from Asian States; six members from Latin American and the Caribbean States; four members from Eastern European States; and eight members from Western European and other States. In resolution 1995/4, ECOSOC invited the Commission on Science and Technology for Development to give consideration to ways and means of taking advantage of the twentieth anniversary of the Vienna Conference on Science and Technology for Development for the formulation of a common vision regarding the future contribution of science and technology for development. The Commission was encouraged to sharpen the focus of its work, devoting particular attention to the issues of transfer of technology and capacity-building, in particular of the developing countries. At its regular session, the Commission selects a theme to be addressed during its inter-session periods. Thus, for the biennium of 1997-1999 CSTD selected a theme "Science and Technology partnerships and networking for national capacity building", for 2001- 2003- "Technology development and capacity-building for competitiveness in a digital world". At the seventh session of CSTD in 2004 the participants confirmed the unique role of the Commission as the only high-level United Nations organ established to provide high-quality advice to the Economic and Social Council and the General Assembly on science and technology for development. The primary role of the Commission remains that of a policy "think tank", which studies the role of science and technology for development, including technology transfer and the opportunities and risks presented by new and emerging technologies. CSTD called for the increasing by the Governments of their research and development expenditure in science and technology to at least 1 per cent of gross domestic product (GDP). Governments were also advised to implement fiscal and other incentives to encourage R&D in the private sector and joint projects between private and public sector. The Commission noted that, in most developing countries, scientific advice still was not in the centre of the decision-making process. The Commission also called for the creation of national science advisory bodies. The importance of a multi-stakeholder and people-centred approach in policy-making was stressed.

At the inter-agency level, the ACC Task Force on Science and Technology for Development was abolished in early 1993. Its coordination functions in the cross-cutting issues of science and technology were split between different inter-agencies committees.

The secretariat of UN dealing with science and technology issues was also reorganized. Most of the functions of the Centre for Science and Technology for Development, including those of providing secretariat to CSTD, have been transferred to UNCTAD.

Certainly, with such profound changes the UN system has lost some effectiveness of its operations in the field of science and technology policy for many years. Nevertheless, its parts have continued to play an important role in the promotion of science and technology policy.

7. Science and technology policy in the work of UN bodies

Most of the arms of the United Nations have been involved in studying various aspects, mainly sectoral, of science and technology policy.

(The UNESCO work in the field of science and technology policy for 1945-1994 is described in detail in a separate article). Since 1994 UNESCO has continued to be the main source of statistics on science and technology in the developing countries. In the period 1994-2000 some unconnected activities in this field have been carried by UNESCO in the regions. Thus, in 1996 UNESCO fielded a consultancy mission, at the request of the Albanian Government, to provide an advice on the formulation of a national science and technology policy and a national science budget. The corresponding missions were arranged for several African countries: Zambia, Burkina Faso and Namibia. The training courses in the fields of science and technology policy were proposed for the African countries. In 1999 UNESCO organized the "World Conference on Science for the twenty-first century: a new commitment" held in Budapest, Hungary. In the Declaration adopted by the Conference, it highlighted the vital importance of science policy. The Conference emphasized the fact that regional and international networking and co-operation can facilitate the exchange of national experiences and the design of more coherent science policies. In response to the demand of Members States and the recommendations of the World Conference on Science, UNESCO has been gradually restoring its science and technology policy programmes. The new mission of UNESCO is to promote a new contract between science and society and provide assistance in the formulation and implementation of science and technology policies at the national, regional and world-wide levels, with a view to increase and mobilize scientific and technological resources in the service of the advancement of knowledge and in the support of sustainable development and peace.

The ILO's Technology and Employment Branch produced a number of studies and reports on a variety of topics related to science and technology policy and played an important role in the debate on the issue of appropriate technology. An ILO study of the secondary sector has considered non-governmental organizations to be same importance as governmental institutions and macroeconomic policy in the accumulation of technological capacity. ILO examined the global distribution of employment as a result of new technologies introduction and proposed socio-economic policies and appropriate investments for their implementation.

Such UN agencies as FAO and UNIDO also have strong science and technology programmes in their particular sectors. FAO, for example, prepared a conceptual framework, guidelines and indicators relating to assessment and transfer of technology for sustainable agricultural and rural development. The publication "Bio-technology in Agriculture, Forestry and Fisheries" set out policy guidelines for the development and use of biotechnologies. UNIDO initiated specialized newsletters and Technology Trends series with a view to sensitizing policy makers to the impacts of new and emerging technologies and stimulating the formulation of policies for the application of advanced technologies. The World Bank programmes had been focused on technology policy with a strong economic perspective. The Inter-Parliamentary Union has, time from time, organized regional conferences devoted to governmental policy, critical to effective utilization of science and technology. According to the IPU, scientific and technological issues should be placed at the centre of governmental decision-making processes, especially through ensuring that scientific and technical institutions have a key role within the machinery of government and through enabling scientists and technologists to be more prominent at higher levels of decision-making.

The Institute for New Technologies of the United Nations University (UNU-INTECH) was set up in Maastricht, the Netherlands, as a specialized research and training centre of the United Nations University. Within the UNU family, UNU-INTECH conducts research and policy-oriented analysis in the area of new technologies and the nature of their economic and social impact, especially in relation to developing countries. Monitoring the emergence of new technologies and assessing their likely socio-economic impacts are key challenges for policymakers today. The UNU-INTECH has always supported national, regional and local level policy-makers and researchers in the developing world to explore and assess the opportunities created by new technologies and to anticipate the potential consequences for their countries. Among the UNU-INTECH projects relevant to policy-making were: Technological change and innovation policy in industrializing countries, Comparative reviews of national innovation systems: a policy perspective (CRONIS), Collection and analysis of all existing innovation surveys, etc.

The regional economic and social commissions of UN supported national efforts to develop national science and technology policies and to increase the efficiency of existing scientific and technological institutions. They encouraged their member states to give a high priority to science and technology in their national policies. For example, since 1971 Senior Advisers to ECE Governments on Science and Technology (SAST) was a subsidiary body of the UN European Economic Commission dealing with issues of science and technology policy. Among the SAST's terms of reference, as revised in 1991, were: exchange of national information in the formulation and/or implementation of science and technology policy and promotion of international cooperation in the field of science and technology policy. SAST carried out biannual review of changes in national policies, priorities and institutions and international cooperation; studied national policies of the countries in transition aimed at promoting innovation and the reorganization of their systems of management of science and technology under new economic and political conditions. For example, the last review was carried out at the 22nd session of SAST held in 1996. Twenty-six governments, including 14 from the countries in transition, contributed to this review. A particular attention was given to financing scientific and technological research activities; diffusion of technology and the problems of brain drain. However, following the general strategy for eliminating science and technology policy from the UN, the 52nd session of the Economic Commission for Europe in May 1997 decided to dissolve SAST and therefore to abolish the Industry and Technology Division of ECE responsible for the work of SAST . Similar situation could be reflected in other regional commissions.

UNCTAD, from the days of its establishing in 1964, has always been actively involved in policy analysis on technology transfer. Its activities included analysis and identification of policies that favour technological capacity-building, innovation and technology flows to developing countries. In 1970 UNCTAD set up the Intergovernmental Group on Transfer of Technology that helped to stimulate intergovernmental discussions on technology-related issues. In 1974 the Group was transformed into the Committee on Transfer of Technology. The discussions during the 1979 Vienna conference subsequently led to the launching of negotiations on an international code of conduct on the transfer of technology. In the 1980-90s, the Committee's work included the identification of policy measures for strengthening the technological capacity of developing countries, leading to the formulation and implementation of a strategy for technological transformation. UNCTAD also monitored trends in international technology flows and their developmental implications, and considered policies to stimulate those flows, to enable developing countries to attract resources and investment. In 1992, the eighth session of UNCTAD laid the foundations of national and international policies aimed at enhancing the development prospects of all countries, particularly those of developing countries. Consequently, the Committee on Transfer of Technology was suspended and an Ad Hoc Working Group on the Interrelationship between Investment and Technology Transfer was created. This Group focused on the interrelated areas of investment, technology transfer, capacity-building and competitiveness. It also addressed different policy-related aspects of technological innovation, including the transfer and development of environmentally sound technologies, university-enterprise cooperation, industrial districts and technology partnerships. The ninth session of UNCTAD held in 1996 called for science, technology and investment policy reviews to be conducted with interested countries in order to identify options for national policies that foster technological capability and innovation. It also called for an exchange of experiences among countries at different levels of development in the formulation of policies to promote technological capacity-building and innovation in developing countries. The new division dealing with investment, technology and enterprise development was created within the UNCTAD. It has taken over the work on science and technology for development. Since that the intergovernmental discussions have been focused on capacity-building in certain fields: biotechnology, information and communication technologies and other emerging technologies. Commercialization of science and technology, including marketing and intellectual property protection, has also been under the UNCTAD's programme of work. It is interesting to note that the wording "science and technology policy" has disappeared from the UNCTAD's lexicon. It is replaced by the term "science advice". For example the new UNCTAD programme on "science and technology diplomacy" was initiated by UNCTAD in 2001. Its title is used to mean the provision of science and technology advice to multilateral negotiations and the implementation of the results of such negotiations at the national level. It therefore covered activities at both international level and national level pursuant to international commitments.

8. Conclusion

In the view of the author of this article, the restructuring of the UN system in the field of science and technology and the shift in its science and technology policy programmes that occurred in the 1990s in favour of technology and innovation policies and commercialization of research should not obscure the multiplicity of problems of science and technology policy that confront the developing countries. We also have some political lessons learned. Transition to the market does not imply the end of state intervention, especially in scientific and technological matters. It calls for a clearer delineation between the sectors where state intervention can be useful and those that have to be left to market forces. Termination or modelling science and technology policy programmes of various UN bodies after the ECOSOC imposed model of the UNCTAD/CSTD programmes could be in conflict with real needs of developing countries. The best available advice is to be incorporated in the various science and technology policy programmes advocated by the leadership of different UN bodies in their particular fields. That advice can be no more authentic than the empirical findings to date. It is of great urgency that these be bolstered by a more robust appreciation of science and technology policy studies and their application in the national policies of the developing countries. The re-establishment of UNESCO science and technology policy programme seems to mount that need with the most urgent priority.

Glossary

UNCAST:	1963 United Nations Conference on the Application of Science and Technology for the Benefit of the Less Developed Areas (the Geneva Conference)
UNCSTD:	1979 United Nations Conference on Science and Technology for Development (the Vienna Conference)
WPA:	World Plan of Action for the Application of Science and Technology to Development
VPA:	Vienna Programme of Action for Science and Technology for Development
PREPCOM:	Preparatory Committee for UNCSTD
IGCSTD:	Intergovernmental Committee on Science and Technology for Development
OST:	Office of Science and Technology of the UN Secretariat
CSTD:	from 1971 till 1979 - ECOSOC Committee on Science and Technology for Development; from 1979 till 1992 - Centre for Science and Technology for Development of the UN Secretariat; from 1992 - Commission on Science and Technology for Development

UNFSSTD:	UN Financing System for Science and Technology
IFSTD:	UN Interim Fund for Science, Technology and Development
ACAST:	UN Advisory Committee on the Application of Science and Technology for Development (1963-1992)
ACSTD:	UN Advisory Committee on Science and Technology for Development (1992-
ACC:	Administrative Committee on Coordination
ACCTF:	ACC Task Force on Science and Technology for Development
ECOSOC:	UN Economic and Social Council
UNESCO:	United Nations Educational, Scientific and Cultural Organisation
FAO:	Food and Agriculture Organization
ILO:	International Labour Organization
UNCTAD:	United Nations Conference on Trade and Development
UNIDO:	United Nations Industrial and Development Organization
UNU:	United Nations University
ECE:	UN European Economic Commission
ECE SAST:	Senior Advisers to ECE Governments on Science and Technology

Bibliography

Anandakrishnan M. (ed.). *Trends and prospects in planning and management of science and technology for development* (1984). Proceedings of Interregional Seminar, Moscow, USSR, 8-27 0ctober, New York, UNCSTD [Discussing follow up of the Vienna Conference].

Baark E., Regis C. and Jamison A.. *Science and Technology for Development in the United Nations System: A preliminary Study* (1988), Research policy studies, Discussion paper no. 183, Lund, Research policy Institute, University of Lund, Sweden.[Analysis of the UN programmes in the field of science and technology].

Contribution to the follow-up to the United Nations conference on science and technology for development (1984), Technical Report, ECE/SC. TECH./R.168, Geneva, ECE [Discusses possible follow-up action of the European countries]

Pati S.C. (1986). *Experiences with the Vienna Programme of Action,* 87 pp, Munchen, Forschungsstelle Gottstein in der Max-Planck-Gesellschaft [This is study on the follow-up activities of developing countries on the recommendations of UNCSTD]

The Vienna Programme of Action on Science and Technology for Development. The United Nations Conference on science and Technology for Development (1979), 36 pp., United Nations, New York [On 31 august 1979 delegates from 142 States adopted by consensus a programme of action and gave it the name of the city that had been host to the UN Conference on Science and Technology for Development]

U.S. Background Preparations and Policy Formulation for the United Nations Conference on Science and technology for Development, (1979), Report prepared by the Science Policy Research Division, Congressional Research Service of Library of Congress for the Committee on Science and Technology, U.S. House of Representatives, ninety-six Congress, First session, 395 pp. Washington U.S. Government Printing Office [It contains background information dealing with U.S. preparation for UNCSTD, summarizes the history of preparatory activity, outlines policy position of the U.S. government and the other countries participating in the Conference]

World Plan of Action for the Application of Science and Technology to Development (1971), Prepared by the Advisory Committee on the Application of Science and Technology for the Second United Nations

Development Decade, 286 pp; United Nations, New York [This United Nations publication consisted of two parts. Part One lists priority areas in which science and technology can make a resounding impact. Part two consists of more detailed proposals selected from those submitted by various stockholders].

Biographical Sketch

Vladislav P. Kotchetkov is a consultant to the UNESCO. He graduated from the Moscow Higher Technological University in 1959 and worked on satellite and missile technology. In 1969 he was awarded by the USSR Laureate State Prize. He received a diploma *cum lauda* in international economics from the USSR Foreign Trade Academy in 1977 and worked as a Director of a Division of International Scientific and Technological Co-operation for the USSR State Committee for Science and Technology. He was involved in the preparation of the 1979 United Nations Conference on Science and Technology for Development(UNCSTD) , was a Secretary of the USSR Preparatory Committee for UNCSTD, represented the country at the sessions of the UNCSTD Preparatory Committee , was elected as a Rapporteur at the European Regional Conference for UNCSTD in Bucharest in 1978 and participated in the Vienna Conference in 1979. From 1980 he has been working for the UNESCO Science and Technology Policy Division, was a Chief of the UNESCO Science, Technology and Society Programme and a UNESCO representative at the UN inter–Agency Task Force on Science and Technology for Development. He also served as the Executive Secretary of the Moscow International Energy Club and the UNESCO International Council for Scientific Communication, represented UNESCO at the IUCN Commission on Education and Communication. He is a member of the UNESCO-EOLSS Joint Committee.

TRANSITION TO SUSTAINABILITY IN THE DEVELOPING COUNTRIES: THE ROLE OF SCIENCE

Carlos B. Aguirre
National Academy of Sciences of Bolivia, La Paz, Bolivia

Keywords : developing countries, transition, sustainability, science policy

Contents

Summary

Developing countries need not only to provide quick responses to the increasing social and economic demands of their societies and improve their participation in the world dynamics, but also create at the same time, conditions for a long term sustained development. For these countries, a response to such a challenge is difficult for several reasons, one of them, because they lack sufficient capacities to generate and use scientific and technological knowledge.

Considering its complexity, a transition to sustainability requires the definition and adoption of a large set of interrelated polices and strategies based on a systemic approach. Here, a simple scenario is described in which sustainable development is considered as composed of five dimensions and their linkages: economic, social, cultural, preservation and adequate use of the environment, and governance. Science plays a key role in understanding and providing solutions to problems in each of these dimensions.

Because science makes part of a non-linear development process, science (or research) policy by itself, does not constitute an appropriate response to the complex challenge of transition to sustainability that the developing countries must face. Such policy must be integrated into an innovation policy. Further, because science policy is principally a social contract, it is in effect an element of the new scenario. It is in this context, that the role which science plays in a scenario of transition to sustainability in the developing countries is discussed, and a set of policy objectives suggested.

While recognizing its importance, the epistemological limit of science is also recognized. In spite of such a limit, it should be borne in mind that no country that has ever developed has done it without conducting an effort in science. Today's developing countries will not be the exception in the new millennium.

1. Introduction: The New World, Present and Future Scenarios

World interdependence and globalization are fundamental forces shaping contemporary society. Two processes have accelerated the world towards these, the development of information and communication technologies, and the dominance and success of the liberal economic ideology. Both forces, while modifying the dynamics of the economy, have created challenges and opportunities for all countries, and are today leading to the emergence of an effective "knowledge economy," knowledge becoming an important organizing principle for society. These processes have been further enhanced by the transit of a large number of countries to a democratic system. Of course, together with globalization, the world has also transited towards the rebirth of nationalisms, as expression of the desire to and need for preserving cultural pluralism.

In spite of the advances triggered by globalization and the opportunities it has created, many developing countries are facing enormous difficulties re: the demands of their societies while adjusting to the new dynamics. As a result of their inability to produce appropriate responses, the economic and social conditions in these countries can be characterized as dramatic.

An understanding of the many edges of globalization is essential to grasp the meaning of such a situation. In particular, it should be clear that among its most powerful agents are transnational conglomerates and corporations which effectively dominate a large fraction of the world's production of goods and services, trade, technology and finances. Such entities have more financial resources under their control than the total GNP of the least developed countries. Their power and influence narrow the range of options and capabilities of weak national governments, when trying to implement consistent autonomous integrated economic, research and innovation policies.

In spite of the dependence and the severe restrictions to the degree of autonomy of national policies, there exists a potentially sufficient space, for local decisions, to influence on the distribution of income, the assignment of resources and the profile of international insertion, in other words, sufficient potential to formulate a valid strategy for sustainable development.

In light of the world situation, several authors have explored the limits of the present patterns of economic growth and its cultural, social and environmental consequences and future scenarios.

Hammond suggests a "global market" scenario, that considers the private sector as the motor of growth and the global economic integration, the motor of development. To reach it, it is then enough to provide the private sector all facilities and break all barriers to trade. This scenario considers that the private initiative and an expanding economy are the best ways to improve life and to provide opportunities and incentives, and estimates that within the next five decades, the world will have expanded its economy six times, providing prosperity without precedent to many social classes.

In this scenario, an excessive materialism, the existence of great inequities and the persistence of poverty are evident. Thurow has pointed out that "capitalism is shortsighted and cannot make the long term investments in education, infrastructure and research that are needed for its own survival." Thus, a strictly market-oriented future, in the absence of important and deliberate interventions to mark the course of human destiny includes a large number of risks.

To improve this scenario, a fundamentally new outlook is needed. Machado considers that governments have overestimated the benefits of free market mechanisms. In his vision, the inequalities that it has generated have created an unbalanced economic process that tends to increase social problems and also the political pressures to reverse the liberalization process. He believes that a sustainable solution is the massive and accelerated economic growth of developing countries, on the base of the production of new environmentally adequate products and services. Such a solution requires creativity, leadership and international consensus, conditions that are considered absent and implies a review of orthodox policies and the international framework that supports them, in other words, a new industrial revolution.

In considering different scenarios, it is important to recognize the political and financial difficulties that a substantive modification of the situation implies, but at the same time recognize that, growth cannot follow the same pattern adopted by the advanced occidental economies in the past, there simply aren't sufficient resources and energy for that to happen and additionally the resulting pollution would destroy the planet.

Just as an example of the latter, climate scientists have predicted that the increase of carbon dioxide in the atmosphere would disrupt weather patterns, and indeed, annual damages from weather-initiated disasters have increased over 40-fold; in 1999 such damages amounted to about 100 billion US Dollars. The consumption of chemicals has exploited, with about three new synthetic chemicals introduced each day. One of the most ominous chronic effects is that as the use of pesticides increased, so has the evolution of pesticide-resistant pests, amounting in 1999, to about 500 pesticide-resistant weeds and crop diseases and because of overfishing, the total world catch of Atlantic Perch decreased from 27 million kilograms in 1970 to about 1 million in 1999.

In the face of difficulties almost inherent to a "global market" type scenario, of responding to growing poverty and the gap between the rich and poor, certain political circles in the developing countries argue that no orthodox economic policy or its variations can satisfy the needs of society and that they would only continue the tendency of concentration of wealth putting at high risk the capacity of governance and therefore democracy. A possible result of the application of such vision, is depicted by a scenario of "isolated fortresses" also constructed by Hammond, in which the countries start to raise barriers to trade and the free circulation of persons. Although this might be a response to the present mismanagement of world affairs, it is a clear step backward in history. In such a scenario, it is possible to foresee that many countries would simply collapse and that the rich will constitute "islands of prosperity in an ocean of poverty."

As an alternative, Hammond constructs a scenario of a "transformed world," constituting an optimistic view of the future, containing a society with power, of illuminated corporate actions and radical political changes. A scenario based on market forces, but not substituted by deliberate social choices, promoting competence but also viewing cooperation and solidarity among societies as equally important forces.

Rattner asserts that such a scenario can be reached by the creation of a more extended and enlightened leadership, thus he considers that a great effort must be made in the development of leaders for sustainable development, taking advantage of the recent evolution of new organizational forms which have created opportunities for the advent of participatory practices, consulting with the population and sharing with them the bulk of political and administrative decisions. In his vision human beings are never more fulfilled than when they are united in a group or community, each one aware and conscious about his/her personal commitment in serving the common endeavor. To this view, it should be added that an important element in motivating human change is what people believe is right and wrong. Questions that arise here are rooted in an ethical, moral, spiritual and a religious system, which must be integrated into development thinking.

Brown et al, construct a scenario for a sustainable world towards the year 2030 that considers an advance toward sustainability based on the collective deepening of the sense of responsibility towards earth and towards future generations. This scenario deals with several basic needs and a new economic approach, and calls for a transformation of priorities and individual values. This view is coincident with that of Guimaraes, which calls for the creation of "sustainable regimes." Both calls are of course fundamental, in view of the present patterns of development and consumption

The dismal forecasts that can be derived from an extrapolation of present trends need not occur. Advances in science and technology, if supported by the necessary political will and international cooperation, mobilized by appropriate policies can produce, substantial progress over the next years towards a sustainable world. As such, this contribution examines the role that science can play in a scenario of transformation to sustainability in developing countries. It suggests a set of policy objectives addressed to the creation and strengthening of scientific capacities that are needed to effectively contribute to such a transition.

2. Sustainable Development

In 1987, the World Commission on Environment and Development recognized that the failure of man to live in harmony with his environment had changed and endangered the planet's life-supporting systems. The overexploitation of natural resources and the methods of production and life styles in industrialized countries have in effect caused severe pollution and degraded the environment. The Commission's report concluded: "We borrow environmental capital from future generations without any intention or prospect of repaying."

As a response to the challenges posed by the World Commission, national and international institutions, started conducting efforts to reverse environmental damage, create awareness, modify technologies, and adopt new economic and social measures. In the process, the concepts and practices of sustainability have been greatly enriched. As a result it has become clear that sustainable development is brought about by the capacities that a society can develop to transform, mobilize its potential, and reaffirm its cultural identity. Thus, sustainable development is not imported "turn key," it is mainly an endogenous effort. It requires fundamental changes in the way of thinking, living, producing, consuming, and forms of human interrelationships, in sum, it needs new political, administrative, social, and economic systems and regimes.

There are three particular dimensions, drawn from the above conditions, in the creation of endogenous capacities for sustainability that are crucial for the developing countries. They relate to human, infrastructure and management capacities. The first, the increase and improvement of the existing system of generation, transmission and use of knowledge, the second, the production of an accelerated development of the physical and telecommunications infrastructure, and the third, an increase of the efficiency of management at all levels, but particularly in the public sector.

3. A Scenario of Transition to Sustainability

Sustainable development can be considered as composed of five closely interconnected dimensions: economic growth, social equity, preservation and adequate use of the environment, culture and governance. None of them by themselves constitute sustainable development. In considering a scenario of transition to sustainability, it should be clear that the problem is not just of saving the planet for future generations, but also of saving the planet's current generation.

Economic growth is the result of competitivity. A country's structural competitivity can be measured in many ways. The World Economic Forum uses eight indicators:

- Internal strength of the economy: the degree of macroeconomic behavior.
- Internationalization of the economy: the degree of the participation of the country in the international flows of commerce and investment.
- Government: the degree in which governmental policies lead to competitivity;
- Finances: the behavior of capital markets and quality of the financial services.
- Infrastructure: the degree that the resources and systems serve the basic needs of business.
- Management: the degree in which firms are administered innovatively, responsibly and profitably.
- Science and technology: representing the technological capacity and the success of research.
- Human factor: the availability and qualifications of human resources

Considering these indicators, competitivity, can then be defined as:

> The capacity of the country, given its economic and institutional structure, of producing a sustained and high per capita economic growth in a given period of time.

An important vision of the paradigm of competitivity is that of Bitar and Bradford. They view development as a structural transformation process. Instead of being characterized as continuous, economic growth is a discontinuous process led by the diffusion of technological change in the economy. Competitivity is conceived as a project originated internally, as a way of moving society towards transformations and economic dynamics. In this way, competitivity leads to a cooperative relationship among all economic and social sectors as the key of a growth process centered in the development of knowledge.

Competitivity-based development, represents a very different direction to the doctrines of the past and the orthodoxy of the 1980s. National strategies of development come first, then they generate global growth and export. Growth is from the inside out and has a certain supply-sided nature. Orthodoxy sees trade from the point of view of demand and stable macroeconomic policies and the liberalization of trade in traditional patterns as sufficient conditions to generate growth.

The orthodox vision is centered in economic efficiency as the criteria for the allocation of resources. The new paradigm of competitivity looks at the initiatives of economic agents that create new rules, thus creating opportunities and different ways of organizations. The market paradigm involves an automatic answer of the economic agents, whereas competitivity involves interactive work between enterprises and other economical agents. The orthodox vision establishes the priority of the economy over policy, while the latter is required in the new paradigm.

Reaching high rates of competitivity is therefore in the basis of economic growth, and a transition to sustainability in the developing countries demands the search of structural competitivity, that is, to make institutions and enterprises and conglomerates competitive, as well as the national system under which they operate. The new paradigm is built on the idea that no market, sector, or industry is isolated from the world and that relative efficiency as well as efficacy are relevant to all the social actors.

The social equity dimension stems from the need of creating growth conditions that respect the individual and at the same time provide him/her with the conditions and opportunities to actively participate in them. In more recent times, the term sustainable human development has been introduced as a better expression of the search of equity in sustainable development, and in the view of some authors, should actually substitute this term, as it clearly identifies the subject of development.

This dimension must be understood in a scenario of transition to sustainability, as including aspects such as inter- and intragenerational equity, access to education and health, and a better quality of life. By attributing priority to education and health, the basic infrastructure conditions are created for the solution of the more complex social problems of society. Without clear and transparent policies and guidelines in view of a rapid transformation of the health and education systems, there are no chances to attain sustainability.

The world economy is in its way destroying the natural systems which support it. Thus the importance of the preservation and adequate use of the environment, as a key dimension of sustainable development. The environmental dimension plays a central role in this scenario, not only in terms of the sustainability of production and development patterns in the medium and long term, but also as a factor that may inhibit the access to specific markets, and thus an element that directly influences the effective competitiveness of nations and enterprises.

The roots of environmental degradation stem from social inequities and greed. In a scenario towards sustainability, the use of natural resources must be linked to a better distribution of wealth and some control over trade, technology and investment flows. At the same time, it must be considered that inequities will diminish if development policies are based on the concept of endogenous capacity building.

A basic dimension of a scenario of transition to sustainability is culture. It not only helps to create self-esteem in a society, but also contains elements that are determinant to create wealth. Landes, Yergin, and Stanislaw, starting from very different considerations come to the conclusion that cultural elements such as hard work, the passion for progress and for the creation of knowledge, are key sources of wealth.

The concept of culture does not only refer to the arts and the aesthetic aspects of life, but also to the ideas and values, the attitudes and preferences, and therefore to the behaviors derived from them. Particularly important today in this conceptualization of culture, is the concept and the practice of ethics. The controversies or perplexities that today invade the discussions on ethics are probably due to the fact that not only values have changed, but also the nature of ethics itself.

On the other hand, the destruction of nature's biodiversity, meets its parallel in the gradual loss of cultural diversity. Thus a scenario of transition to sustainability must include the defense and development of ethnic minorities, their traditions, including languages, religion, ways of life and music. It is important that individual and collective identity be nurtured, leading to a feeling of "belonging" to a community.

Today's democratic processes are exposed to the social pressure of growing poverty. They find themselves between the need of advancing in the institutional consolidation of democracy, and at the same time, promote growth with social equity, that is, to reconcile political democracy with market economy and social justice. Globalization and interdependence have particularly contributed to the complexity of the problems that governments must face and demand an improved understanding and management of complex systems.

This situation constitutes without doubt the central problem of governance, as in the framework of globalization and interdependence, in front of the risks of the lack of legitimacy of the political system, the deepening of socio economic problems, and the deficiencies of the state management, it must make democracy respond to the dangers and conflicts that tend to undermine the bases of legitimacy and the institutional operation of the political system, and simultaneously increase the institutional capacity of action and management of the state, that is, strengthen its capacity of articulating social interests and the adoption of appropriate decisions to solve problems of growth and social equity.

The concepts and practices of adjustment and governance, in a scenario of transition to sustainability, are related to those of modernization and the results of adopted policies. If modernization is accepted as that which covers an ample opening of the local economies to external trade and to international capital flows, a government reduced in its functions to only the essentials, but capable of maintaining price stability, and a structure favorable to the private activity, then, there should be no doubt that with the adoption of adjustment policies, fundamental steps have been taken in many developing countries.

If however, a definition of modernization includes the existence of effective mechanisms of democratic participation, coordination among private economic agents and of these with government, the streamlining of extreme social conflicts and the preparation of a sustainable long term growth, there is still a long way to go.

4. Innovation in a Scenario of Transition to Sustainability: Challenges and Opportunities

4.1 The Context

The economic and social situation of the developing world is appalling, when comparison is made to other nations. The average annual GDP per capita in the OECD countries varies between US$ 12 000 and US$ 25 000, while in the developing countries, it is around US$ 5000. In the least developed countries between US$ 150 and US$ 2000. The number of people living below the official poverty line (US$ 1/day), has increased from 1.2 billion in 1987 to 1.5 billion in 1999, and the estimate for the year 2010 is 2 billion, that is approximately one third of the world population. About 2.8 billion people do not have access today to basic sanitation, clean water and health services. 20% of the world's children do not go to school and have no chance to learn a trade or profession.

The situation is further aggravated when countries themselves are analyzed. The gap between rich and poor, in them is even wider. Such a situation of poverty degrades the dignity of all human beings, no matter where it occurs. Following Landes: "Now the big challenge and threat is the gap in wealth and health that separates rich and poor. ... Here is the greatest single problem and danger facing the world millennium."

From the perspective of this presentation, it should be recognized that the neo-classical model for economic development has not worked in most developing countries, because of the effect of the global-fast, science-based technological development, which has increased the economic benefits of a highly trained work force and higher educational standards, conditions which are absent in them. According to Sagasti, "We are in the middle of an exploitation of knowledge that is dividing the world between rich societies, of rapid transformation, that use knowledge efficiently, and poor societies, of slow transformation, that do not use it with efficacy."

In the above context, the achievement of a scenario of transition to sustainability in the developing countries, confronts a set of enormous challenges while also provides new opportunities. To face such challenges and take advantage of the opportunities, these countries require, among many others, the adoption of science and innovation policies which would provide an appropriate environment to the strengthening of "national innovation systems," and of the "science system" within it.

4.2 Innovation in a Scenario of Transition to Sustainability

Innovation is the result of a social process. It facilitates both individual progress and that of the whole of society. Innovations are thus not only of a technological nature, but also of the conditions of the context that favor economic, social and cultural growth. Technological innovation is not a linear process, it is cumulative and interactive. Various factors make it interactive. Consumers demand more and more high quality and faster delivery times. Science and technology are interdependent and technological development requires complex experience in many activities. Interdependence is crucial and it is proven that the innovative enterprises are the ones that dominate all the interactions and combine the market demands with the technological supply.

The cumulative nature of innovation is due to the fact that technology is today more dependent of science and scientific progress because of the accumulation of knowledge. Those enterprises that dominate complex technological systems will have to go through a long process of learning in order to develop their technical competence. At the moment that knowledge and abilities combine together, various mechanisms start playing a role, producing economic returns of the investment made in research and to the technological capital in the enterprise. In this way, the experience won in each stage receives a mutual feed-back and this makes the acquisition of higher innovative abilities possible. It is within this context that innovation policies are today recognized as a crucial factor for competitivity.

The different characteristics of innovation are crucial for understanding the role of science and the design of science policy for sustainability. In fact, today, it is necessary to abandon the linear model that assumes that societal benefits are directly proportional to the support of basic research. The search for a new scenario of transition, presupposes a new type of interaction between capital and work, government and enterprise, between society as a whole and innovation, and thus basic and applied sciences.

The creation of knowledge, innovations, and technical progress is based on the adequate operation of a "national innovation system," defined in a scenario of transition to sustainability as:

> The network comprised of institutions in the private and public sectors whose activities and interactions, generate, import, modify, and diffuse new and traditional knowledge, and educate innovators and entrepreneurs, transferring the benefits of science and technology, adapted to the demands and requirements of greater competitivity in the economy and society.

In this definition, the concept of traditional knowledge has been incorporated to allow for the idea of minor and incremental innovations that make part of any development model with a certain degree of autonomy. Under it, science and technology policies are designed in such a way that the concepts of technology gaps and obsolescence are referred mainly to the needs of each society and to its capacity and social objectives that are determined democratically, and not only to the technologies and research lines dominant in more advanced countries or in the international market.

The introduction of the idea of traditional technologies in the definition of the "national innovation system," responds also to the need of introducing a mentality change that predominates in governments, firms and even academia of many developing countries, that perceive themselves as mere technology receptors and not as creators or even adaptors. In this situation, capacities must be built by a change of attitude and new forms of administration and urgent measures taken for institutional change in order to apply policies, public and private institutions capable of interacting in new and creative ways are needed.

4.3 Challenges

Developing countries must in the shortest term possible, find responses to the increasing social demands of their societies, in general, overcome poverty, while at the same time preserve their national cultures and make part of a globalized world. Such challenge is immense and must be faced by all social and economic actors.

Governments must adopt well-defined policies, exempt from particular interests, in order to reach an adequate distribution of wealth. This goal will not be obtained only through technology transfer or foreign investment. An important challenge is to the enterprises which must adopt innovation policies. In particular they need to modify their vision of technology as a static solution for production and produce technical change to reach competitivity goals. Enterprises must build competitive capacities in two directions, the first through the accumulation of knowledge, abilities and technology which will have a fundamental value for its clients. The second, through a clear understanding of the business context, anticipating and accompanying its evolution, while establishing an advanced position in the productive chain. For all of these, it is crucial to detect technological discontinuities and markets and act quickly on the business opportunities that are created through the use of accumulated competitive capacities.

A third important challenge is to the education system, particularly in universities. The crisis of the 1980s and adjustment policies adopted as a consequence, have seriously affected their capacity to respond to the world dynamics. The situation at the postgraduate level is especially delicate. In the view of Gibbons:

> It has long been recognized that higher education institutions, particularly universities, are among the most stable and change resistant social institutions to have existed during the past 500 years. Based on the model of the physical campus, residential students, face-to-face student-teacher interaction, a lecture format, and ready access to written texts, these institutions have effectively developed and transmitted the store of knowledge from one generation to another. They have fulfilled this responsibility in the midst of political and social upheaval, social development, and technological advancement while remaining

essentially unchanged in structure and method. Will this proven model retain its resilience and relevance in the 21st century?

Finally, it must be recognized that in most of the developing countries, there is consensus, both in the public and private sectors, that technology is a requirement for competitivity and productivity growth, and that creativity and efficiency are important ingredients for a dynamic economy. Consensus fails however, when the discussion arises on how far market forces induce sufficient investment in research and innovation to reach development goals.

Experience in all countries of the world and at all times, is that the state plays a key role in the definition of promotion policies and that the market forces are not sufficient to initiate accelerated processes in science, technology and innovation. The challenge then rests on the forms and ways that policies and investments in the public sector must take.

4.4 Opportunities

In spite of the problems that globalization poses for the developing countries, it should also be clear that it creates objective conditions for the structural transformation of the present system into a solid, pacific, and cooperative world society.

Globalization stimulates the social awareness of political actors and institutions and their perception of environmental and social problems. Increasing consciousness avoids the risk and threats to survival caused by environmental degradation, engenders favorable conditions for public oriented democratic policies in search for solutions for problems that transcend national boundaries. At the same time, globalization creates opportunities for approximation, contacts and communication between social movements and NGOs plugged into on-line networks, receiving and transmitting information, knowledge and practical experiences.

Such networks, are open to new ideas, proposals and messages without censorship, allowing for the free association of interest groups without discrimination of color, age, gender, ethnic, religious origin, or social status. In these contacts, it is possible to find the roots of an emerging identity, a sense of belonging to a global community, quite different from formal democratic systems, constantly under pressure by interest groups.

On the other hand, as a result of the application of the recent structural transformation measures, new conditions that can be utilized in the generation of innovation capacities have been created. Some of these measures have to do with the improvement of the infrastructure that is useful for innovative processes. For example, the strengthening and better operation of the telecommunications system or the more export oriented production system being put in place, are all opportunity factors of importance.

There are at least three other factors that should further encourage innovation in the developing countries. The first is the need of maintaining and perfecting the emerging models of sustainable development that have been put into place. Only the creation of larger competitive capacities based on the production of innovations and technical progress, will allow, as time passes by, to reach the established objectives and maintain the sustainability of the model itself.

The second factor is related to the need for creating competitive enterprises within the regional economic integration arrangements. Certainly, the possibility of success of enterprises in these markets will depend on the development of its innovative capacities. A third factor is related to international trade and in particular to the new established international trade rules. The results of the Uruguay Round on intellectual property and investments in GATT, have a deep impact on the acquisition and development of technology. As markets become liberalized, competition inside, including the domestic markets, will be each time more intense and it will be defined in favor of those who have the greatest technological superiority.

Also in the last years a still fragile movement from the consumers side has initiated in some countries. They start to see quality and not only the price of a product as an element of their buying decision. This movement, together with those that have been defined to protect the consumers, including environmental considerations, is an additional encouragement factor to innovation. It should also be considered that in spite of the traditional lack of linkages among the different elements of the national innovation system, there exists a new and slowly growing perception of the need to approximate research to production. Such phenomena are certainly a positive factor in favor of the adoption and implementation of innovation policies in developing countries.

4.5 The National Innovation Systems in Developing Countries: An Overview

The national innovation system of any country plays a definite role in the production of knowledge, innovations and technical change. In many developing countries, in spite of the conceptual and operative efforts that have been put forward to understand sustainability and in some cases define policies and strategies, the conceptualization and operative roles that the national innovation system can play to accelerate and facilitate the integration of the different dimensions of sustainable development and their direct contribution to them has not been fully understood and thus its build up has not been given a priority. As a consequence, the "national innovation system" in the developing countries cannot produce the necessary knowledge base and the products that a scenario of transition to sustainability requires as a key input.

Studies on enterprises clearly reveal the weaknesses of the national innovation system of developing countries: the absence of leadership in the firms, which maintain a traditional organizational and managerial style, that does not value creativity or innovation behaviors; undervalue given to human capital; poor functioning of the capital markets; inefficient information flows, lack of linkages between research and production and the absence of local scientific and technological developments.

A key element of the national innovation system is science, which in the developing countries faces several limitations that have been amply described in the literature. Shortage of funds, lack of critical masses and infrastructure, isolation of the scientific communities, are among the most common. In many countries the situation is more difficult because of the presence of a large number of weak research institutions, showing an enormous dispersion of efforts. Such a situation contributes to low production and poor quality of projects.

It is important to discuss some of the causes for the limitations. The first is the still extended vision among political and economic decision makers, and entrepreneurs, that while recognizing the importance that scientific and technological knowledge has for the acquisition of competitive capacities and the advancement of culture, they feel it is a luxury that only richer countries can support, and that even if science is done locally the results can hardly compete with those obtained in larger countries and therefore that knowledge obtained through research should be sought abroad. Further, this view is exacerbated by that which considers that any transfer of technology must be conducted strictly by the private sector firms under any kind of investment mechanism. This situation is leading to the loss of the capacities that have been built through many years of efforts in universities and other research centers, which has led in turn to abandon benchmarks of excellence.

Also, a vision has been developed throughout the years, that only science appropriate to developing environments should be developed. This has led to the belief that there is science for the poor and science for the rich, and has been one of the most damaging visions to the scientific enterprise in many of the developing countries.

In later years, new arguments have been raised that these countries must concentrate strictly in satisfying basic social needs, giving priority to basic education. While this argument has a strong basis, it has unfortunately led many countries to decrease public support to university education and in consequence to research. Further, non-university public research organizations have also been dismantled, with no substitute from the private sector emerging. It should be mentioned however that views on this issue are changing as signaled by recent World Bank documents recognizing the role of knowledge generation and higher education in reaching sustainable goals.

To complicate matters, as the science enterprise in the developing countries has traditionally depended on "science aid" from more developed countries, there is, as discussed by Oldham, a disappointment of donors of the little impact, with few exceptions, that such "aid" has had in building up strong capacities so as to bring the benefits of science and technology to impact on the development efforts of these countries. This disappointment, coupled to the visions already mentioned above, have certainly made the situation much more difficult in building up science.

Besides specific limitations affecting these countries, the future of science itself is rather complex, as the scientific communities and institutions throughout the world are being pressed by economic, political and cultural concerns. This perception is accompanied by a lack of confidence on how can the institutions and scientists respond to a changing world. Some dominant factors will be a larger political interest in the organization and results of science and a lesser participation and more critical attitudes of the general public on the value of scientific research.

In spite of this situation, there are many experiences in developing countries that point to the existing potentials for the development of science. Several centers of excellence and individual scientists have made contributions that are of value to the development of knowledge. A science policy can probably reproduce some of the conditions under which such contributions have been made.

5. The Role of Science in a Scenario of Transition to Sustainability in Developing Countries

"The advancement of civilization has, in more than one way, been driven by the generation of knowledge through scientific and technological research. The world scientific community has unraveled many of the secrets of nature, and the behavior of many of its life forms as well as understood the human dimension. Science … has allowed entire civilizations to reach forward into the future, and to stretch innovation." As such, investment in science is crucial for sustainable development and learning and using science, greatly benefits the life of each single member of the world community.

Spectacular advances have made science based technology a key of today's globalized economy and new scientific and technological knowledge will continue to be an essential element of a transition to sustainability, its handling cannot be put aside if a competitive economy is desired. For this to happen, although new technology might be available globally, it has to be modified or adapted to local conditions, if it is to serve its purposes. This adaptation cannot be done without an insight in the scientific background of the technology, an insight that often must be based on active research in relevant areas, most typically in the basic sciences. In fact, science has a definite role to play in the way technological progress is achieved and technology trajectories are defined. The latter is a key issue in a scenario of transition to sustainability.

Today it is also recognized that science must be seriously concerned, not only with its contribution to the sole advancement of knowledge, thus to culture, but also with the many unsolved problems of humankind, which includes an understanding of the physical world and the social problems that hold today's civilization in the grip of numerous contradictions. Science is then called to face up to the natural and social complexity of today. Science is an important contributor to social equity, through improvements in education and health, and the creation of new and more productive employment. It can explain conditions for human development, including the impediments that result from economic inefficiencies, social inequalities and gender biases. It can also provide a better basis for transitions to economies that provide increased human welfare with less consumption of energy and materials.

Education is an essential element of a scenario of transition to sustainability as it relates to all of its dimensions. One important example is literacy of women on which, choices on size and timing of families and therefore demographic growth, depends. Education in the sciences is the basis for production and innovative economic activity, and thus closely related to the creation of jobs, improvement of income and quality of life. Advanced education, critically dependant on the degree of development of science, provides the cutting edge for technological advances, and provides the basis for wise and informed choices in many critical policy areas.

The contributions of science to the use and preservation of the environment has been extensively discussed in the past years. Issues such as the understanding of global and environmental change, its causes, and policies to mitigate their effects have greatly advanced. At the same time in its linkage with the economic dimension of sustainability, such aspects as the access to environmentally friendly technologies, the improvement of management capacities, which consider the environmental dimension and the dissemination of innovations successfully coping with the environmental impact of their production activities, have a very strong scientific base. In a long-term view, it is certain that science can continue to respond to question of critical importance for the welfare of coming generations.

For some authors the key question for sustainability is, "With what technologies will economic growth occur in the future?" Only the population explosion rivals this question in fundamental importance to the planetary environment. Particularly important at this point is the problem of human consumption patterns. The forces that drive consumption are multiple and complex; they include economic activity, the distribution of wealth and income, technological choices, social values, institutional structures, and public policies. Human consumption involves the transformation of materials and energy, and as more people in the developing countries aspire to a materials-intensive lifestyle, and as affluence increases consumptive behaviors, and challenges will continue to grow. Science, combined with other human endeavors, is key to produce an improved and qualitative change in present consumption patterns.

Science also contributes to governance in several ways. The vitality of democracy depends on an educated public, and here scientists have a role to play—for example, in developing curricula. Another contribution has to do with communications. Sustainability depends on the capacity to gather groups of people of diverse origin and provide them with a structure where they can communicate their perceptions on reality and interconnected visions of the future. Science based technologies have and can provide the means to improve this communication for example through the quite obvious development of information technologies. The latter, however, as discussed by UNCSTD, is not exempted of risks, due in part to the fragile social and economic capacity of many developing countries.

A long-term vision is another element of governance. That means, among others that, the scenario of transition to sustainability is characterized by foresight. Governance requires that each member of society be able to examine carefully and systematically long-term issues. Two main characteristics of foresight contribute to that: as an instrument for decision making at all levels and as a tool for creating shared visions of the future, integrated with competitive intelligence mechanisms, built up from a scientific basis.

Sustainability is also based on the possibility of creating consensus and establishing development priorities. Science and technology dialogues destined to the construction of agendas and the definition of investments, make a fundamental part of a scenario of transition, because they lead to the development of common and concerted visions among stakeholders. In this context, the contribution of science is unique as it provides the means for a more autonomous development, based on the perceptions of society.

Throughout time, science has constituted a valuable and indispensable tool for understanding and dominating nature, strengthening culture and creating new cultural values. The role of science in a scenario of transition to sustainability, stems from the character of science itself as it includes not only a catalog of facts and theories on different aspects of nature, but also its philosophical basis, the history of its development, the social structures in which it takes place and expresses itself, the laws that regulate it and the policies that favor or limit its development. It can be said, in fact, that because of these characteristics, "science is a way of living life."

In this conception, it can be thought that when developing science, its spirit is incorporated in culture. That means, among others, to allow access of society to the challenges and the satisfaction of understanding the universe and above all to image and collectively construct possible worlds. Certainly culture could not imagine more ambitious goals. Maybe for this reason science is referred to sometimes as the "star" of culture. This is the most beautiful face of science, the other, less humane, but equally important is that directed towards the economy.

Both technology on one hand, and society and culture on the other, affect each other on a mutual basis. Science and science based technologies have changed society's lifestyles, including its value system. It has affected management and organizational behavior, the way of decision-making and of work, including interpersonal relations. Culture in turn, influences technology in a multitude of ways, including the definition of policies, attitudes of society towards the individual scientist, the educational system, and so on. In this case, science serves to reach a critical conscience of society and sets the basis for a local technological development.

In the above context, science is not only an instrument, or an exogenous element or an externality to development, it is in fact part and source of human progress and its objectives are strongly based in its cultural dimension. Science is a force absolutely fundamental to the well being of human kind and in fact to its survival.

As the new millenium opens, there are several challenges facing the science system in general:

- The increasing complexity of the global problems and of the social organizations that govern human interactions and cooperation. An approach to the complexity problem, be it of nature or of social organization, requires international cooperation.
- The growing need for sustainability in science itself, that is the need to commit resources overtime to the science enterprise, and to create a new social contract between science, its patrons and the general public. The time frame for science is long. This transcends, by a considerable margin the attention spans of politicians and the financial supporters of science. Commitment is essential at three levels; spending, university based research, and international collaboration.
- The burgeoning diversity of human scientific resources, which calls to make the best use of the scientific community by strengthening it. Also, the breakdown of disciplinary barriers, the challenge of communicating science to a broader audience, taking opportunity of the creativity and contribution of citizens who are not scientists and the need to bring science closer to the political leaders and the public.

- The "massification" of higher education in most countries, i.e. the sharp increase in both the number of students and also the proportion of young people enrolled in higher education institutions.

- A flagging support on the part of national and international agencies for the organization and continuation of adequately-resourced programs at the national level and of international scientific cooperation. There are budget restraints combined with the increasing marginal cost of scientific progress.

- The demand for increased accountability, i.e. for an organization and management of the science system that leads to the most effective production of scientific results, also, a more critical public perception of science and the emergence of an increasingly wide spectrum set of ethical issues.

- The demand for increased economic relevance and wealth creation, i.e. for a stronger, more direct, more visible and more rapid contribution to economic competitiveness and growth, and related to it, the increasing need for multidisciplinary and interdisciplinary research due to economic and social demands, for example for a stronger, more direct, more visible and more rapid response to societal and environmental problems.

- The burgeoning scientific opportunities across the entire spectrum of scientific disciplines and the increasing role of megascience, including the need in big and small science projects for obtaining, processing, interpreting, storing, and later accessing very large amounts of data.

- A reduced long-term research effort in many industrial enterprises.

- The definition of policies aimed at developing partnerships between scientists in different sectors as well as across scientific disciplines.

- The development of a political and economic climate that is proving less hospitable to the international organization of new initiatives that involve long-term programs of scientific inquiry.

- The development of new electronic communications networks.

Facing up to these challenges by the active participation of scientists of developing countries, can provide key inputs to the construction of a scenario of transition to sustainability in the countries themselves and at the world level.

6. Science Policy for a Transition to Sustainability in Developing Countries

6.1 Science Policy or Research Policy: Why a Policy at All?

It is increasingly difficult to separate science from technology, following Lane: "To my mind, the question is not, where the dividing lines are between science and technology, or between basic and applied research, but rather how do we take better advantage of the interrelationships in order for the nation to reap the full benefits of its integrated investments in science and technology"

Branscomb has argued convincingly that if instead of debating the boundaries of science and technology policy, the term research policy be used and its requirements for creating technical knowledge found, and use the term innovation policy, to cover the important subject of incentives for innovation. In this concept, research may be viewed as creating both new understanding of nature and new technical opportunities, some of which may be immediately available while others mature in a longer term. Thus, both scientific and technological research contribute to public value and are natural companions.

> Technological as well as scientific research aimed at building national capability, creating new opportunities, and guiding technological decisions should motivate the government's investment in research. When science itself is the driver to create new opportunities and new understanding, the research community should have the primary voice in setting goals and priorities. When, on the other hand, it is public needs that drive the research, the political process, informed by research, provides the funding and the overall goals. But in both cases, researchers need an environment that favors risk-taking and allows them considerable latitute in setting research strategies.

The centerpiece of any strategy to achieve a scenario of transition to sustainability, is precisely the accelerated build up of capacities in science, technology, and innovation. Here, while recognizing the importance of using the term "research policy" as a much better approximation to the concept of policy for capacity building, the term "science policy" will still be used, as it is widely disseminated in the developing countries. Nevertheless, "science policy" in this paper, will be used almost as a synonym of "research policy" as defined above.

In the developed world, explicit public innovation policies have been adopted, to promote the development of a trade oriented environment for research, legislative and regulatory measures considered obstacles to the diffusion of knowledge have been eliminated, new rules have been established to create in the scientific communities more interest in the commercial exploitation of their work, incentives have been multiplied to promote science based industrial activities and important investments are being made in fundamental research and in new and enhanced approaches to international collaboration. Such efforts have been accompanied by a reorientation of investment to attend areas of major economic relevance. The latter has become a way of measuring and evaluating research proposals.

In most of the developing countries, particularly after the crisis of the 80's, the environment for the development of science, technology and innovation has been adverse. The adjustment policies, the drive towards privatization, and the political agendas, did not consider them. As a consequence, cutbacks in government funds have forced many technological institutions to close down, and weakened universities, and the already fragile institutional structures became more vulnerable. The technological change that occurred was the result of foreign investment in the privatized enterprises and other chance factors, not necessarily the product of determined policies. Such change, while contributing to promote further investment and modernization mainly of the service sectors and exploitation of non renewable natural resources, did little to contribute to the manufacture or transformation sectors, thus introducing limitations to the generation of employment, increasing the negative commercial balance of payments and other phenomena, resulting in the dramatic situation, already discussed.

Such situation cannot continue. Present challenges (and opportunities) urge the search for solutions that have an important research and innovation component. The continuity and acceleration of the important structural transformation policies, require their substantial contribution. Without it the effort and investments made are in danger of being lost. It is necessary to build new structures for the ongoing changes and particularly important, build in society, visions of the future based on knowledge.

In the context of globalization of the economic and political relationships and in consideration of the specific internal challenges that the developing developed countries face, the acceleration of science, technology and innovation is clearly a priority. A transition to sustainability, requires a long term vision to create abilities to anticipate and respond to change. It is one of the purposes of a policy to develop such abilities. No country can expect to succeed in the application and attainment of its development goals unless it defines a policy addressed to these key elements.

A science policy should be destined to contribute to the revitalization of the national innovation systems, through a set of specific actions, conceived as a way to mobilize the national intelligence. This purpose implies the creation and strengthening of national capacities of science, and preserving the accumulated capacities of the past; and reorienting the strategies and instruments that impede necessary changes. Science constitutes the base of social welfare and cultural and economic development. Its progress constitutes, therefore, a new and renewed priority. Any nation that lives in the "era of knowledge," needs science as a key element for its growth and welfare.

While the importance in the definition of science policy should be understood, it should also be taken into account the negative experience, wide in many developing countries, of formulating it, isolated from other human endeavors. Thus, a science policy in the developing countries must make part of an innovation policy, which is at the same time oriented towards all dimensions of sustainability. As such, science must be included as the generator of knowledge that will contribute to all faces of sustained development.

6.2 The Role of the State

Today no modern state, interventionist or liberal, abstains from adopting a policy addressed to science, technology and innovation. Such policy is key to articulate the knowledge sources with the economic and social policies. Even though the creation or acquisition of technological and innovation capacities are primarily a matter of competence of the enterprises in a free market economy, the responsibility to create favorable conditions to build strengths is a main function of the state. Enterprises compete on a platform created by the state. It has become in recent years the main promoter and stimulator of creativity and innovation in firms. The development of knowledge as a key resource of the economic base cannot be left simply to the operation of market forces.

Particularly important to consider here is the fact that an enormous number of national and international meetings, studies, declarations have indicated the need for state intervention through policy formulation. The recipes of what should be done in science and technology are enormous. In spite of these, the results so far have been very limited and new strategies are needed to improve the situation. One such strategy is to lobby governments. There should be no reason, for example, why developed countries and multilateral organizations, impose on developing countries to fight against poverty, drugs, preservation of the environment, illiteracy, open markets, but at the same time do very little to pressure them to adopt strategies leading to the generation and advancement of knowledge.

6.3 Main Science Policy Issues

Science policy, as part of a broader innovation policy, must, in the developing countries, consider a set of issues which are pivotal in the search for sustainability. Without attempting to be exhaustive, some of these issues will be discussed here.

6.3.1 Policy Objectives

Science policy in the developing countries must aim to expand endogenous capacities, that is a "home grown" capacity to make reasonable and independent decisions. The creation of such capacities require a serious, concerted and resilient effort for the development of human resources and of a minimum indigenous research capacity. Further, once acquired and expanded it must be maintained, through a persistent response to changes in the social, economic, political, and physical environments.

In the above framework, science policy in the developing countries must, more specifically, aim to:

- Sustain long-term research linking it to societal goals and develop a system by which science can detect demands from society, and in turn translate societal problems into researchable scientific questions, integrating disciplinary knowledge in locally focused problem-driven research and applications efforts.
- Through fundamental research, create a solid foundation of basic knowledge, and promote the development and enhanced use of information and communications technologies.
- Put its scientists at the same level as those of any other country and increase and expand the training of human resources.
- Link S&T capacities with government and private sector in collaborative research partnerships.
- Facilitate the use of local capacities and transform private and public institutions into "learning organizations."
- Promote cooperation with more developed countries, that is, to link the country's scientific activity to the mainstream of world knowledge generation, and thus contributing to couple global, national and local institutions into effective research systems.

As noted in the preparatory phases and the UNESCO-ICSU World Science Conference Declaration and Plan of Action, the world community should also introduce into their policy objectives, specific goals aiming at creating a culture of peace, to improve knowledge and further critical analysis that contribute to the harmonization of the complex interrelationships between science, technology and society, and pursue a process of democratization of science itself.

6.3.2 Research Projects

The acquisition of scientific and technological capacities requires the execution of a set of high quality research projects and successful results. The size of the country and the availability of resources require a high degree of selectivity. To implement a selective policy, three criteria for the execution of research projects should be in their base:

- To profit from "national comparative advantages," for example geographic situation, history and cultural traditions and natural resources, so that the research results constitute a real and original contribution to the advancement of knowledge.
- To participate in international scientific projects and networks, to which, because of existing local capacities, and interest of further developing knowledge for the benefit of society at large, scientists can make a contribution of relevance.
- To contribute with results to the creation of the necessary technology for the production that social advances and the economy require. This is, a series of projects that contribute mainly to the modernization of the productive structure, raise productivity and competitivity, adjusting itself to the size and the existing productive capacities; create new capacities and satisfy specific demands of internal and export markets.

The adoption of such criteria, will certainly contribute to maintain quality of research projects. The first and second criteria are more related to science and the third to technology. All are equally important. In the more specific case of the third, it should be clear that research should answer to a long-term vision, giving priority to new technologies to contribute to the modernization of production.

The above criteria should be linked to the need of regarding the plans and priorities of governments which already have defined investments. Additionally, the projects that already count with human resources and installed infrastructure should be further developed initially, looking for results that will forge a new attitude of relationship of its actors with its surrounding. This will allow the improvement and consolidation of the current system and will produce a qualitative change, necessary for a later diversification.

The economic convenience of strengthening local research in the developing countries must be discussed when defining the best mix between resources dedicated to the external transfer of scientific and technological knowledge and those assigned to internal production. Particularly important is the need of the continuous upgrading of the foundations for technology applications. In this way, training and research in the basic sciences become key activities. Such research is best supported by provision of incentives for efficient work and good performance. It is important to understand that once capacity has been established within a field, the probability will increase that scientists working in other countries in that field may find it not only possible but also attractive to cooperate with the newly created research environment. From the point of view of brain drain, once science capacity has been created in a field it may become possible to benefit from the return of nationals, provided of course that other incentives can be put in place.

Finally, technology development depends on the ability of training institutions to transfer scientific knowledge, which means that teachers must not only have the opportunity of following and understanding the latest developments an be able to relate them to the society's in which they live, but also, through research be prepared to teach scientific methodologies and views. A local research capacity in the basic sciences is essential in this respect.

6.3.3 Independence of Research

The independence of research constitutes an issue that must be confronted in developing countries. It is certainly at the center of any science policy. For some it is a critical goal, essential for the future of science and of a civilized democratic society and for others it should be in a necessary, although maybe uncomfortable, relation with the economy or political factors. In the case of the developing countries, the main argument is that because of very limited resources, science can only follow definite social and economic demands, and that there is no room for scientists pursuing knowledge for other purposes. What is clear in today's debates is that independence cannot be taken as a norm that will not take challenges, its existence in an interdependent and complex system means that it can be affected by very diverse pressures and processes.

To attempt a response, it should be considered that a science policy in the developing countries must include the principle of independence as it is the only way to liberate creativity and produce innovations. Further, it is a way by which scientists create and strengthen their self-esteem, especially when choosing to execute research at the frontiers of science, collaborating with scientists from other more advanced countries.

Some guidelines for policy should include the fact that if a scientist in a least developed country, freely chooses a research topic, he or she should do it, in general in conjunction with groups of more developed countries, who will secure excellence and eventual impact. The country should not then judge only for the scientific merit of a proposal but also look at the way the scientist will interact with more developed environments. In the end, the results will always have a positive impact. Science policy in the developing countries should therefore include a specific incentive for scientists participating in international science, under the concept of independence of research.

It should also be considered, in contextualizing policy, that a tide of anti -science that has appeared in the world, created because of the lack of training of citizens and decision makers, has led to diminished freedom and independence of science. This is an issue that must be taken into account in deciding policies leading to a scenario of transition to sustainability.

6.3.4 Image of Science

Science has always had an impact on society. For this reason its image is key to its own development. In many of the developing countries, the image of science remains positive which is an advantage with the more developed ones where the image of science is sometimes questioned and thus is connected to the wishes of the general public to provide funds for research. Image, nevertheless needs to be further improved. Decision makers lack interest in the support of science because of a diffused image. Few actually understand the ways of science and in particular its fragility and complexity, peculiarities that have been shown so vividly by the works of Kuhn.

A number of practical solutions can be taken to improve the image of science in the developing countries and these should make part of any policy leading to a scenario of transition to sustainability:

- Scientists must recognize the need for transmitting honest information to the public. All opinions must be based on carefully balanced scientific visions. Policy must facilitate means for transmitting such information. In the developing countries this is not trivial. It is important to note that as public perception of science grows, scientific institutions need to better define their roles and contributions. This is closely related to the issue of accountability, as it requires that scientists work the best they can and stay within socially determined boundaries, for example as to what is ethically acceptable.

- To maintain and improve credibility, scientists must distinguish between those problems that require of their opinion and to which they can really contribute, from those in which a scientists cannot contribute more than any other person. Further, it is important to recognize the existence of problems to which science contributes only very little.

- A stronger link must be sought with the media. Popularization of science, the advancement of scientific journalism and similar activities, must be supported by the scientific community and individual scientists. Complexity must be transmitted in simple terms.

- Activities with government and private industry support to promote the greater public awareness, such as science weeks and fairs must be sought. Such efforts among others, will help to open or improve dialogues among scientists, policy makers, and the informed public, which are required if changes of attitude with respect to the scientific enterprise are expected to come with the support of a larger social base.

The teaching of the history of science in schools and universities is recognized as one of the most effective ways to improve the public understanding of scientific methods, results and the problems of science. Further, educating children in the basic sciences, so that they become better informed and appreciate both the power and limits of science, is essential if they are to be sufficiently equipped to survive in the present millennium. An important part of the issue of image, is the need of establishing means by which the prestige and self esteem of scientists in the developing countries, can be recuperated.

6.3.5 Human Resources for Science

A scenario of transition to sustainability requires the participation of the most qualified human resources. Well-trained, motivated and creative individuals are essential to its construction. The training of such individuals is probably one of the biggest challenges facing the developing countries.

Higher education institutions, in general universities, have a fundamental responsibility in the training of such individuals. In the case of the developing countries, particularly important are the state universities, as they hold the largest percentage of installed infrastructure in certain key disciplines, such as those related to the basic sciences.

Universities in most of the developing countries, must develop a new model, one which recognizes the different ways knowledge is generated today, so they can harmonize the imperatives of academic excellence with the different demands of service to society and the economy; the local with the global conditions; the short and long term needs; education with the labor market; and science and technology with society. This process means the critical incorporation to the national culture, of the knowledge being generated outside, in all science and technology fields, allowing that the opening to the outside world, be done with understanding, confrontation and critique. This will in turn allow the creation of propitious renewal and permanent updating of plans and study programs and research.

Science policy must be addressed to facilitate the insertion of the universities in the international transformation process, and at the same time, contribute to the acceleration of changes that these require in the organizational and management forms.

For the development of science it is absolutely necessary to create a strict postgraduate training program, which is also one of the ways to motivate the overcoming of the innumerable deficiencies that are found today in the lower levels of education. Policy must facilitate the creation of networks and cooperative postgraduate training between universities in the same country, abroad and particularly those in the more developed world. Postgraduate training cannot be considered, as it is today, simply as a remedial step in education and what is worst simply a easier way of obtaining financial resources.

The training of human resources at the postgraduate level, principally the doctorate, should make a substantive part of any policy. In the case of the developing countries, it is necessary to overcome a severe shortage of individuals at this level. Such a scientific elite is required by industry, government and the universities, and should be capable of maintaining the accumulation process with the rhythm required to accelerate competitive and innovative capacities. This elite is the instrument to advance a science-based transformation to sustainability.

To support advanced training, science policy must contemplate two obvious directions, research training abroad and locally, creating for each, conditions that do not exist today. Two general guidelines must be at the base of such policy:

- Take advantage of all high quality research activities going on within the country, and also those outside universities, for example in industry.

- Benefit for research training opportunities abroad, making sure that the projects in which researchers from developing countries participate will contribute to strengthen their national and individual capacities.

In fields where domestic research activities are insufficient, either in quantity, quality or both, for the necessary support of domestic research or further, because it is considered important that a certain group participates in an advanced scientific international program, at least part of the training must take place abroad. In practice, today, research training is mostly organized on an individual basis, with little institutional involvement, reducing benefits for the country and some times leading to brain drain, thus the importance of introducing such issue explicitly as a part of a national science policy, which can consider as objectives:

- Upgrading of the standards in the national institution.
- Lower costs of training.
- Identify relevant choices of subjects for the research training.
- Future research and educational cooperation.
- Reduce brain-drain.
- Improve negotiating skills and research training strategies based on experience.
- Obtaining financial support from a local source, even if at a small percentage.

In the case of local training, unless there is sufficient competence in some specific area, which is not the case in most of the smaller developing countries, training can take the form of cooperative programs (sandwich programs) where trainees perform their studies both at home and a foreign institution. It has been pointed out, that such North–South collaboration requires some ingredients to be effective, and these should be part of the policy considerations, amongst them:

- Each cooperating group should include a substantial number of researchers.
- Partners should meet regularly to review ongoing work and plan activities.
- Research papers should be written jointly.

A cooperative system can involve a limited number of universities in the country or neighboring countries, and one or more universities of developed countries that secure the standards and provide a quality seal to the mechanism. In general, most of the developing countries have some centers of excellence where research work can be undertaken by doctoral candidates, these can be complemented by those centers belonging to the universities of the more developed countries.

6.3.6 Centers of Excellence

Scientists conduct their research at scientific institutions. Unless these exist, not only in number but also in quality, science will never be able to develop, impact and contribute to any scenario of transition to sustainability.

Scientific institutions of excellence must be at the forefront of the search for a scenario of transition. Such institutions must be built or preserved in the developing countries. Each one already has such an institution or the potential to build one, from the Desert Research Institute of Namibia to the Cosmic Ray Laboratory of Chacaltaya in Bolivia, the range is ample.

From a policy perspective, no matter in which discipline of science they work, these centers must be protected, allowed to grow and diffuse their work. In time others can be added. The first step is to identify the institutions and try to preserve and reproduce the contextual conditions that have permitted their growth and provide further incentives for their work.

Such conditions show that centers of excellence normally have a scientific career among their researchers, thus policy can establish the need to create such career at national or at least university levels. Full dedication of researchers to universities is an absolute requirement. In many universities of the developing countries, no more than 10% top 20% of the teaching and research staff are dedicated full time to them. More recently many countries have taken important steps to allow their scientific communities to access international data transmission networks and participate in research networks. This should be further encouraged, and incentives must be put in place for this to happen. Some times just a lower telephone tariff for Internet connections may be sufficient.

On the other hand specific support actions should be developed for research groups of excellence to allow their insertion in the dynamics of international scientific development. National policies for science, must explicitly take into consideration financial and technical support of regional centers of excellence. Many of these do exist. The developing countries have in these centers valuable sources of know how and training.

Regional centers as well as "big science" (large international facilities) projects can contribute to break the isolation to which many scientists in the developing countries find themselves. In time and without contacts such scientists cannot contribute to the generation of knowledge and to the national development efforts. The participation of scientists in such endeavors can be facilitated at times with small investments. This participation, in turn, attracts investment in local efforts. Scientists of the developing countries could join their colleagues of larger countries in their research groups in all fields of science, the large astronomical telescopes in Chile, the particle accelerators in the USA and Switzerland, the new cosmic ray facility AUGER in Argentina and the USA, and so on, just to mention a few examples from physics.

For both scientific advance and education, the participation of scientists of the developing countries, in large experimental facilities around the world is of key importance. This is not explicitly recognized by these countries. In fact, the leave of absence system, or other means of allowing scientists to travel for a short period of time is extremely complicated by bureaucratic procedures and worst, the return might be a nightmare for a young scientist. It is then necessary that advanced countries provide opportunities for scientists from the developing countries to work at their installations, but it is first necessary that the countries themselves adopt and actually implement explicit policies allowing their nationals to participate without problems.

6.3.7 Internationalization of Science

The progress of science and therefore its contribution and impact on development, both in its economic and cultural dimensions, depends on the linkages and cooperation that can be established among the world scientific communities and of these with other actors of the "national innovation system." The relevance of scientific cooperation has been of course recognized for many years. Several countries have made it an important landmark of their development policies and have created a series of instruments for its application.

The spirit of scientific international collaboration is to ensure that the world functions as one giant community, not just to promote the development of science, but also to fulfill the needs of society for scientific and technological knowledge.

Today and probably more each day, dramatic new opportunities abound in the international landscape for science. How can developing countries partake in this opportunity while at the same time respond to short term pressures imposed by social needs, in particular poverty? This question must be approached from a policy perspective in a systematic way.

Oldham has pointed out many reasons for international scientific collaboration:

- Scientific problems and disciplines are a function of geography, for example environmental issues.
- From the economic perspective, the cost of building equipment or computer tasks, may be too large for a single country to carry. Cost sharing is then the response, such is the case of "big science."
- In the commercial dimension, to enhance economic competitiveness of countries and regions, collaboration may mean new types of relationships between government laboratories, universities and private sectors.
- Politically, some meetings of heads of state or bilateral agreements, originated in political interests, have led, in several occasions to successful collaboration.
- In this global view, scientific collaboration, in the interest of the developing countries, face some problems that need to be addressed:
- Scientific colonialism, that is the use of scientists of the concerned countries used almost strictly to collect data, leading to a fragile effort of capacity building.
- The commercial exploitation of research results carried out mostly by the enterprises of the developed countries.
- The increase of brain drain.

Much has been accomplished in the past years to overcome these problems. The PUGWASH Guidelines have argued for real scientific collaboration, and its application has led scientists in the developing countries to actively and effectively participate in international scientific collaboration. Much however remains to be done, and the answer to the question of how involved should the small scientific communities of these countries be in large international science projects requires reflection and policy definitions of importance.

Whatever the motive for cooperation, it is clear today that no country, let alone the smaller ones, has the resources capable of sustaining science and the development of human resources for science. But scientific collaboration is not hands out or hands down, it is building up of knowledge and capacity to enable a country to create its own wealth. Scientific cooperation must allow and make sure that each nation will have the capability to use scientific knowledge, by improving science education, or improving the public understanding of science or overcoming isolation of scientists from one another and from society. The power of a program for mobilizing society for helping to inform policies to governments, and for helping developing countries is immense; international scientific collaborations must be built.

International scientific cooperation represents the opportunity for scientists in developing countries to partake in work being carried out at the frontiers of science. This represents, not only a way of advancing and transferring knowledge to their own environment, but also a way to contribute to the advancement of knowledge at world level and thus feel an active part of the world scientific community, thus providing self-confidence and much needed self-esteem.

The internationalization of science is the way of reaching high standards and optimal results. The developing countries require a policy for internationalization if it wishes that their scientific communities will actually make a serious contribution to a scenario of transition to sustainability. Policy under this issue maybe concerned with:

- International mobility at all levels.
- The development and wider use of electronic communication networks.
- The establishment of "research networks" and "virtual centers," interconnecting research teams.
- The education of scientists themselves with regards to opportunities and challenges abroad.
- The support of teachers and others to implement changes needed in school systems, including work to develop curricula and use www resources.
- The achievement of a better understanding of the global production and distribution of scientific information. Particular attention must be drawn to new developments such as electronic journals, publications and libraries.
- Identifying the advantages and consequences of interdependencies among the nation's various research programs and policies.
- Examining the technological opportunities that exist, or are likely soon to emerge for supporting greater integration and collaborative use of shared facilities.

More specifically, a policy for the internationalization of science must consider different levels, four of which are important to be considered here. The first level refers to the development and implementation of global agendas, which should address major unsolved problems, adhering to the sustainability criteria. Developing such a complex agenda is surely a great challenge to the world community, but its contents and the results of its application will have a definite impact on a scenario of transition to sustainability in the developing countries.

In science and technology an impressive number of agendas have been defined at the international and regional levels. In the first case, for example, the first UN Conference held in Geneva in the 1960s, later the UN Conference held in Vienna in 1979, and the more recent UNESCO-ICSU Conference in Budapest in 1999 are of importance, not to mention the adoption of several others at the different UNCTAD meetings.

The UNESCO-ICSU World Science Conference, built on a set of preparatory activities, has formulated a declaration and a plan of action, containing a large set of recommendations, many of which are included in some way or the other in this contribution. However, as is the case of all UN conferences, much care has been taken not to leave out any issue, in order to satisfy the political view of over 180 countries members of the system, and thus making such recommendations extremely general.

The application of the recommendations contained in conference documents are the responsibility of all countries and the UN system itself. In the context of the policy environment of the developing countries, the recommendations, even if general, fill important gaps and must be adapted to each country situation. Whether this actually happens or not with the latest UNESCO-ICSU plan is uncertain. The experiences of both Geneva and Vienna show that the declarations and recommendations remain at that, in spite of their rich conceptual and operative contents. On the other hand, world organizations require revamping if they will serve the purposes of their own declarations and plans. Under this context, policies in developing countries must consider in an explicit way the wealth contained in such declarations and act accordingly so that these institutions will actually deliver responses.

A second level of internationalization refers to the undertaking of international science projects, in which developing countries can participate. Policy must include some specifics, for example how to use, ICSU's strongly international network which is very well suited to bring together scientists who can generate solutions, for example, to global threats of disease, poverty, food shortage and a changing environment.

Decisions-makers, even in science, believe that the developing countries should pursue "small science" alone. This is a mistaken approach, as scientists from these countries can well contribute to "big science," as well. The problem is of emphasis not of choice. Of course if all the available resources are put into "big science" projects, it is possible that investment is not well planned. But certainly it is possible and should be the policy of a country to dedicate a fraction of the national investment in science and technology into large projects of international reach. It should also be considered, that excellence in "big science" depends on excellence in "small science," and that international cooperation with rigorous competition is essential for both.

At the level of international science, the developing countries and their developed counterparts must define policies aiming to:

- Advance understanding of the laws of nature and to promote peace, development

and prosperity.
- Provide an international forum where scientists and governments can meet to discuss, consult and plan regional and international projects on science.
- Promote dialog and cooperation between different disciplines and fields of science.
- Invigorate science education at all levels.
- Reinforce links between science and technology, between science and society and between science and development.
- Broaden communications links among and within scientific communities.

A third level of internationalization is that of regional cooperation. Regional groupings, sharing particular interests, are important to promote collaboration. These normally have common economic, political, and cultural interests, which can demand on science and scientific institutions. Experience has shown that agendas built on these interests can be very effective and actually implemented. Further, many regional networks are connected to world networks, and hence participating in world efforts.

Regional "knowledge networks" must be built quickly. Globalization in its economic forms is going faster than the creation of knowledge; "it is as if the thinking head of the region is getting behind, while the material body moves faster and with greater imagination in the plane of investments, commerce and entrepreneurial alliances." Along this line, policy fora should be regionally promoted, which could:

- Provide guidelines to regional and national science development efforts.
- Collaborate with governments and regional institutions in the identification of opportunities for cooperation, through information exchange between government officials and members of the scientific communities.
- Identify and explore policy issues, which are relevant to the successful planning and implementation of multilateral projects.
- Secure long term and stable financial arrangements.

A fourth level of internationalization is science aid, whose main purpose should be to strengthen the capacity in the developing countries to conduct relevant and high quality research.

For this objective to be accomplished, there are many conditions. A key issue in policy is the need of redressing the imbalance in the North–South research partnerships. Unless conscious efforts are made to incorporate this issue in policy, collaborative programs stand little chance of achieving their development objective, and to some extent also their scientific objectives. To this end, the dual objective of development and scientific relevance must be introduced. The Southern context needs to be taken into account in a more rigorous way than is the case now. Several recommendations have been made for redressing imbalances, that refer to transparency in financial matters, the participation of researchers in the South in the preparatory phases of projects, efficient communications channels and project monitoring and evaluation, among others.

Other issues that need to be considered, especially when the objective is capacity building. One is certainly that the first capacity required in any country is that to make decisions on what type of science and technology is needed to meet its needs. The second capacity is that required in a cumulative way to generate, absorb, transfer and use knowledge. For each there is need of an appropriate institution and persons. It is not possible to think that science aid will be successful unless such conditions exist.

At the same time, donors' policies are important and in many cases they must pass through permanent review and follow with internationally agreed set of norms and practices. This includes a coordination of donors. In the past, such aid has at times created collections of institutions unrelated to one another, or investments on one type of research activity at the expense of others. The result is an unbalanced system with many unconnected parts. "It is a system capable of absorbing large sums of money with little noticeable impact on the lives of poor people in these countries"

Science policy to foster aid, must then consider conditions, demands, opportunities and potentials, which can be summarized as follows:

- There exists capacity in most developing countries to absorb external research funding, and they constitute, with the exception of the least developed countries, only a small percentage of total domestic investment in research. National governments should then ensure that the scientific and technological infrastructure serves the development of research and innovation.
- Many problems facing developing countries are economic, social and political rather than scientific and technological, and by no means can such problems be solved through research.
- Research alone is not sufficient to guarantee development and close links must be sought between researchers and the production system. It is necessary to create innovative skills.
- Social sciences must be involved in the design of scientific research, in order that technical change has no adverse economic or social consequences.
- There is a need for persistent and long-term research. Short-term projects are not conducive to any results of impact, either intellectual or practical. It is not easy to measure benefits when projects are evaluated because of their capacity building impact.
- Dissemination of research results must be taken very seriously.

6.3.8 Institutional Structure

The present world situation requires that all the institutional components of the national innovation system have to adapt to the new context. In the case of the science system, research councils, learned institutions and universities are faced with significant challenges.

Research councils in most of the developing countries are the most fragile. They were never able to attain the level of importance of their counterparts in more advanced nations. The causes for this situation have been amply analyzed, but the main one is that science and technology have never been an important part of the political agendas, and thus, the institutions managing them lack political support. Innovative ways need to be searched for making these institutions effective in their promotional and investment roles.

Learned institutions, such as academies of science, when they exist, are also under pressure by the scientific communities, which demand a more active role in the absence or weakness of research councils and to become a more democratic mechanism to allow further participation by all scientists. Academies in most countries are private institutions, created by government charters, and because of their weak financial and political positions, they can hardly cope with their own business, let alone consider outside pressures.

In spite of the above, there are examples of academies of science, that besides their more traditional activities related to the organization of meetings, exchanges, publications and the representation of the interests and points of views of science, have rendered valuable advisory services, and led debates on science, technology and innovation, thus responding to societal demands. Through the provision of independent advice on policy matters with important technical content, academies can increasingly be a force for wise decision-making, enhanced further by international cooperation which builds a collective capacity for understanding and meeting global challenges.

The institutional structure must be aware that the laboratory—mainly in the university—is the centerpiece of the science system, but the centerpiece of the national innovation system is the enterprise; it must organize itself accordingly. This poses a difficult challenge to policy, as each element of the system has specific requirements, but at the same time it is their synergy, that actually create research and innovation capacities. What cannot be done, as is the case today, is to define policies and strategies in the science system and suppose that these will work on the whole of the national innovation system. Experience shows that in the developing countries, whatever advances have been obtained in the development of technology capacities, have been due in great part because of the formulation of industrial policies, and a much smaller part because of the application of an independent science and technology policy.

6.3.9 Management of Science

The management of science and innovation has become a key issue in today's globalized world. In the developing countries, the needs for managers of scientific institutions and projects is enormous. A rather limited set of reasons, will illustrate this point.

Organizations need to undergo change at a much faster pace that is the case today if they wish to respond to increased local and global competition and rapid advances in the generation of knowledge. Managers are thus forced to understand the way information systems that support research change, while at the same time preserving experience-based parts of their businesses. Policy must then be concerned with training of managers that understand for systems solutions.

Science managers in the developing countries must also grasp the concepts and practices of innovation. A development in this line, will allow the establishment of benchmarking practices and the redefinition of research methods. Today, very few developing countries can actually design, promote and implement innovation projects of their own. Managers knowledgeable of the interface research-innovation, will definitely contribute to the development of the management of intellectual capital for innovation.

In today's world, multidisciplinary research has become of extreme importance in the search for a scenario of transition to sustainability. More so, solution to several interdisciplinary problems can be faced with a multidisciplinary approach. Managers in the developing countries must be ready to face this phenomenon with great urgency.

6.3.10 Investing in Science

Two fundamental conditions are needed in order that the developing countries define and implement a science policy for a transition to sustainability, the political will to undertake an effort in science, as a mechanism for endogenous capacity building, and the commitment of financial resources to that purpose. Many national and international policies and declarations, made up mainly by the scientific communities, already contain several of the recommendations discussed in this paper, but for their implementation, these two conditions are absent.

The political will, that includes both public and private sectors, can be constructed in a process of dialog among stakeholders, which is slowly starting to take place in many countries. It can also be strengthened by the commitment of science institutions to renovate and start facing up to the new challenges posed by globalization, showing that their work is relevant to a transition to sustainability. There are several other ways that maybe thought of for improving the environment for science, that will contribute to a better social base and thus political will. Many of these depend at the same time on improving investments.

The financing of science in the developing countries has been traditionally small, and the trend is that the situation will not improve unless dramatic changes are made at the level of policies and compromise of political and economic decision makers.

The first question that is asked by decision makers when defining investments in science is, "what are the relative costs and benefits to society of the application of existing knowledge, produced elsewhere, mostly in the developed countries versus generating and even adapting new knowledge?" More specifically, resources for science and higher education must compete with resources for poverty alleviation and other pressing short-term social needs.

As discussed in this paper, decision-makers would rather access existing knowledge to satisfy the different social and economic requirements and demands. While this would seem to be an appropriate approach in times of shortage of fiscal funds, it simply does not satisfy the most urgent need for long term development and much less a sustained one, the creation of autonomous decisional and operative response capacities. There is an immense number of examples of the economic failure of transferred knowledge into a developing environment, because of the lack of strong local capacities.

The final result of this situation is quite evident: poverty and the extreme economic and political dependence of developing countries' governments and enterprises on their equivalents in the developed countries and transnational corporations, and in the case of the least developed countries extreme dependence on multilateral and bilateral credit and aid. If the latter would be cut, these countries, which today are "symbols" of the success of the liberal ideology, would simply collapse.

The improvement of an investment climate requires first an understanding of the complexity of science. For example, if pure and applied mathematics (essential in today's information society) cannot fully merge, they cannot flourish for long without each other. The exchange of ideas between these two aspects are an essential component of the intellectual enterprise into the twenty-first century, and this merits accelerated research and therefore investment. This example is brought forward, as mathematics is a discipline that can extensively grow in the developing countries. The understanding of science by decision makers and the general public is as already discussed elsewhere a key issue facing the scientific communities of developing countries.

New concepts and practices need to be built if the present situation of under funding can be overcome. Existing science and higher education institutions in the public sector, which comprise by far the largest group in the majority of the developing countries, must lead the effort by applying internal reforms. It is not possible to conceive the present atomization and duplication of research projects, many of low quality. These institutions must develop new budget concepts. Those which will not jeopardize institutional pluralism, but at the same time, should lead to a trade off, the transfer of funds from less relevant projects to ones of higher quality or relevance to cultural, social and economic needs. Such type of reform does not require any additional funding and will rationalize the existing ones. A full set of needed reforms can be thought of, only as examples, the definition of rules to allow consulting and advisory services, the adoption of administrative mechanisms by which resources produced by research or service are actually used to strengthen the institutions that produce them, and not the bureaucratic structures of the parent institution.

Measures promoting the participation of enterprises in research are necessary. National firms are slowly realizing the importance of innovation for competitivity and thus incentives must be created. Tax and other incentives must be introduced. In the past, such instruments were used and abused by government, and today these have gone to the other extremes, in many developing countries, simply no incentives at all. Of great importance is to involve foreign firms setting up offices and other activities in the developing countries, in the local research and innovation efforts. In many cases it is possible to think that resources from privatization and capitalization should be destined to these endeavors, for which special legislation might be necessary, although in some cases, simply a good negotiation, under clear policy guidelines, can produce the desired effect.

On the other hand, international organizations should adopt the policy to destine a part of their grants and loans to build up research and innovation capacities. In turn they should press local governments to establish the appropriate rules to use these funds efficiently.

At both the national and international levels, new mechanisms for raising resources must be found. Some are currently under discussion and should make part of a new system, for example taxes on international financial transactions or on international flows on information or international air transportation. The search for such mechanisms, imply the need of diversification of financial institutions and activities, trying to shift the center of gravity from automatic subsidies to the supply of private and public modalities, more attuned to demands and based, in the case of fiscal resources on benchmarks established in contracts, evaluations, project competitions, competitive bidding, or in formulas that promote the reach of public interest objectives.

Because North - South cooperation will continue to be of importance, particularly for the least developed countries, it is necessary to find the best ways for improving its intensity and quality. The latter must be part of policy considerations in all parts involved.

7. Conclusion

Science plays a fundamental role in a scenario for transition to sustainability. The better knowledge of nature, its links with new technologies and the process of innovation, its direct relation to teaching and learning, the constant deployment of new methods and views of research, and its impacts on culture, society, environment and governance, are just examples of its contributions. Science is not in the instruments in the laboratory, but mainly in the minds of people and it is through these minds that sustained development will actually be reached.

While recognizing the epistemological limit of science, it should be borne in mind that no country that has developed in the past has done it without an effort in the build up of science capacities. The developing countries of today will not be the exception.

Acknowledgments

This paper has drawn and taken many of the ideas and specific proposals of several authors, mainly from those listed in the bibliography.

Glossary

GATT:	General Agreement on Trade and Tariffs.
GNP:	Gross national product.
ICSU:	International Council for Science.
OECD:	Organization for Economic Cooperation and Development.
R&D:	Research and development.
R&I:	Research and innovation.
S&T:	Science and technology.
UNCSTD:	United Nations Commission for Science and Technology for Development.
UNCTAD:	United Nations Conference on Trade and Development
UNESCO	United Nations Educational, Scientific, and Cultural Organization

Bibliography

Bitar S. and Bradford C. I. Jr. (1992). Strategic options for Latin America in the 1990s. In C. I. Bradford Jr., ed. *Strategic Options for Latin America in the 1990s*, 287 pp. París: BID/OECD. [An in-depth discussion on the paradigm of competitivity and orthodox policies.]

Branscomb L. M. and Keller J. H., eds. (1998). *Investing in Innovation: Creating a Research and Innovation Policy that Works*, 516 pp. Cambridge, MA: MIT Press. 516 pp. [Review and discussion of recent policies adopted in developed countries to enhance innovation capacities.]

Caswill C. (1992). *Academies, Research Councils and Universities: Their Role in Modern Europe*, 66 pp. London: Academia Europaea. [A review of critical issues about the role of scientific institutions in a globalized and market-oriented world.]

Clark W. and Kates R. W. (1999). *Our Common Journey: A transition toward* sustainability, 528 pp. Board on Sustainable Development, Policy Division, Washington, DC: National Academy Press. [An in-depth study on the way science can be used in order to respond to sustainability challenges in specific economic and social sectors.]

Gibbons M. (1998). *Higher Education Relevance in the 21st Century*, 60 pp. Paris: Contribution to the UNESCO Conference on Higher Education, 5–9 October. [A discussion of the role of universities in a world where the production of knowledge is taking many different ways and involving a diversity of actors.]

Hammond A. (1998). *Which World: Scenarios for the 21st. Century*, 306 pp. Washington, DC: Island Press/Shearwater Books. [Construction and discussion of three scenarios for the 21st century, from the perspective of a free market economy.]

Machado F. (1998). Administración eficiente de la innovación tecnológica en los países endesarrollo. *Comercio Exterior*, **48**(8), 12. [An overview of what could be a new industrial revolution, based on the development of new products, attuned to world environmental demands, produced by the developing countries.]

Oldham J. (1994). International scientific collaboration, 25 pp. ORSTOM/UNESCO Conference: *20th Century Science: Beyond the Metropolis*, September. [A critical review of the state and results of international cooperation and science aid.]

Rattner H. (1999). Leadership for sustainable society, 18 pp. Paper prepared for the W. F. Kellog Foundation Leadership Forum, Leading Change in the New Millennium: A Call to Action, Washington D.C., November. [A discussion of the needs of creating a new leadership in the world, based on principles of sustainability and of the need of strengthening social cohesion through the traditional and new roles of non-governmental organizations.]

Thulstrup E. W. (1999). Promoting science driven economic development, 15 pp. Paper presented at the Bolivian Science and Society Symposium, Cochabamba, September. [A critical review of issues related to the need of promoting science and cooperation in developing countries.]

Biographical Sketch

Professor Carlos B. Aguirre is a Bolivian physicist, President of the National Academy of Sciences of Bolivia, and also professor of policy and management in the Universidad Mayor de San Andres, La Paz. His present activities include membership or directive posts in several national, regional and international organizations of science. Between 1995 and 1996, he was the Executive Secretary of the Bolivian National Science and Technology Council. In the early 1990s, he directed a project for the monitoring of new technologies for the Andean Countries and was editor of the Andean R&D Newsletter. In the 1980s, he was Head of the Technology Department of the Andean Community Secretariat in Lima, Peru. In this period, he was also the Secretary of the Andean Council for Science and Technology and Secretary to the Andean-European Commission for Science and Technology Cooperation. Before joining the Secretariat, he was between 1978 and 1979, the first Director of Science and Technology, in the Bolivian Ministry of Planning. Between 1973 and 1978, he was the Director of the Physical Research Institute of the Universidad Mayor de San Andres, where he also held the positions of Director of the Research Planning

Center and Secretary General. In 1969, he was appointed full professor at the Physics Department of the Faculty of Sciences. He has been an Associate Member of the International Centre for Theoretical Physics, in Trieste, Italy and visiting professor in Japan, Argentina, Colombia, Peru, Venezuela, the former Soviet Union, and Germany. Professor Aguirre is a member of (and has held a number of directive posts in) international and national scientific organizations, and professional societies. He has conducted consultancy assignments for UN organizations, bilateral cooperation agencies, international NGOs, and others, in the fields of education, science, technology and innovation policies; he has worked in Austria, Bolivia, Brazil, Colombia, Costa Rica, Ecuador, Namibia, Mexico, Indonesia, Spain, Sweden, Switzerland, Tanzania, and Venezuela. He has received several distinctions and awards for his services to science and development. Professor Aguirre has published extensively in the fields of cosmic ray physics, science and innovation policies and planning, higher education, and environmental issues.

SCIENCE AND SOCIETY: AFRICA'S PERSPECTIVE

Shem O. Wandiga and Eric O. Odada
Kenya National Academy of Sciences, Nairobi, Kenya

Keywords : science, society, Africa, toxic chemicals, sustainable development, land degradation, water resources

Contents

Summary

Uplifting food production to the level of self-sufficiency in Africa remains a major challenge for science in the next century. It calls for leveling of high population growth, economic development of rural areas through relevant education in science, improvement of human health and environment and sustainable use of natural resources. Achievement in these areas will result in changed consumption pattern and reduced waste generation that may be scientifically managed. Science and scientific knowledge application hold the key for Africa's socioeconomic improvement. Mobilization of resources should put women, youth and the elderly as priority groups. International co-operation and assistance directed to disadvantaged societies in ways that improve their socioeconomic status holds promise.

1. Introduction

African society derives its roots in its customs and cultural perception of nature, supernatural powers and humanity. Many African cultures use the environment to support life. Multiple cures were and are still found from plants, animal parts, water and the living organisms within, like fish, and water plants. Traditionally, destruction of environment or life is prohibited in many customs and cultural laws. The soil has been, for decades, left alone to yield plants that are used for food. Modern large plantation agriculture was introduced by the 'white settlers.' Through them, modern cash crops like maize, wheat, barley, tea, coffee, cocoa and rubber plants were introduced.

African science evolved from the use of plants and animals for the cure of diseases. In many herbal prescriptions, plants roots, barks, and leaves were boiled and the extract drunk or chewed, and leachates swallowed or mixed with other animal parts to expel evil spirits. There are evidences that surgery or bone fracture treatment were practiced by some societies in Africa long before the advent of modern medicine. Africans lived with animals and culled them for food. Taming of animals evolved with time. Some tamed animals were imported by foreign settlers and quickly adopted by pastorialists.

Modern western science was introduced to Africa by the missionaries and colonial governments. Some of the first African degree holders in chemistry, mathematics, physics or biology are still alive. Similarly, some of the first African medical doctors to graduate from medical school are still living. Modern African science adopted the British, French, Belgian or German syllabus and system. The missionaries looked down on African traditional science as primitive and/or inferior. Great efforts were made to modernize the African to the extent that most traditional African science has been abandoned by the present society. Indeed, many modern Africans look down on African herbal medicine and would prefer to swallow western drugs at any cost.

Through colonization, the African society has been transformed into a western society aping either the British, French or American (due to their global cultural influence) societies. The present African lives a dual personality, one indigenous and the other western. Neither shall the two be reconciled. The dual cultural heritage of the African creates development agenda and issues that may be unique to Africa. From a cultural tradition that used resources fully aware of their limits evolved a self-centered person who accumulates wealth and glorify riches; a person who destroys the environment and all living things in it. The self-serving African of today would delight in destroying the source of livelihood for others, as long as individual gains are maximized. This is the African that we need to transform back to the original roots in the twenty-first century.

Present public images of Africa is that of a continent with emaciated children, uncontrolled population growth, chronically suffering from food scarcity, has a plethora of diseases, and with abundant tribal conflicts. Children are taught to learn that bad leaders, who practice misrule and plunder of resources, are found mainly in Africa. Political coups repeat themselves year in year out. Above all, poverty is widespread in the continent. Poverty according to *Webster's Seventh New College Dictionary* is defined as "lack of money or material possessions; want; scarcity, dearth; debility due to malnutrition; lack of fertility". Accordingly, the World Summit for Social Development and Beyond has focused the attention of the world community to Africa, as one of the priority continents for socioeconomic developmental assistance. Today, almost half of Africa's population is living in absolute poverty, with about 30% classified as extremely poor (living on less than US$ 1/day). Africa is noted for its falling per capita income during the last 20 years. For example, in 1980 the GNP per capita was US$ 770, in 1998 it dropped to US$ 480. The challenge of poverty reduction is dependent on broad-based and sustained economic growth, complemented by the more efficient provision of social services such as education, health care, improved standards of nutrition, and access to clean water and sanitation.

On the other hand, sustainability implies the use, preservation and conservation of nature, with its life-support systems, including human, natural resources, physical and biological environment and all activities that help keep them beautiful, clean and healthy. Development emphasizes improvement of people's lives, economy and society. Often, reference is made to economic development, which comprises improvement of productive sectors for increased employment and desired consumption and wealth. Therefore, no matter whether one is desirous of reducing poverty or improving sustainable development, one cannot achieve much without the use of scientific knowledge.

The following environmental problems affect all nations but have most impact in Africa: climate change, loss of biological diversity, land degradation and desertification, deforestation and forest degradation, pollution of fresh and marine waters, depletion or increase of stratospheric ozone, and accumulation of persistent organic pollutants. The manifestation of these problems except for the last two are observed through droughts or floods, permanent disappearance of animal and plant species, soil erosions, enlarging desert areas, declining tropical forest areas (only about 2 million square kilometers of frontier and non-frontier forests remain), and uncontrolled growth of macrophytes in fresh water lakes, as well as non-useable river waters and polluted coastal areas.

Given these concerns, it is essential that any national policy on development that does not take into consideration an integrated approach to environmental issues will fall short of its goals. Sustainable development can only be achieved through incorporating environmental issues into all sectoral decisions, including management of agriculture, land resources, forestry, energy, and water resources. Concerted efforts should be made to correct market, policy, institutional, and knowledge failures. Both management and administrative mechanisms, 'command and control' policies, market mechanisms, and voluntary agreements should be used in various combinations to effect improvement.

Exploring the use of scientific knowledge in Africa for poverty eradication, sustainable development forms the basis of this paper.

2. Missed Opportunities

The twentieth century has been a lost period for Africa. During this period, the continent evolved from being colonies to full independent states. It has now taken approximately three to four decades of self-rule for most African countries, since the end of colonization. During the period of full independence rule, the Cold War divided the political allegiance of most of the countries into Eastern or Western spheres. With the collapse of the Union of Soviet Socialist Republic, many African nations found their support for economic and political authority changed. Old regimes had to quickly abandon their ways and adopt democratic principles. The end result of the Cold War and the democratization process taking place have seen the emergence of internal conflicts, the deterioration of infrastructure, economic stagnation and increased prevalence of poverty for most of the population.

Nature has not spared Africa and the Africans either. Global climate change has brought disasters associated with unpredictable weather. Old diseases like yellow fever, dysentery, typhoid, and malaria have re-emerged with a vengeance. New diseases, like ebola, human immuno-deficiency virus/acquired immuno-deficiency syndrome (HIV/AIDS) have become pandemic. Concurrent with these changes has come the burden of financial debt repayment. Repayment of squandered debts has stifled any meaningful economic growth. As a result, educational institutions, health services, social services and infrastructure have collapsed or have been left to decay for lack of financial resources.

At the beginning of the twenty-first century, democratic systems have taken root in most African nations. Liberalization of the economy has brought competitiveness in trade. All current forces such as population, gross domestic product (GDP) per capita, world gross domestic product, food, energy, water carbon dioxide, except hunger, show trend scenarios of increases ranging from 1.5 to 4.5 times the 1995 level.

Africa is unfortunate that food production lags behind population growth, with population levels projected to reach 1454 million in 2025 and 2.050 million in 2050. On the other hand, Africa is possibly the only continent that may realize a faster growing economy in this century. An economic growth scenario of up to 4% is projected during the coming century. Despite these projections, it will take at least a century for developing countries to catch up with the countries belonging to the Organization of Economic Cooperation and Development (OECD). Therefore, given that economic as well as energy growth is primarily to be in developing countries, coupled with debt relief, wealth reallocation by developed countries to developing countries, Africa will most likely see a reversal of its situation. In order for these projections to hold, we discuss here some of the issues that require the attention in order to maximize the possible positive trends.

3. Population and Science

Feeding the fast-increasing population of Africa remains the single most important challenge to science and scientists in the continent. At over 744 million people and growing at an average annual rate of 2.93, stabilization of population growth to a replacement level poses a second challenge to the continent. Some slight decrease in fertility rate (6.6 in 1970–1975 to 6.0 in 1999–1995) has been recorded. However, such a decrease remains small, though significant. UN projections indicate that replacement level will not be reached in Africa until about 2045–2050. In the meantime, feeding over one billion people in the next century, the majority of whom are already born, remains a problem. Despite the earlier indication that life expectancy at birth was improving (46.1 in 1970–1975 to 53.0 in 1990–1995) such gains have been wiped out by epidemic disease like HIV/AIDS, malaria, typhoid and many other enteric water-borne diseases. The crude death rate per 1000 population, although showed signs of improvement (19 in 1970–1975 to 14 in 1990–1995) in the early 1990s, might have actually reverted to higher numbers in the late 1990s. However, infant mortality rate (95 per 1000 births) is still above the average world rate; 62 in 1990–1995.

Figure l. Feeding the increasing population, many of whom are already born, is a problem

Stabilization of population depends firstly on provision of quality education especially to the girl- child. Such education includes better knowledge of human body and its function, delayed entry into procreation and use of birth control methods. Some of the proven population stabilization methods include abstinence, the rhythm system, and use of contraceptives and condoms. Science has a contribution to population stabilization. Some of the science challenges include development of male contraceptives, safe and improved female contraceptives, as well as prophylaxis devices. Population has been shown to decline with economic prosperity. Africa's economic growth depends on the use of scientific knowledge.

Africa has seen the highest number of displaced people, both internally and internationally. In the 1990s, there have been many internal wars like the Rwanda tragedy, and those in Burundi, Somalia. Democratic Republic of Congo, Republic of Congo, Sierra Leone and Sudan. These wars and many others have forced thousands of people to immigrate. The total number of internally displaced persons in Africa stood at 1.9 million in 1994 and 1.3 million in 1995. Pastoralist migration has always been a tradition in Africa. Those migrating normally return to their original places at the end of a drought or flood. However, today, permanent migration in search for jobs, escape from economic hardship, wars, political or religious persecution, have become a daily occurrence. In 1999, there were 21.4 million refugees in all regions of the world out of these Africa had 6.28 million with 1.6 million being displaced persons (Table 1).

	1994	1995	1996
Refugees	6 752 200	5 692 100	3 270 860
Asylum seekers	–	–	63 350
Returnees	3 084 000	2 085 400	1 296 770
Internally displaced persons and others of concern	1 973 100	1 344 000	1 653 700

* UNHCR's funding mandate defines refugees as those who are outside their countries and who cannot or do not want to return because of a well-founded fear of being persecuted for reasons of their race, religion, nationality, political opinion or membership in a particular social group. Displaced persons are people who live in refugee-like situations within their own countries.

Table 1. Persons of concern to UNHCR*

Rural-urban migration is another feature of movement of Africans. In average, urban population is increasing at an average rate of 4%, but may be as high as 7% in some countries. Investment in human capital in Africa is one of the lowest. A large percentage of adult African women are illiterate (54% in 1990) compared to men (30% in 1990). Gross primary school enrollment as a percentage of age group stands at 89 and 105 for female and male in 1990.

Gross enrollment percentage is obtained by dividing of the age group that should be enrolled in school according to the rules of a country. Numbers greater than 100 reflect inaccuracies in the data. However, at tertiary level of education only 5.6% of the cohort age group are enrolled; of those enrolled only 4.3% are women. Statistics for health services are no better and show that as low as 15% of Mali people have access to health facilities while in Mauritius, 100% of its population receive such services. Other countries among the 48 African states have figures between these two countries with only 20 countries having a percentage greater than 50.

Human conflict arises primarily from insecurity, scarcity of resources or economic opportunities and/or human greed. Scientific understanding of the bases of these factors improve tolerance and understanding amongst people. Furthermore, scientific knowledge improves natural resources' exploitation, add value to resources and improve production processes. Wealth creation lies with those who possess knowledge. Increasing the knowledge base in African societies and full utilization of the human resources with that knowledge would enhance socioeconomic development of such societies, stabilize migration and improve educational standards.

4. Combating Poverty

Poverty is a measure of voidness of material wealth. It implies that a person has not accumulated financial savings. It denotes non-possession of land, housing, clothing and most importantly, food, or an inability to buy any of these things. A fair reflection of how poor we are is also measured by our dependence on charity, social security system or any other organized system for looking after destitute persons. Morally, the poor may be blessed, destitute, indigent, working, deserving or voluntary. The working poor are often referred to as deserving poor. Through the ages the poor have always received sympathy. In the African tradition they were looked after by the well-to-do families. They were never rejected. The continent does not practice a systematic system of discrimination or exclusion based on race or caste system. There were few inherited kingdoms and wherever they were, they did not build walls against sections of their subjects based on tribes. This historical background sets apart the African poor from, for instance, those found in some parts of Asia or the gypsies in Europe. Geo-political considerations associated the poor masses with those exploited and manipulated by the rich, the upper classes. Poverty may also result from structural systems in the society that create unemployment or under-employment in the labour market, exclusion of certain groups within a society or community from capital accumulation or access to fair share of economic benefits. This new form of poverty may be emerging in Africa.

The discussion in this paper confines itself to the first definition.

According to World Bank categorization, there are two poverty groups found in Africa today. The first is the urban poor who are persons born of poor parents and have lived in a city slum all of their lives, or have been attracted to urban centers to look for jobs, but drifted to urban slums for shelter; and the second is the rural poor. The first category of urban slum dwellers have lost their ancestral roots. They neither know their ancestral land nor own one. They have cut social and cultural ties with their tribe, or ethnic group. The 'new' urban poor have ancestral land and strong social and cultural ties with their ethnic group. They may have children and a wife living at ancestral home. All urban poor are willing to work, except that they cannot find a decent well-paying job. They can move out of their poverty situation, provided jobs and decent housing can be provided. A second common factor is that the majority of urban poor are uneducated or have no professional skills.

Provision of education and labor skills is a ladder that enables any of them to climb out of poverty.

A second poverty group forms the majority of persons living in the rural areas. They have land, housing, though of some temporary structure. Some farm, fish or engage in some economic activities. There may be a natural resource of economic potential in their areas but they lack the knowledge for its exploitation. As a general rule, they have no roads, running water, electricity, neighborhood hospitals and other essential amenities. The majority of the rural poor have a low education level.

Poverty in most parts of Africa has not become systematic. Provision of hope, restoration of dignity and teaching of the means to fight poverty can allow reversal of the situation. Therefore, extension of education, teaching of labor skills, provision of roads, electricity and other means of communication, such as telephone, would greatly transform these rural communities. Economic and/or cultural exploitation of the communities is unethical. Fair compensation for labor and products should be observed if the fighting spirit, ability and willingness to work are to be recognized. Science that helps these communities exploit their natural resources need to form a basis of education curricula.

5. Improved Human Health

Many of the diseases that were considered conquered have re-emerged in the continent. It is true that some diseases like smallpox have been eliminated and soon, polio may also be eliminated. However, water-borne and other enteric diseases have become a nuisance. Malaria continues to be one of the major killers. To this list has been added HIV/AIDS. Progress is being made at eliminating river blindness and leprosy. However, re-emergence of yellow fever all point to the tortuous battle that is yet won. The spread of some of these diseases may be worsened by climate change. Diseases like malaria, yellow fever and dengue, which are spread by insects, may penetrate to higher plateaus that were hitherto inhabitable by insect vectors due to warmer temperatures.

Most of the pandemic diseases could be prevented by observance of public health principles. Others await advancement through research for the discovery of cures. A paradox that faces the sick in the continent is that medical help and hospitalization has become unaffordable for the majority. Since most of them are poor, they cannot afford drugs or fees for a doctor. Restructuring of government services through economic transformation imposed by the World Bank and International Monetary Fund (IMF) has worsened the medical care of most patients. These institutions have imposed payment of fees for health services, eliminated subsidies by government to patients including the disadvantaged social groups in the society. The major international pharmaceutical companies are unwilling to invest in diseases of the poor for fear of economic losses. Above all, enforcement of public health regulations together with public education would constitute corrective initiatives that need to be restored. Education that promotes the public understanding of cause of diseases and their prevention include emphasis of basic sciences. The present science curricular teaches one to be competitive in the world economy. This in itself maybe advantageous for an African who may find work outside his/her immediate environment. The taught African science curricula are structured along the European or North American universities system. Such structures make science teaching very expensive as most inputs are imported. Often resources and infrastructure to maintain academic standards and quality have been lacking. Elite class education has evolved, which excludes majority children of peasant farmers from education at the primary, secondary or tertiary levels. The examination system models those found in the west. Adopting and adapting local resources to both the examination system and curricula has not taken place.

Therefore, there is need to change the present science curricula and introduce use of locally found educational materials. There is need to emphasize at local community level, preventive measures for disease control. Most importantly, there is a need to look for traditional cures for some or all of the diseases. The possibility exists that both western medicines could work hand in hand or as alternative to traditional medicine. The Africans need to restore pride and respectability to their herbal medicines. A science curricula that teaches the rural communities how to grow more food, prevent and cure common diseases, use the natural resources as well as conserve their environment, industrialize and trade in goods, shall certainly improve the human health.

It is necessary that the government assist such communities by building hospitals, roads, providing potable water and other essential infrastructures. Until these communities are able to have skilled personnel, it is essential that the government pays for teachers, doctors, administrators who would work with them, educate them and improve their lives.

6. Promotion of Sustainable Agriculture and Rural Development

There still exists in Africa communities that rely on gathering and herding as means of survival. Alongside these communities exist others that practice modern agriculture through cultivation of cash crops like tea, coffee, tobacco, sisal, maize, pyrethrum, flowers and other horticultural crops. A third group of community practice subsistence agriculture, producing only what is needed for survival. The first and last category of communities needs relevant and pertinent science to teach them how to be self sufficient in food. It is in these communities that re-orientation of science curricula is needed most. A science curriculum that has community development objective would not only teach students how to grow food crops but would also teach agricultural practices that enhance environmental protection.

In promoting agriculture, there is need to de-emphasize the widespread use of synthetic fertilizers and pesticides. The reasons for this are: first, most people cannot afford their cost. Second, environmental damage associated with fertilizers has resulted in proliferation of invasive plants in receiving waters. Added nitrogen fertilizers have been proven to be sources of greenhouse gases like N_2O, NO and NO_2. Some of these gases promote depletion of ozone in the stratosphere and formation of ozone in troposphere. Third, pesticides are not only expensive, but persistent organic pesticides have been proven to adversely affect both human health and the environment. The non-persistent pesticides cause more deaths to uneducated, unsophisticated users, than persistent pesticides.

	Total cases	F(%)
Drugs		
Chloroquine	196	72.4
Alcohol	25	24.0
Others	293	73.6
Household chemicals		
Kerosene	254	4.5
Carbon monoxide	25	46.7
Others	293	52.7
Pesticides		
Organophosphates	514	39.1
Organochlorine	3	33.3
Carbarmates	4	25.0
Rat poison	84	64.3
Fungicides	11	45.4
Herbicides	47	48.9
Others		

Food poisoning	89	45.6
Plants	48	50.0
Venoms	208	47.6
Unclassified	40	45.0

Distribution of poisoning as percentage of age groups

	Age unknown	0–5	6–14	15–20	Over 25
Percentage	7.2	29.9	10.2	19.1	33.6

Table 2. Cases of human poisoning (1888) in 19 Kenyan hospitals between 1991 and 1993 by various poison agents and percent gender distribution (Source: Maitai et.al. 1998. A retrospective study of childhood poisoning in Kenya)

The second community group needs science for use and management of chemical fertilizers and pesticides in order to increase crop yields. However, in some communities, extensification of cropland by conversion of wild lands cannot be avoided. Sustainable land use and food sufficiency can only be achieved through science, whatever agricultural practice is adopted. Elimination of the 'bads' of agriculture, such as soil erosion, land degradation, fertilizer run-offs, water pollution, pesticide contamination of foods, land and water systems, and production of greenhouse gases can only be achieved through scientific knowledge and application of appropriate eco-technologies.

While it is not possible to completely eliminate the use of fertilizers and pesticides in agriculture, it is argued that the use of alternative organic manure; the application of appropriate amount of synthetic fertilizers to be absorbed by plants; the application of integrated pest management schemes as well as limited use of pesticides, offer advantages to present farming methods. These are farming techniques that can only be acquired through scientific research and education. More needs to be done in these areas to improve agriculture in the continent.

7. Promoting Human Sustainable Development

Modern economy not only depends on mass production of goods and services but also on increased consumption. Africa's consumption pattern is far below the developed and newly industrialized countries. However, with modernization of Africa's economy, the consumption pattern is changing in many countries. As these societies move from subsistence economy, there is urgent need to educate the people on how to reduce wasteful use of natural resources. Application of cleaner technologies, recycling of wastes, conservation of water, energy and environment all extend the life span of available resources. A firmer application of these technologies can only arise from improvements made in industrial production through science and technology. Using only what is needed and necessary has become the goal of future society. Therefore, improvement of processing technologies, installation of eco-friendly industries require that African societies train scientist that can adapt and adopt some of the developed countries technologies UN agencies and assistance by developed countries, NGOs in building capacities of Africans in these areas should be one of the priority programs.

Assistance by developed countries, UN agencies, and NGOs in building capacities of Africans in these areas should be one of the priority programs.

8. Atmosphere is Becoming Polluted

The clean air of Africa is fast deteriorating. Many African nations have polluted indoor and outdoor air. The major pollutants are sulfur dioxide, ozone, lead, manganese, cadmium, volatile organic compounds, oxides of nitrogen and carbon, particulates and mercury. These are products of impartial combustion of fossil fuel burning, industrial manufacturing, photochemical reactions, evaporation of solvents, cooking and break-down of fungicides, paints and spills of mercury- containing products as well as gold mining activities. Domestic biomass burning for cooking and heating may be the single most important source of health threatening air pollution. At present about 70% of energy source comes from biomass in the continent. Health risks associated with this source include chronic obstructive lung diseases, lung cancers and adverse pregnancy outcomes in women, and respiratory problems in children. Recent studies indicate the level of greenhouse gases to be increasing as a result of biomass burning. These emissions have impacts on both regional and global photoxidation processes and tropospheric ozone. Table 3 gives our calculated contribution by bio-fuel burning in Kenya regional and global emission budgets of trace gases.

Although the amounts of trace gases contributed by Kenya and Africa are small compared to global contributions, their significance should not be overlooked, given the fact that deforestation is taking place at a fast rate in the continent.

	Kenya	Africa		Global	
	Biofuel use	Biomass burning[a]	All sources[b]	Biomass burning[a]	All sources[b]
C (CO_2)	9.70	337	565.9	1901	8351
C (CO)	0.92	96	112	287	482
C (SCH_4)	0.10	6.2	18.1	24	380
C (NMHC)[c]	0.06	13.1	123	52	534
N ($N2_X$)	0.02	3.1	7.1	8	40
N (N_2O)	0.001	–	–	0.1–1[c]	13–16[c]
N (NH_3)	0.002	–	–	0.48[d]	4.4[d]

Unless otherwise stated, all other values in columns under Africa and global are derived from Marufu L. (1999). *Photochemistry of the African troposphere: The influence of biomass burning.*

[a] Includes: biofuel, savanna, agricultural waste and deforestation fires.
[b] Includes: biomass buring, industrial, biogenic, ocean and lightning.
[c] Bouwman A. F., Hoek K. W. V. D., and Oliver J. G. J. (1995*). Uncertainties in the global source distribution of nitrous oxide.*
[d] Andreae M. O. (1993). *The influence of tropical biomass burning on climate and the atmospheric environment.*

Table 3. Contribution by Bio-fuel Burning in Kenya to Regional and Global Emission Budgets of Trace Gases (all values are in Tg C or N per year)

Additional research is needed to definitely quantify these problems. Improvement in the energy ladder scale from biomass to charcoal, to liquefied propane/methane gas, to electricity, reduce the health risks. Improvement in the energy ladder occurs through socioeconomic development, propelled by use of scientific knowledge. Use of new energy sources such as renewable organic oils, gas turbines and gasification of coal will improve the environment, but technology must be adopted through the use of scientific knowledge.

9. Protecting Water Resources

Africa with the Indian Ocean in the East, the Atlantic Ocean in the West and Mediterranean Sea in the North, has one of the longest coastal lines. Several major rivers like the Niger, Volta, Congo, Zambezi, (Limpopo), and Nile drain the hinter-lands into one of the surrounding seas. In addition, there are several minor rivers and major fresh-water and brine water inland lakes, all creating a criss-cross of water system that sustains life in the continent. The number and water quantities of ground water aquifers have not been accurately established. It is possible that this source of water when quantified may add, as has been experience by Libya and Botswana, a major source of water for agriculture and life support system.

An increasing demand for water by an increasing population has created scarcity for water. Already 14 African countries experience water stress, and 22 will do so by 2050—a situation when 20 to 40% of available fresh water is already being used. Future agriculture will depend more on irrigation than on rain. Taking the world average of 70% of water withdrawn from lakes, rivers and groundwater aquifers used for irrigation on agriculture, it becomes evident that Africa will be experiencing acute water shortages. Scarcity of water may create conflicts between neighbors and nations. Such conflict hinders sustainable development.

Traditionally, Africans have managed water with great respect. Water was known as a 'no respector of an individual' in some societies. It gave equal treatment to all. As such it was not known as a source of disease in the years past. Today, drinking water has become a major source of disease in the continent. The majority of people use untreated water. The numbers affected may be as high as one half of the population. Water-borne diseases have re-emerged with a vengeance, mainly due to water pollution by municipalities, industry and people. Providing safe drinking water to the population remains a pivotal priority. Treated water provision is mainly found in urban centers. Most of Africa's rural areas use untreated or bore-hole water. Water treatment is mainly and solely done by municipal water departments or the ministry of water. The dwindling financial resource base on most councils or governments place into question the sustainability of provision of quality water. Water research has been mainly conducted at higher institutions and often relates to hydrological and limnological studies. Today there are no research instructions in Africa, south of the Sahara, that handles basic research, water management and treatment studies. Yet, water availability, quality and quantity, vulnerability of natural and socioeconomic systems to global change, changes in land use and water use are issues that beg scientific solutions in the continent. There is therefore a need for capacity building and integrated studies about water sources, management, climate and land use patterns in order to evolve not only sustainable natural, but also social systems. Public education on sustainable water-use should be immediately started in every African country. Unless this is done, both economic productivity and social development in the region will be hampered. Sustainable use of ground water, surface water and rivers need study and propagation. Human activities that may lead to water reduction, land degradation and desertification should be discouraged.

Future water use strategies should take into account alterations caused by climate change. It is predicted that climate change may reduce the over-all amount and frequency of precipitation and increase its variability in many countries of Africa. Learning to live with less in the future is a sure way of adaptation. Pollution of water resources (both marine and fresh water) comes mainly from high eroded soils, run-offs of fertilizers and pesticides used in agriculture, and indiscriminate disposal of chemicals. Agriculture followed by municipalities is therefore the most polluting source of water resources. Application of agricultural practices such as terracing, minimum application of fertilizers and pesticides, integrated pest control and crop rotation improve water quality. Legal enforcement of necessary protective laws is required to reduce the risks associated with water pollution from municipalities and other sources.

10. Desertification and Land Degradation

The continent has one of the most sensitive ecologies whose land productivity depends mainly on rainfall. Loss of forests leads to decrease in conventional rains, since mostly evaporation over forest cover results in about 75% of tropical rains.

Conversion of forests and grasslands to farm lands also leads to high rain water run-offs and soil erosion. Decreased soil fertility and high loss of biodiversity result from intensive use of land, over-grazing and clearing of land cover.

Desertification and land degradation are results of poor land management. Their effects are heightened by climatic variations and over population. Both forces are at play in the continent. Control of desertification and land degradation improve chances for achieving sustainable development through poverty reduction. Research is needed on the traditional forest and land use management. Deforestation can only be slowed through public education based on scientific data that relate effects to results. Application of results of scientific research to land management halts desertification and land degradation.

11. Energy for Sustainable Development

Major energy sources are biomass, charcoal and coal, hydro, geo and thermal energy. Though solar and wind energy have potential, their uses are undeveloped. Reliance on biomass energy has serious consequences for the environment. Biomass burning is the major cause for deforestation, emission of greenhouse gases and consequently, acceleration of climate change. In addition, combustion of fossil fuels in thermal generator, coal and charcoal burning add considerable amounts of carbon oxides and volatile hydrocarbons to the atmosphere. Energy technologies that use renewable sources of energy are or have been developed in institutions outside Africa. Research that aims at adopting, copying these twenty-first century technologies need to be started. Supply and management of solar panels for rural homes' lighting has great promise for Africa. Therefore there is a need for study and proper use of energy saving devices, combination of energy sources, improved housing architecture that utilizes less energy to heat or cool, and enhanced use of renewable energy materials. Such strategies would contribute considerably to energy conservation.

12. Toxic Chemicals and Hazardous Wastes Management

The continent is an importer of toxic chemicals and a generator of hazardous wastes from use, storage and disposal of the same. The bulk of imported toxic chemicals include pesticides, especially persistent organic pesticides, as well as biodegradable pesticides, herbicides, acaricides, oils (refined fossil fuel products), bulk manufacturing chemicals, and pharmaceuticals. Classification of toxic chemicals and hazardous wastes and their registry has been done in a handful of African countries. In general, Africa is a free Bazaar for Chemicals manufacturers.

Apart from countries like Uganda and Republic of South Africa, there are very few countries in black Africa with prescribed regulations for use, storage and disposal of toxic chemicals and hazardous wastes. Installed high temperature incinerators for disposal of toxic chemicals, are non-existent in most countries. The mode of disposal of these compounds is either through landfill, dumping into the sewer or ocean, or open air combustion. Such disposal processes expose both human and environment to risks that could be avoided. Open sea incineration is only available to countries with financial resources to hire such services. The science of toxic chemicals and hazardous wastes management from cradle to grave awaits exploitation and development in the continent. There is a need for the study of toxic chemicals and their effect on human and environment, disposal and detoxification mechanisms, management, use and storage of agricultural chemicals as well as appropriate cleaner chemicals technologies.

13. Solid and Liquid Wastes Management

Collection and disposal of solid wastes has become a major problem for most African cities and urban centers. Per capita generation of solid wastes in urban centers is around 1 kg/d. Over 60% of this waste is wet and about 90% of solid wastes is combustible. Open surface dumping is the prevailing mode of disposal. Very few cities have scientifically managed landfill sites. Often open burning is the mode of incineration as cost for modern incinerators are far beyond the reach of most cities. Artisan collection and selection of recyclable wastes has become a means of livelihood for many people. Recycling is done for newspapers, bottles, aluminum, and other metals. Composting of wastes is also practiced by some women's groups, though sales of organic manure is a problem due to low consumer education.

Figure 2. Open Waste Dump Burning

There are very few operational functional sewage plants in most cities. Raw sewage wastes are often dumped into rivers, lakes or the ocean. Urban centers with well-maintained open-lagoon sewage treatment plants have better treatment of their wastes than those with modern, aerated and mechanically-driven treatment plants. The cost of spare parts and replacement parts of sewage plants tend to be high for most urban centers, since operation and maintenance costs have not been harmonized with charges for sewage treatment. As a result of financial insufficiency, most installed capacities have deteriorated or become non-functional. The technology for solid/liquid wastes treatment is available and most are in the public domain. Innovation and adaptation of these technologies remain potential for African cities. However, there are needs for study of solid waste characteristics and scientific methods for disposal. The bacterial load of water systems is very high and there is need for clean-up. Treatment of municipal and industrial liquid waste using affordable technologies needs investigation. Public education on waste disposal is required if sustainable water use is to be assured.

14. Management of Radioactive Wastes

The establishment of the Basel Convention on the Control of Trans-boundary Movement of Hazardous Wastes and their Disposal (Basel, 1989), was a major catalyst for the establishment of the Bamako Convention on the ban on the import into Africa and the control of trans-boundary movement and management of hazardous wastes in Africa, by the African nations.

Fear of import of radioactive wastes by some unscrupulous leaders for disposal in African soil catapulted the Organization of African Unity to ban all such imports through the Bamako Convention. However, it is unfortunate that since its establishment, less than 10 countries have ratified it. It is only through moral force and public condemnation that Africa has not yet become a dumping ground for radioactive wastes. It is also fortunate that the amount of radioactive materials used in the continent is low and consists mainly of research and medical chemicals. To date, no African country has installed a fully functional radioactive wastes processing plant. As the need for energy increases and the use of nuclear fuels for power generation becomes more of a possibility, it is essential that manpower training for radioactive wastes management is intensified. Second, disposal of radioactive and mining wastes should be studied with a special consideration for their disposal. Data should be collected regarding sites with radioactive substances such that their soil is not used for road construction.

15. Biotechnology and the Future

Advances in genetic engineering have led to mass production of genetically modified plants. The majority of robust plants are food crops like wheat, maize, potatoes and other cash crops. Through manipulation of genes of these crops, a new crop which is resistant to a particular disease or which has greater seed yield is produced. Advances have also been achieved in modifying plant genes to produce vaccines, or medicines against some common diseases. A few countries in Africa, such as Republic of South Africa, Zimbabwe, Nigeria, Cote D'Ivoire, Cameroon, Kenya and Ethiopia have capabilities of applying biotechnology for crop breeding. In fact, some of these countries are presently experimenting with field plantation of genetically modified food crops. Government approval of mass production of such crops has not been given. However, genetic improvement of certain food crops through the study of their species and selection of the most productive and disease resistant species have already resulted in improved coffee, tea, maize, rice and banana species. Farmers already benefit from the high yields of such genetically modified crops. On the other hand, some farmers in some countries have raised doubts about yield repeatability with time. Similar advances have also been achieved in animal breeding through improvement of animal species. High meat or dairy yielding cattle, sheep and goats have been reared. Improvement of poultry species have benefited from genetic engineering. Gene selection and genetic modification of crops and animal husbandry have great potential for Africa. Application of these technologies vary from country to country depending on the availability of trained persons and government policies. There is need for capacity building in these areas, proper public policy formulation and consumer education if Africa is to benefit from genetic engineering.

Cloning of species has now become a regular feat of most advanced laboratories. It was first achieved in the United Kingdom, with the sheep 'Dolly'. Cloning of human species, though theoretically possible, has not been attempted. However, *in vitro* fertilization of human eggs is possible and has been regularly performed. It is also possible to produce human parts like the heart and limbs in other animal species, like the pig.

Moral and ethical considerations have prevented mass application of gains achieved in science through genetic engineering. In the area of plants, two moral and ethical issues arise. The first concerns biological selection. The plant species that are currently being used have gone through natural selection process of many centuries. The source of carbohydrates, proteins and fruits in human diet are mainly plants. These plant nutrition ingredients help in propagation and regeneration of human tissues and cells. It is possible that through genetic manipulation, new variant genes may be introduced into humans through such food crops. The period for observation and experimentation with genetically modified foods has been considerably shorter than the biological selection period. It may be that these fears will be dispelled with time. However, at present a precautionary principle is necessary.

Advances in biotechnology have been achieved at high research cost. Most of genetic engineering research has been conducted by the private sector, who patented their results. In addition to patenting research results, some corporations have hinted at including terminator genes ensuring that replication of seeds is not possible. If genetically modified foods are widely grown, it is possible that they would cross pollinate similar plant species. In the case of plants with or without terminator genes, such cross pollination would wipe out all known similar species of foods crops. Therefore, questions are being asked as to whether it is socially and ethically correct to subject rural farmers to total dependence on multinational companies for seeds each season. Much more important, is it morally, ethically and scientifically necessary to reduce or eliminate all the genetic pool of such food crops that have gone through centuries of selection?

Human cloning is morally and ethically unacceptable to most cultures. The moral and cultural practices of most Africans do not allow borrowing of human parts. Much more importantly, is the possibility that in wrong hands, human cloning could be used for race cleansing. Growing of human parts in animals like pigs may also introduce unknown diseases through gene transfer. Many citizens of Africa are non-Christians, Christians and Muslims. Their spirituality conflict with reproduction of humans in the laboratory.

The above examples illustrate the complex moral, ethical, spiritual and scientific issues that advances in biotechnology have opened. Open and widespread discussion of these issues has not taken place in the continent. However, the momentum of progress being made in these areas by the developed countries is not waiting for Africans to be involved. It is necessary that these issues be discussed by scientists and policy makers. That there be a moral consensus in each country on how best to apply biotechnology. Training of a critical mass of scientists lags behind in most states. Establishment of centers of excellence for biotechnology, research and development has also not been achieved in the continent. International cooperation in capacity building is essential if Africa is to be given opportunity to benefit humanity from its vast resources of biodiversity of species.

In the meantime, time is ripe for the United Nations special agencies like UNESCO to take up the issues of establishing instruments that would safeguard food crops germ plasm and experimentation with human cloning. The UNESCO's Universal Declaration on Human Genome and Human Rights is one such positive instrument that has been established. Other similar instruments should be drawn to cover areas and aspects not covered by Human Genome Convention.

16. Mobilizing Resources for Africa's Development

Sustainable development for Africa cannot be achieved while 11 million of its women citizens remain illiterate, two million children miss places for primary education each year, less than 30% of university students are women and less than 15% of university students read science courses.

Empowerment of women and youth through appropriate education form the first task for achieving the goals for sustainable development. A science curriculum that has no relevance to society disinterests the learners. As the needs of science vary from community to community, there is a need to decentralize the science syllabus in order to solve such needs. A syllabus that relates to poverty alleviation, job creation, crop production, and animal husbandry has a chance of making a society overcome its shortcomings.

Provision of essential infrastructures like roads, potable water, electricity, telephone, medical centers, and schools should become the priority preoccupation of central and local governments. Availability of such infrastructures has the ability to multiply several folds the work output of peoples. Furthermore, utilization of local resources as inputs for commercial goods are improved enormously when electricity and roads for transport and communication are available. The ability to train local communities on skills that improve the available resources can only succeed when the labor products can be transported to markets for sale. Transformation of local resources into marketable goods motivates communities and allows application of science in a much more transparent way.

Concurrent with the training of the rural communities should be creation of awareness about governance, environment and economy. It is important that the political, moral, social and ethical sophistication of the rural communities be enhanced to allow practice of democracy to prevail. Utilization of culture to improve the common good of society and to enhance its development should be the focus of development. Sustainable use of local resources can only be achieved by a society that is aware of environmental issues. Continuing education on environmental conservation is very important if we are to develop.

We live in a world which is becoming more and more dependant on trade. Training of local communities to climb the economic ladder from barter trade to free trade is a necessary component of rural development. Viable and stable markets for community produced goods should be identified. Methods of satisfying the market through improved products and market communication should be studied and imparted to the community.

The role of women groups, youth, and retired civil servants, organized through cooperative groups or into non-governmental organizations, is very important in achieving the objectives of rural communities' development. However, many such community development groups fail because of poor leadership and lack of management skills. Training of community leaders is essential before any action is taken for mobilization. Any rural development project should start with the goal of meeting demands for foods. Once this is satisfied, other activities will follow naturally. For example, once demand for food has been satisfied, the community could aim at meeting energy demands, housing demands, health demands, fiber demands and many more, depending on the locally available resources.

A second role of local and central government in this process of development would be to make policies that would enable scientific understanding of the nature links between rural development and human need demands: Identification of innovative technologies that would improve product quality without raising the cost; formulation of policies that promote both political will and public commitment to the objectives of rural development; improved coordination between national and international organizations that are willing to assist the community. The public commitment shall enable mobilization of both individuals and the private sector into becoming more socially responsible.

The history of official development assistance to developing countries has created a culture of dependency on aid. The social, cultural and moral value for hard work and savings have also been eroded. There is a need to review the whole concept of development assistance. It is fair to say that development assistance should first of all be directed to rural communities. Assistance should not be given as a hand-out, but should enable a community to do what it could not do by itself. A second role for development assistance should be training, continuing education and capacity building. Assistance to government for mega projects should only be considered when clear objectives for the project have community support and there are safe guards for close monitoring. Assistance for mega projects that serve the political elites as show-off case projects have been used to siphon project funds to private accounts. It is for this reason that involvement of stakeholders in any major projects should be promoted.

The success of any rural development is measured by reduction of tensions. Quite often, such tensions are created by scarcity and conflict of human needs and demands. Satisfaction of human needs through creation of employment, self-sufficiency in food, conservation of environment, self-fulfillment through education, leisure, recreation are all ingredients of a peaceful secure society. Denial of progress in development is a good breeding ground for wars. Our future is inextricably linked to one another and to resources available. Cooperation at all levels in these areas assures a better future for all of our children.

Glossary

GDP:	Gross domestic product.
HIV/AIDS:	Human immune deficiency virus/acquired immune deficiency syndrome.

NGO: Non-governmental organization.
OECD: Organization of Economic Cooperation and Development.
UNHCR: United Nations high commission for refugees.
WSSD: World Summit for Social Development and Beyond.

Bibliography

Anddreae M. O. (1993). The influence of tropical biomass burning on climate and the atmospheric environment. (ed. R. D. Oremland). *Biogeochemistry of Global Change: Radiatively Active Trace Gases.* New York: Chapman and Hall, pp. 113–150. [This paper highlights the significance of biomass burning in the atmosphere.]

Bamako Convention (1991) on the Import into Africa and the Control of Trans-boundary Movement and Management of Hazardous Wastes in Africa. [This is a convention for the control of hazardous wastes into Africa.]

Basel Convention (1998) on the Control of Transboundary Movement of Hazardous Wastes and Their Disposal. [This is a manual for the implementation of the Basel Convention Series ISBC No. 94/004, UNEP, Geneva, 1994.]

Bouwman A. F., Hoek K. W. V. D., and Oliver J. G. J. (1995). Uncertainties in one global source distribution of nitrous oxide. *Journal of Geography Research* **100,** 2785–2800. [The paper reviews uncertainties in nitrous oxide estimates from the literature sources.]

Friedman J. (1996). Rethinking poverty. *International Social Science Journal* **148**, 161–172. Blackwell Publishers/UNESCO. [This paper reviews various concepts of poverty.]

Gash J., Roseburg H.C., Odada M., Oyabande E. Lekan O. and Schulz, R. E. (eds.) (2001). *Freshwater Resources Research in Africa*, 146 pp. Workshop report, 26–30 October 1999, Nairobi, Kenya. Potsdam: BAHC International Project Office. [The workshop papers include water and human health, hydrology , integrated land and water management, capacity building and issues.]

Kituyi E. S., Wandiga O., and Helas G. (2000). Biomass burning in Africa: An assessment of its role in atmosphere change. In *Climate and Change: Science and Policy for Africa* (ed. Pak Sum Low) (in press). [The paper reviews Africa's contribution to the greenhouse gases debate.]

Maitai C. K., Kibwage I. O., Guantai A. N., Ombega J. N. and Ndemo F.N. (1988). A retrospective study of childhood poisoning in Kenya. *East and Central African Journal of Pharma. Science* **1,** 7–10. [This is a study of chemical poisoning in Kenya using hospital reports.]

National Research Council (1999). *Our Common Journey: A Transition Toward Sustainability*, 363 pp. Washington, DC: National Academy Press [This is a book that reviews world trends and forces for the twenty-first century. It gives scenarios for 2025, 2050 and 2100.]

Population of concern to UNHCR: A statistical overview (1995-1999). [This is statistical information at UNHCR. http://www.unhcr.ch/refworld/refbib/refstat/1996/table04 html and http://www.unhcr.ch/statist/main.htm]

Royal Society of London, U.S. National Academy of Sciences, the Brazilian Academy of Sciences, the Chinese Academy of Sciences, the Indian National Science Academy, the Mexican Academy of Sciences and the Third World Academy of Sciences (2000). *Transgenic Plants and World Agriculture*, 40 pp. [This is a monograph reviewing the science of genetically modified foods.]

Salati E. and Vose P. B. (1985). The water cycle in tropical forests with special reference to the Amazon. (ed. G. B. Marini-Bettolo) *Proceedings of the Pontifical Academy of Sciences*, Vol. 56, pp. 623–648. Pontifical Academy of Sciences Press. [This paper reviews the water cycle in tropical forests.]

Scholes, M. and Andreae, M. O. (2000). Biogenic and pyrogenic emissions from Africa and their impact on the global atmosphere. *Ambio*, **29**, 23–29. [The paper reviews sources of greenhouse gases in Africa.]

Serageldin I. and Collins W., eds. (1997). *Biotechnology and Biosafety*, 214 pp. Washington, DC: The World Bank. [The workshop report reviews the scientific biasis of genetic engineering, application, uses, and biosafety issues and policy framework for research, application and biosafety.]

United Nations Educational Scientific and Cultural Organization (UNESCO) (1998). *Statistical Yearbook*, 687 pp. UNESCO Publishing and Bernan Press. [This is a yearly statistics report on education, science and technology, culture and communication.]

United Nations Environment Programme (UNEP) (1999). *Global Environment Outlook 2000*, 398 pp. Earthscan Publications Ltd. [This is a book that gives the status of the global environment.]

United Nations, *World Summit on Social Development and Beyond* (1995 and 2000), 132 pp. [These are conference reports that review consensus on world community action on poverty reduction.]

World Bank (1992). *Welfare Monitoring Survey I Kenya, 1992*. Washington, DC: The World Bank. http://www4.worldbank.org/afr/poverty/databank/surnav/database/ShowSurvey.cfm?CFGRIDKEY=18 [This provides poverty measurement by conducting welfare assessment for Kenyan population.]

World Resources Institute, United Nations Environment Programme, United Nations Development Programme (1995). World Resources 1994–1995, 400 pp. Oxford, New York. [This is a biannual report that gives statistics on people and the environment, resource, consumption, population growth and women.]

Biographical Sketches

Shem Oyoo Wandiga has a Ph.D. in Chemistry from the Case Western Reserve University, US. A professor of Chemistry in the University of Nairobi, Kenya. Professor Wandiga was also Deputy Vice Chancellor (Administration and Finance) of the University. He was subsequently appointed as the Coordinator of the Policy and Planning Task Group of the Ministry of Education (1991), elected as Kenya's Representative to the Executive Board of UNESCO (1995), President of the Program and External Relations Commission of Executive Board (1997), and as a member of the General Committee and Advisory Committee on Environment of the International Council of Science. Currently Chairman of the Kenya National Academy of Sciences, Professor Wandiga is the author of a large number of publications in scientific and educational fields. He has also chaired several national committees on university education and been a consultant on World Bank, UNESCO and United Nations Environment Programme projects undertaken in Kenya and East Africa.

Eric Onyango Odada is an aquatic geochemist focusing on the environmental quality assessment of African lakes and rivers and the study of human impact on these natural systems. He is active in science-driven policy initiatives; currently, leading a consortium of African, European and North American scientists interested in promoting the investigation of African Great Lakes as archives of records of environmental and climatic dynamics in East Africa. He is a Consultant to the Western Indian Ocean States on design and implementation of Integrated Coastal Zone Management; Program Leader of the Global Change regional research and capacity building initiatives in Africa; Coordinator and Chief Scientist in the UN-IAEA Isotope Techniques in Lake Dynamics Investigations in Africa and is Global Environmental Facility (GEF) Strategic Adviser on International Waters. He has over 60 peer-reviewed publications.

INDEX

A

B

C

P

Peer Review, 4, 35, 42, 44-47, 53, 56, 62-63, 116
Performance Indicators, 9
Policies Recommended By Outside Bodies (Especially the World Bank), 72
Policy
    By (Means of) Science, 26
    Capacity, 106, 129
    Cultures, 82-83, 85, 88
    For Science, 26, 86-90, 238
Policy-Making Processes and Evaluation Tools: S&T Indicators, 1-21
Political-Bureaucratic Culture, 84-85, 93
Population and Science, 286, 289
Possible Extreme Options for the Use of Indicators in Decision-Making, 16
Post-War Catch Up and S&T Policy, 167, 169
Private Sector Share of GDP, 188
Procedures for the Evaluation of Socio-Economic Programs, 28
Production and Use of S&T Indicators in Practice – The Question of the Data Sources, 1, 11
Promoting Human Sustainable Development, 286, 295
Promotion Of
    Common Systems of Scientific Policy and Technical Issues, 162
    Sustainable Agriculture and Rural Development, 286, 294
Protecting Water Resources, 286, 297
Public Policy, 1-2, 11, 19, 36, 40, 103-104, 129, 153, 167, 169, 176, 183, 301

R

R&D
    Expenditures, 8, 102, 170-172, 191, 195, 199, 202
    Funding, 102, 116, 133, 171
    Programmes, 22, 94
    Projects, 133, 143, 152, 196
Reaching Beyond the North American Sphere, 153, 163
Regional
    Institutions, 16, 65, 78, 278
    Ministerial Conferences, 207, 218, 221, 224, 233
    Networks, 207, 223, 226, 278
    Research Space, 153, 158
Relationship between Science Policies, Promotion and Management of R&D Activities, 22, 26
Relative Impact Indicators, 42, 59
Republic Of South Africa, 65, 299, 301
Research
    And Development (R&D) Programs, 29
    Diaspora, 153
    Evaluation, 22, 25, 38-40, 43, 45-47, 53, 57, 62-64, 180
    Groups, 42, 44-45, 48, 53, 55-56, 59, 63, 206, 210, 216, 274
    Institutions, 5, 11, 16, 32, 78, 97-98, 111, 135, 141, 147, 162, 168, 173, 186, 200, 261
    Performance, 23, 42, 45, 48, 50, 53, 63, 207, 209, 222
Role Of
    Bibliometrics in Institutional Evaluation, 47, 53
    Science in a Scenario of Transition to Sustainability in Developing Countries, 249, 262

S

S&T
    Indicators
        Definition, Terms of Reference and Categories, 1, 5
        Production Activity, 11

A source of knowledge for sustainable development and global security to lead to fulfillment of human needs through simultaneous socio-economic and technological progress and conservation of the Earth's natural systems

It is a virtual dynamic library with contributions from thousands of scholars from over 100 countries and edited by well over 300 subject experts, for a wide audience: pre-university/university students, educators, professional practitioners, informed specialists, researchers, policy analysts, managers, and decision makers.

This archive is now available at **http://www.eolss.net** as an integrated compendium of twenty component encyclopedias

1. Earth and Atmospheric Sciences
2. Mathematical Sciences
3. Biological, Physiological and Health Sciences
4. Biotechnology
5. Land Use, Land Cover and Soil Sciences
6. Tropical Biology and Conservation
7. Social Sciences and Humanities
8. Physical Sciences, Engineering and Technology Resources
9. Control Systems, Robotics and Automation
10. Chemical Sciences Engineering and Technology Resources
11. Water Sciences, Engineering and Technology Resources
12. Energy Sciences, Engineering and Technology Resources
13. Environmental and Ecological Sciences, Engineering and Technology Resources
14. Food and Agricultural Sciences, Engineering and Technology Resources
15. Human Resources Policy, Development and Management
16. Natural Resources Policy and Management
17. Development and Economic Sciences
18. Institutional and Infrastructural Resources
19. Technology, Information and System Management Resources
20. Area Studies (Africa, Brazil, Canada and USA, China, Europe, Japan, Russia)

Within these twenty on-line encyclopedias, there are about 235 Themes, each of which has been compiled under the editorial supervision of a recognized world expert or a team of experts such as an International Commission specially appointed for the purpose. Each of these 'Honorary Theme Editors' was responsible for selection and appointment of authors to produce the material specified by EOLSS. On average each Theme contains about thirty chapters. It deals in detail with *interdisciplinary* subjects, but it is also *disciplinary*, as each major core subject is covered in great depth by world experts.

EOLSS Presentations are in gradually increasing depth.

- **Level 1 (Theme Level)**: Presentations of broad perspectives of major subjects

- **Level 2 (Topic Level)**: Presentations of perspectives of the special topics within the subjects
- **Level 3 (Article Level)**: In-depth presentations of the subjects in various aspects

Some themes have a Three level structure and some Two level structure. The EOLSS web of knowledge is woven over a hierarchical structure through cross reference links

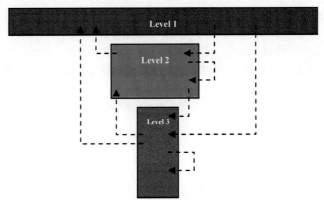

The three level structure

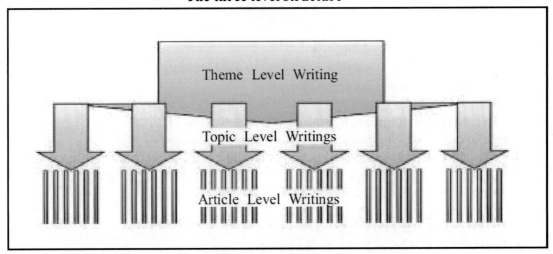

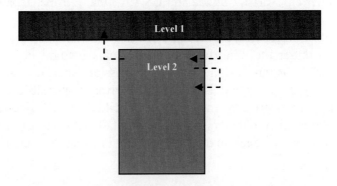

The two level structure

EOLSS is the result of an unprecedented global effort involving the collective expertise of nearly 400 Subject Editors and 10000Authors from over 100 countries.

The EOLSS project is coordinated by the UNESCO-EOLSS Joint Committee (http://www.eolss.net/eolss_international.aspx) and sponsored by Eolss Publishers, which is based in Oxford, United Kingdom. Through many and diverse consultation meetings around the world, the EOLSS has benefited immensely from the academic, intellectual, and scholarly advice of each and every member of a large International Editorial Council (http://www.eolss.net/eolss_international.aspx), which includes Nobel and UN Kalinga Laureates, World Food Prize Laureates, and several fellows of academies of science and engineering of countries throughout the world.

From 1996 thousands of scientists, engineers and policy-makers began meeting just to define the scope of the project, before discussing the details of the contributions. Regional workshops were held in Washington DC, Tokyo, Moscow, Mexico City, Beijing, Panama, Abu Sultan (Egypt), and Kuala Lumpur, to develop a list of possible subjects and debate analytical approaches for treating them.

A UNESCO-EOLSS Joint Committee was established with the objectives of (a) seeking, selecting, inviting and appointing Honorary Theme Editors (HTEs)/International Commissions for the Themes, (b) providing assistance to them, (c) obtaining appropriate contributions for the different levels of the Encyclopedia, and (d) monitoring the text development (See UNESCO-EOLSS Joint Committee below).

Motivation for Publication in Electronic Media

Print projects for such large publications are highly material and energy intensive, a consideration that is very important in the case of the EOLSS since it advocates minimal use of natural resources.

Another great advantage of the electronic medium for large publications, in addition to portability, is the search facility. For instance, information on any chosen aspect can be found by search with the click of a button from the huge body of knowledge. Parts of the information may be downloaded, organized, printed, or adapted to one's own academic needs such as research studies, course-packs, etc. This is not the case with the print versions.

EOLSS Updating and Augmentation

Efforts to update and augment the EOLSS body of knowledge: Editors and Authors are given free access to access their works and forward updates if any. New themes are added covering many additional subjects. Updating/augmenting is frequent- almost every month

EOLSS in Official UN Languages

The Universal Networking Digital Language Foundation (UNDL) is now active on the translation project. The 1st International Workshop on Natural Language Processing Using the Universal Networking Language (UNL), took place during 4-7 May 2007 at Alexandria, Egypt. UNDL started to work with the six official UN languages. Twelve

UNL language centers situated in France, Russia USA, Cairo, Sao Paulo, Madrid, Tokyo, China, etc. are participating in this work.

Unite Nations DECADE OF EDUCATION FOR SUSTAINABLE DEVELOPMENT: 2005-2014

EOLSS is dedicated to contribute to the implementation of Unite Nations DESD: 2005-2014

A sample of opinions on the EOLSS project

"Scientific curiosity and technological endeavors are deeply rooted aspects of human nature. They are responsible for the development of human society and welfare. But they are also responsible for much of the environmental problems we are facing today. Solutions for these shortcomings are inconceivable without full scientific and technological support. EOLSS has the goal to provide a firm knowledge base for future activities to prolong the lifetime of the human race in a hospitable environment." **Richard R. Ernst, Nobel Laureate- Chemistry**

"The EOLSS is not only appropriate, but it is imaginative and, to my knowledge, unique. Much of what we can write about science, about energy, about our far-ranging knowledge base, can indeed be found in major encyclopedias, but as I understand your vision, never as a central theme; the theme of humanity, embedded in nature and constrained to find ways of maintaining a relationship with nature based upon understanding and respect." **Leon M. Lederman, Nobel Laureate in Physics**

"The population of our planet and its development over the ages sets the scene for considering all global problems and it is reasonable to begin their discussion with population growth. ... Thus we are dealing with an interdisciplinary problem in an attempt to describe the total human experience, right from its very beginning. But without this perspective of time it is not possible to objectively assess what is happening today and provide an objective view of the present state of development, the challenge now facing humanity." **S.P. Kapitza, UNESCO Kalinga Prize Winner**

"Pursuit of knowledge and truth supersedes present considerations of what nature, life or the world are or should be, for our own vision can only be a narrow one. Ethical evaluation and rules of justice have changed and will change over time and will have to adapt. Law is made for man, not man for law. If it does not fit any more, change it.... Some think that it is being arrogant to try to modify nature; arrogance is to claim that we are perfect as we are! With all the caution that must be exercised and despite the risks that will be encountered, carefully pondering each step, mankind must and will continue along its path, for we have no right to switch off the lights of the future.... We have to walk the path from the tree of knowledge to the control of destiny." **Jean-Marie Lehn, Nobel Laureate in Chemistry**

"Ecotechnology involving appropriate blends of traditional technologies and the ecological prudence of the past with frontier technologies such as biotechnology, information technology, space technology, new materials, renewable energy technology and management technology, can help us to promote global sustainable development involving harmony between humankind and nature on the one hand and tolerance and love of diversity and pluralism in human societies on the other. We need shifts in technology and public policy. This is a challenging task to which the Encyclopedia of Life Support Systems should address itself." **M.S. Swaminathan, First World Food Prize winner**

"EOLSS is concerned with the Life Support Systems... Each of these systems is a very complex one. ...we have to think of all these "systems" as closely related "subsystems" of the Planet Earth System. ...Rational decisions will be more and more possible to envision if one will be able to couple the physical modeling to economic and financial models and to human factors..." **J.L. Lions, Japan Prize winner in Applied Mathematics**

"In the coming decades, the need for working together, across countries, is essential. We need to bring 3 billion poor into society and make it possible for them to lead healthy lives, to get the health interventions that we all take as something as self-evident. It can be done. It is within reach. We need to invest in the future. We need to invest in the health of all people worldwide." **Gro Harlem Brundtland, Director General, WHO,** in an interview with Patricia Morales

"We must learn to live at one with nature. Nature does not bear grudges, but it must not be brought to the point where it can no longer sustain human society and the continuance of humankind on Earth. I believe that one of the most important things is the shaping of a new value system, because nature can live without us, but we cannot live without nature. Instead of a hedonic approach, we should promote an approach that reasonably limits consumerism and which promotes the virtue of enoughness. If we insist on consumerism as the new utopia, nature will reject such a system, as surely as cultural diversity rejected the totalitarian system. Our generation has to face a difficult challenge, but as recent history has proven, walls of difficulty, like the Berlin Wall, can fall." **Mikhail Gorbachev, Nobel Laureate in Peace,** in an interview with Patricia Morales

"We live in an increasingly interdependent world. As a result, we must cooperate together across all boundaries of nation, culture, faith, and race, and at all levels—locally, nationally, regionally, and globally— if we are to achieve our basic environmental, economic and social goals. Furthermore, if we are to make wise choices and to cooperate together effectively, we urgently need a shared vision of fundamental ethical values to guide us. In other words, the development of global ethics is essential. Our very survival as a species is in doubt if we cannot clarify our ethics and develop common values around such basic issues as environmental protection, justice, human rights, cultural diversity, economic equity, eradication of poverty, and peace." **Steven C. Rockefeller, Earth Charter Commissioner,** in an interview with Patricia Morales

EOLSS e-Books and their full color print versions in subject categories

For the convenience of users' special interests the EOLSS body of knowledge is being made available in e-book (pdf) format with the provision for Print on Demand. There is a wide gallery of e-books making up a virtual library and the sets of volumes classified under different component encyclopedias are outlined below. **The ISBN numbers are respectively eISBN for e-books and ISBN for print versions. Forthcoming items are without these numbers.**

Prices of e-Books and their full color print versions

Each e-book (Adobe Reader-single user) volume: **US$78/Euros60/UKP50**
Each print volume (Full Color Library Edition): **US$234/Euros180/UKP150**
Orders from the USA, Europe, and UK will be charged in US$, Euros, UK Pounds respectively.
Orders from all other regions will be charged in US$.
e-Book orders for multiple users will be entitled to get a discount: Five users-10%; Ten users-20%
To place your order visit **http://www.eolss.net/**

DESWARE: Encyclopedia of Desalination and Water Resources

This is a unique publication, the only one of its kind in the filed of desalination and water resources which puts together contributions from leaders from all over the world. For details visit **(www.desware.net)**

OUTLINES OF THE COMPONENT ENCYCLOPEDIAS OF THE EOLSS IN TERMS OF e-BOOKS AND THEIR FULL COLOR PRINT VERSIONS

(The bulleted items in the following pages represent Theme/Section/Topic titles. The size of a Theme may vary from about 10 Chapters to about 240 Chapters. The size of an entry (Chapter) may vary from about 5000 words to about 30000 words.)

EARTH AND ATMOSPHERIC SCIENCES

Environmental Structure and Function: Earth System

Editors: Nikita Glazovsky, *Institute of Geography, Russian Academy of Sciences, Russia*
   Nina Zaitseva, *Department of Earth Sciences, Russian Academy of Sciences, Moscow, Russia*
   eISBN:
   ISBN:

Environmental Structure and Function: Climate System (Vols.1-2)

Editor: George Vadimovich Gruza, *Institute of Global Climate and Ecology, Russian Academy of Sciences, Russia*
   eISBN: 978-1-84826-288-1, 978-1-84826-289-8
   ISBN: 978-1-84826-738-1, 978-1-84826-739-8

Natural Disasters

Editor: Vladimir M. Kotlyakov, *Institute of Geography, Russian Academy of Sciences, Russia*
   eISBN:
   ISBN:

Geography (Vols. 1-2)

Editor: Maria Sala, *University of Barcelona, Spain*
   eISBN: 978-1-905839-60-5, 978-1-905839-61-2
   ISBN: 978-1-84826-960-6, 978-1-84826-961-3

Geology (Vols. 1-5)

Editors: Benedetto De Vivo, *Universita di Napoli "Federico II", Italy*
   Bernhard Grasemann, *University of Vienna, Austria*
   Kurt Stüwe, *Institut für Geologie und Paläontologie, Universitaet Graz, Austria*
   eISBN: 978-1-84826-004-7, 978-1-84826-005-4, 978-1-84826-006-1, 978-1-84826-007-8, 978-1-84826-008-5
   ISBN: 978-1-84826-454-0, 978-1-84826-455-7, 978-1-84826-456-4, 978-1-84826-457-1, 978-1-84826-458-8

Geophysics and Geochemistry (Vols. 1-3)

Editor: Jan Lastovicka, *Institute of Atmospheric Physics, Academy of Sciences of the Czech Republic, Czech Republic*
   eISBN: 978-1-84826-245-4, 978-1-84826-246-1, 978-1-84826-247-8
   ISBN: 978-1-84826-636-0, 978-1-84826-656-8, 978-1-84826-662-9

Oceanography (Vols. 1-3)

Editors: Jacques C.J. Nihoul, *University of Liege, Belgium*
   Chen-Tung Arthur Chen, *National Sun Yat-Sen University, Taiwan*
   eISBN: 978-1-90583-962-9, 978-1-90583-963-6, 978-1-90583-964-3
   ISBN: 978-1-84826-962-0, 978-1-84826-963-7, 978-1-84826-964-4

Geoinformatics (Vols. 1-2)

Editor: Peter M. Atkinson, *University of Southampton, UK*
   eISBN: 978-1-905839-87-2, 978-1-905839-86-5
   ISBN: 978-1-84826-987-3, 978-1-84826-986-6

Advanced Geographic Information Systems (Vols. 1-2)

Editor: Claudia Maria Bauzer Medeiros, *Universidade Estadual de Campinas (UNICAMP), Brazil*
eISBN: 978-1-905839-91-9, 978-1-905839-92-6
ISBN: 978-1-84826-991-0, 978-1-84826-992-7

Astronomy and Astrophysics

Editors: Oddbjørn Engvold, *University of Oslo, Norway*
Rolf Stabell, *University of Oslo, Norway*
Bozena Czerny, *Copernicus Astronomical Center, Bartycka 18, 00-716 Warsaw, Poland*
John Lattanzio, *Monash University, Victoria Australia*
eISBN:
ISBN:

Tropical Meteorology

Editor: Yuqing Wang, *Department of Meteorology and International Pacific Research Center, School of Ocean and Earth Science and Technology University of Hawaii at Manoa*
eISBN:
ISBN:

MATHEMATICAL SCIENCES

Environmetrics

Editors: Abdel H. El-Shaarawi, *National Water research Institute, Burlington Ontario Canada,*
Jana Jureckova *Charles University Sokolovska, Czech Republic*
eISBN: 978-1-84826-102-0
ISBN: 978-1-84826-552-3

Biometrics (Vols. 1-2)

Editors: Susan R. Wilson, *School of Mathematical Sciences, Australian National University, Australia*
Conrad Burden, *The Australian National University, Australia*
eISBN: 978-1-905839-36-0, 978-1-905839-37-7
ISBN: 978-1-84826-936-1, 978-1-84826-937-8

Mathematics: Concepts, and Foundations (Vols. 1-3)

Editor: Huzihiro Araki, *Research Institute of Mathematical Sciences, Kyoto University, Japan*
eISBN: 978-1-84826-130-3, 978-1-84826-131-0, 978-1-84826-132-7
ISBN: 978-1-84826-580-6, 978-1-84826-581-3, 978-1-84826-582-0

Probability and Statistics (Vols. 1-3)

Editor: Reinhard Viertl, *Technische Universität Wien, Austria*
eISBN: 978-1-84826-052-8, 978-1-84826-053-5, 978-1-84826-054-2
ISBN: 978-1-84826-502-8, 978-1-84826-503-5, 978-1-84826-504-2

Mathematical Models of Life Support Systems (Vols. 1-2)

Editors: Valeri I. Agoshkov, *Institute of Numerical Mathematics, Russian Academy of Sciences, Russia*
Jean-Pierre Puel, *Universitetde Versailles St. Quentin, Versailles, France*
eISBN: 978-1-84826-128-0, 978-1-84826-129-7
ISBN: 978-1-84826-578-3, 978-1-84826-579-0

Mathematical Models (Vols. 1-3)

Editor: Jerzy A. Filar, *University of South Australia, Australia*
Jacek B. Krawczyk, *Victoria University of Wellington, New Zealand*
eISBN: 978-1-84826-242-3, 978-1-84826-243-0, 978-1-84826-244-7

ISBN: 978-1-84826-695-7, 978-1-84826-696-4, 978-1-84826-697-1

Computational Models (Vols. 1-2)

Editors: Vladimir V. Shaidurov, *Institute of Computational Modeling of the Russian Academy of Sciences, Russia*

Olivier Pironneau, *Universite de Paris 6, UFR France*

eISBN: 978-1-84826-035-1, 978-1-84826-036-8

ISBN: 978-1-84826-485-4, 978-1-84826-486-1

Optimization and Operations Research (Vols. 1-4)

Editor: Ulrich Derigs, *University of Cologne, Germany*

eISBN: 978-1-905839-48-3, 978-1-905839-49-0, 978-1-905839-50-6, 978-1-905839-51-3

ISBN: 978-1-84826-948-4, 978-1-84826-949-1, 978-1-84826-950-7, 978-1-84826-951-4

Systems Science and Cybernetics (Vols. 1-3)

Editor: Francisco Parra-Luna, *Universidad Complutense de Madrid, Spain*

eISBN: 978-1-84826-202-7, 978-1-84826-203-4, 978-1-84826-204-1

ISBN: 978-1-84826-652-0, 978-1-84826-653-7, 978-1-84826-654-4

History of Mathematics

Editors: Vagn Lundsgaard Hansen, *Technical University of Denmark, Denmark*

Jeremy Gray, *Centre for Mathematical Sciences, Open University, UK*

eISBN:

ISBN:

Mathematical Models in Economics (Vols. 1-2)

Editor: Wei-Bin Zhang, *Ritsumeikan Asia Pacific University, Japan*

eISBN: 978-1-84826-228-7, 978-1-84826-229-4

ISBN: 978-1-84826-678-0, 978-1-84826-679-7

Mathematical Physiology

Editor: Andrea de Gaetano, *CNR IASI Laboratorio di Biomatematica, Roma, Italy*

eISBN:

ISBN:

BIOLOGICAL, PHYSIOLOGICAL AND HEALTH SCIENCES

Global Perspectives in Health (Vols. 1-2)

Editor: B. P. Mansourian, *Research Policy & Cooperation, WHO, Switzerland*

eISBN: 978-1-84826-248-5, 978-1-84826-249-2

ISBN: 978-1-84826-698-8, 978-1-84826-699-5

Water and Health (Vols. 1-2)

Editor: W.O.K. Grabow, *University of Pretoria, South Africa*

eISBN: 978-1-84826-182-2, 978-1-84826-183-9

ISBN: 978-1-84826-632-2, 978-1-84826-633-9

Physical (Biological) Anthropology

Editor: P. Rudan, *Institute for Anthropological Research, University of Zagreb, Zagreb, Croatia*

eISBN: 978-1-84826-226-3

ISBN: 978-1-84826-676-6

Psychology (Vols. 1-3)

Editor: Stefano Carta, *University of Cagliari, Italy*

eISBN: 978-1-905839-65-0, 978-1-905839-66-7, 978-1-905839-67-4
ISBN: 978-1-84826-965-1, 978-1-84826-966-8, 978-1-84826-967-5

Genetics and Molecular Biology

Editor: Kohji Hasunuma, *Yokohama City University, Japan*
eISBN: 978-1-84826-123-5
ISBN: 978-1-84826-573-8

Physiology and Maintenance (Vols. 1-5)

Editors: Osmo Otto Päiviö Hänninen and Mustafa Atalay, *University of Kuopio, Finland*

Volume 1: Physiology and Maintenance

General Physiology
eISBN- 978-1-84826-039-9
ISBN- 978-1-84826-489-2

Volume 2: Physiology and Maintenance

Enzymes: The Biological Catalysts of Life Nutrition and Digestion
eISBN- 978-1-84826-040-5
ISBN- 978-1-84826-490-8

Volume 3: Physiology and Maintenance

Renal Excretion, Endocrinology, Respiration
Blood Circulation: Its Dynamics and Physiological Control
eISBN- 978-1-84826-041-2
ISBN- 978-1-84826-491-5

Volume 4: Physiology and Maintenance

Locomotion in Sedentary Societies
eISBN- 978-1-84826-042-9
ISBN- 978-1-84826-492-2

Volume 5: Physiology and Maintenance

Neurophysiology
Plant Physiology and Environment: A Synopsis
eISBN- 978-1-84826-043-6
ISBN- 978-1-84826-493-9

Medical Sciences

International Commission:

President: B.P. Mansourian, *WHO (ret.) Switzerland*

Vice President: A. Wojtczak, *Institute for International Medical Education, NY, USA*

B. McA. Sayers, *Imperial College of Science, Technology and Medicine, London UK*

S.M. Mahfouz, *Cairo University, Egypt* Members

A. Arata, *USAID, Alexandria, Virginia, USA*

B P.R. Aluwihare, *University of Peradeniya, Kandy, Sri Lanka*

G. W. Brauer, *University of Victoria, Victoria, BC, Canada*

M. R.G. Manciaux, *Nancy University, Vandoeuvre-les-Nancy, France*

J. Szczerban, *Medical University of Warsaw, Warsaw, Poland*

Y. L. G. Verhasselt, *Vrije Universiteit Brussels, Brussels, Belgium*

Alberto Pellegrini, *Ipanema, Rio de Janerio, RJ, Brasil*

Assen Jablensky, *The University of Western Australia, Perth, Australia*

R. Kitney, *Imperial College of Science, Technology and Medicine London, UK*

Arminee Kazanjian, *University of British Columbia, Canada*

Tomris Turmen, *WHO, Turkey*

R. Leidl, *Ludwig Maximilians-University, Munich and GSF-National Research Center for Environment and Health, Institute for Health Economics and Health Care Management, Germany*

K. Sorour, *Cairo University, Egypt*
eISBN: 978-1-84826-283-6, 978-1-84826-284-3
ISBN: 978-1-84826-733-6, 978-1-84826-734-3

Biological Science Fundamentals and Systematics (Vols. 1-4)

Editors: Alessandro Minelli, *University of Padova, Italy*

Giancarlo Contrafatto, *University of Natal, South Africa*
eISBN: 978-1-84826-304-8, 978-1-84826-305-5, 978-1-84826-306-2, 978-1-84826-189-1
ISBN: 978-1-84826-754-1, 978-1-84826-755-8, 978-1-84826-756-5, 978-1-84826-639-1

Extremophiles (Vols. 1-3) (Life under extreme environmental conditions)

Editors: Charles Gerday , *University of Liege, Belgium,*

Nicolas Glansdorff, *Vrjie Universiteit Brussel, Belgium*
eISBN: 978-1-905839-93-3, 978-1-905839-94-0, 978-1-905839-95-7
ISBN: 978-1-84826-993-4, 978-1-84826-994-1, 978-1-84826-995-8

Ethnopharmacology (Vols. 1-2)

Editors: Elaine Elisabetsky, *Universidade Federal do Rio Grande do Sul, Brazil*

Nina L. Etkin, *University of Hawai'I, USA*
eISBN: 978-1-905839-96-4, 978-1-905839-97-1
ISBN: 978-1-84826-996-5, 978-1-84826-997-2

Pharmacology

Editor: Harry Majewski, *RMIT University, Victoria, Australia*
eISBN: 978-1-84826-180-8, 978-1-84826-254-6
ISBN: 978-1-84826-630-8, 978-1-84826-704-6

Biophysics

Editor: Mohamed I.El Gohary,*Al Azhar University, Cairo, Egypt*
eISBN:
ISBN:

Phytochemistry and Pharmacognosy

Editors: John M. Pezzuto, *University of Hawaii, Hilo, 96720, Hawaii*

Massuo Jorge Kato, *University of Sao Paulo, Sao-Paulo, 05508-000, Brasil*
eISBN:
ISBN:

Neuroscience

Editor: Jose Masdeu, *National Institutes of Health (CBDB-NIMH), Bethesda, MD*
eISBN:
ISBN:

History of Medicine

International Commission:

President: Athanasios Diamandopoulos, *Renal department, St. Andrew's regional Hospital, Patras, Greece.*

Vice Presidents:

Carlos Viesca, *UNAM Mexico.*

Alain Lellouch, *DIM de l'Hopital, Saint-Germain-en-Laye, France*

Members:

S. Kotteck, *Israel*

C. Bergdold, *Cologne, Germany*

F. Alecbari, *Atjerbaizan*

B. Fontini, *South America*

T. Sorokina, *Moscow, Russia*

J. Pearns, *Australia*
eISBN:
 ISBN:

Reproduction and Development Biology
Editor: Andre Pires da Silva, *Department of Biology, The University of Texas at Arlington, USA*
eISBN:
 ISBN:

BIOTECHNOLOGY

Editors: Horst W. Doelle, *MIRCEN-Biotechnology, Australia*

Edgar J. DaSilva, *Section of Life Sciences, Division of Basic and Engineering Sciences, UNESCO, France*

(The bulleted items in the following represent Topic titles. The total number of chapters is 240. The size of an entry (Chapter) may vary from about 5000 words to about 30000 words.)

Volume 1: Fundamentals in Biotechnology
eISBN 978-1-84826-255-3
ISBN 978-1-84826-705-3
Volume 2: Methods in Biotechnology
eISBN 978-1-84826-256-0
ISBN 978-1-84826-706-0
Volume 3: Methods in Gene Engineering
eISBN 978-1-84826-257-7
ISBN 978-1-84826-707-7
Volume 4: Bioprocess Engineering - Bioprocess Analysis
eISBN 978-1-84826-258-4
ISBN 978-1-84826-708-4
Volume 5: Industrial Biotechnology - Part I
eISBN 978-1-84826-259-1
ISBN 978-1-84826-709-1
Volume 6: Industrial Biotechnology - Part II
eISBN 978-1-84826-260-7
ISBN 978-1-84826-710-7
Volume 7: Special Processes for Products, Fuel and Energy
eISBN 978-1-84826-261-4
ISBN 978-1-84826-711-4
Volume 8: Agricultural Biotechnology
eISBN 978-1-84826-262-1
ISBN 978-1-84826-712-1
Volume 9: Marine Biotechnology
eISBN 978-1-84826-263-8
ISBN 978-1-84826-713-8
Volume 10: Environmental Biotechnology - Socio-Economic Strategies for Sustainability
eISBN 978-1-84826-264-5
ISBN 978-1-84826-714-5
Volume 11: Medical Biotechnology - Fundamentals and Modern Development – Part I
eISBN 978-1-84826-265-2
ISBN 978-1-84826-715-2
Volume 12: Medical Biotechnology - Fundamentals and Modern Development – Part II
eISBN 978-1-84826-266-9
ISBN 978-1-84826-716-9
Volume 13: Social, Educational and Political Aspects of Biotechnology - An Overview and an Appraisal of Biotechnology in a Changing World – Part I
eISBN 978-1-84826-267-6
ISBN 978-1-84826-717-6
Volume 14: Social, Educational and Political Aspects of Biotechnology - An Overview and an Appraisal of Biotechnology in a Changing World – Part II
eISBN 978-1-84826-268-3
ISBN 978-1-84826-718-3
Volume 15: The Role of microbial resources centers and UNESCO in the Development of Biotechnology

eISBN 978-1-84826-269-0
ISBN 978-1-84826-719-0

TROPICAL BIOLOGY AND CONSERVATION

(*The bulleted items in the following represent Topic titles. The total number of chapters is 240. The size of an entry (Chapter) may vary from about 5000 words to about 30000 words.*)

ISBN 978-1-84826-731-2
Volume 11: Case Studies
eISBN 978-1-84826-282-9
ISBN 978-1-84826-732-9

SOCIAL SCIENCES AND HUMANITIES

Peace, Literature, and Art (Vols. 1-2)

Editor: Ada Aharoni, *Technion - Israel Institute of Technology, and Communications Commission of IPRA, Israel*
eISBN: 978-1-84826-076-4, 978-1-84826-077-1
ISBN: 978-1-84826-526-4, 978-1-84826-527-1

Global Security

Editors: Pinar Bilgin, *Fellow, Woodrow Wilson International Center for Scholars, Washington, DC USA*

Paul D. Williams, *George Washington University, USA*
eISBN:
ISBN:

Environmental Laws and Their Enforcement (Vols. 1-2)

Editors: A. Dan Tarlock, *Illinois Institute of Technology, USA*
eISBN: 978-1-84826-113-6, 978-1-84826-293-5
ISBN: 978-1-84826-563-9, 978-1-84826-743-5

Linguistic Anthropology

Editor: Anita Sujoldzic, *University of Zagreb, Zagreb, Croatia*
eISBN: 978-1-84826-225-6
ISBN: 978-1-84826-675-9

Archaeology (Vols. 1-2)

Editor: Donald L. Hardesty, *University of Nevada, Reno, USA*
eISBN: 978-1-84826-002-3, 978-1-84826-003-0
ISBN: 978-1-84826-452-6, 978-1-84826-453-3

Culture, Civilization and Human Society (Vols. 1-2)

Editors: Herbert Arlt, *Research Institute for Austrian and International Literature and Cultural Studies (INST), Austria,*

Donald G. Daviau, *University of California, Riverside CA, USA*
eISBN: 978-1-84826-190-7, 978-1-84826-191-4
ISBN: 978-1-84826-640-7, 978-1-84826-641-4

Literature and the Fine Arts

Editors: Herbert Arlt, *Research Institute for Austrian and International Literature and Cultural Studies (INST), Austria,*

Donald G. Daviau, *University of California, Riverside CA, USA*

Peter Horn, *University of Cape Town, Cape Town, South Africa*
eISBN: 978-1-84826-122-8
ISBN: 978-1-84826-572-1

Philosophy and World Problems

Editor: John McMurtry, *University of Guelph, Canada*
eISBN:
ISBN:

Fundamental Economics

Editor: Mukul Majumdar, *Cornell University, USA*
eISBN:
ISBN:

Law

Editors: Aaron Schwabach, *Thomas Jefferson School of Law, USA,*
Arthur J. Cockfield *Queen's University, Canada*
eISBN: 978-1-905839-68-1
ISBN: 978-1-848269-68-2

Government and Politics (Vols. 1-2)

Editor: Masashi Sekiguchi, *Tokyo Metropolitan University, Japan*
eISBN: 978-1-905839-69-8, 978-1-905839-70-4
ISBN: 978-1-84826-969-9, 978-1-84826-970-5

Journalism and Mass Communication (Vols. 1-2)

Editor: Rashmi Luthra, *University of Michigan, USA*
eISBN: 978-1-905839-71-1, 978-1-905839-72-8
ISBN: 978-1-84826-971-2, 978-1-84826-972-9

Unity of Knowledge (in Transdisciplinary Research for Sustainability) (Vols. 1-2)

Editor: Gertrude Hirsch Hadorn, *Federal Institute of Technology, ETH Zurich Zentrum, Switzerland*
eISBN: 978-1-905839-82-7, 978-1-905839-83-4
ISBN: 978-1-84826-982-8, 978-1-84826-983-5

Comparative Literature: Sharing Knowledges for Preserving Cultural Diversity

International Commission:

President: Lisa Block de Behar, *University of Republica, Montevideo, Uruguay* (Former President to the commission was Tania Franco Carvalhal, *Federal University of Rio Grande do Sul, Porto Alegre, Brazil* who has passed away in 2006)

Vice Presidents: Paola Mildonian, *Universita Ca Foscari di Venezia, Venezia, Italy*

Jean-Michel Djian, *University of Paris 8, France*

Djelal Kadir, *Pennsylvania State University, University Park, PA, USA*

Alfons Knauth, *Ruhr University of Bochum, Germany*

Dolores Romero Lopez, *Universidad of Complutense de Madrid, Spain*

Márcio Seligmann Silva, *University of Campinas, Sao Paulo, Brazil*
eISBN:
ISBN:

History and Philosophy of Science and Technology

Editors: Pablo Lorenzano, *Universidad Nacional de Quilmes, Argentina*

Eduardo Ortiz, *Imperial College, London,, UK*

Hans-Jorg Rheinberger, *Max Planck Institute fur Wisenschaftsgeschichte, Berlin, 14195, Germany*

Carlos Delfino Galles, *FCEIA-UNR, Pelligrino 250, Rosario 2000, Argentina*
eISBN:
ISBN:

Linguistics

Editor: Vesna Muhvic-Dimanovski, *University of Zagreb,*

Lelija Socanac, *Department of Language Research of the Croatian Academy of Sciences*

eISBN: 978-1-84826-287-4
ISBN: 978-1-84826-737-4

Religion, Culture, and Sustainable Development
Editors: Roberto Blancarte Pimentel, *El Colegio de Mexico, México*
eISBN:
  ISBN:

World System History
Editors: George Modelski, *University of Washington, Settle, Washington WA, USA*
  Robert A. Denemark, *University of Delaware, USA*
eISBN: 978-1-84826-218-8
ISBN: 978-1-84826-668-1

World Civilizations
Editor: Robert Holton, *Trinity College, University of Dublin, Dublin, Ireland*
eISBN:
ISBN:

Historical Developments and Theoretical Approaches in Sociology
Editor: Charles Crothers, *Auckland University of Technology, New Zealand*
eISBN:
ISBN:

Nonviolent Alternatives for Social Change
Editor: Ralph V. Summy, *The University of Queensland, Australia*
eISBN: 978-1-84826-220-1
ISBN: 978-1-84826-670-4

Demography (Vols. 1-2)
Editor: Zeng Yi, *Duke University, Durham, USA; Peking University, China*
eISBN: 978-1-84826-307-9, 978-1-84826-308-6
ISBN: 978-1-84826-757-2, 978-1-84826-758-9

PHYSICAL SCIENCES, ENGINEERING AND TECHNOLOGY RESOURCES

Development of Physics
Editor: Gyo Takeda, *University of Tokyo and Tohoku University, Japan*
eISBN: 978-1-84826-201-0
ISBN: 978-1-84826-651-3

Fundamentals of Physics (Vols. 1-3)
Editor: José Luis Morán López, *Instituto Potosino de Investigación Científica y Tecnológica, Mexico*
eISBN: 978-1-905839-52-0, 978-1-905839-53-7, 978-1-84826-223-2
ISBN: 978-1-84826-952-1, 978-1-84826-953-8, 978-1-84826-673-5

Physical Methods, Instruments and Measurements (Vols. 1-4)
Editor: Yuri Mikhailovitsh Tsipenyuk, *P.L. Kapitza Institute for Physical Problems, Russian Academy of Sciences (RAS), Russia*
eISBN: 978-1-905839-54-4, 978-1-905839-55-1, 978-1-905839-56-8, 978-1-905839-57-5
ISBN: 978-1-84826-954-5, 978-1-84826-955-2, 978-1-84826-956-9, 978-1-84826-957-6

Mechanical Engineering, Energy Systems and Sustainable Development (Vols. 1-5)

Editors: Konstantin V. Frolov, *Mechanical Engineering Research Institute, Russian Academy of Sciences, Russia*

R.A. Chaplin, *University of New Brunswick, Canada*

Christos Frangopoulos, *National Technical University of Athens, Greece*

eISBN: 978-1-84826-294-2, 978-1-84826-295-9, 978-1-84826-296-6, 978-1-84826-297-3, 978-1-84826-298-0
ISBN: 978-1-84826-744-2, 978-1-84826-745-9, 978-1-84826-746-6, 978-1-84826-747-3, 978-1-84826-748-0

Materials Science and Engineering (Vols. 1-3)

Editor: Rees D. Rawlings, *Imperial College of Science, Technology and Medicine, UK*

eISBN: 978-1-84826-032-0, 978-1-84826-033-7, 978-1-84826-034-4
ISBN: 978-1-84826-482-3, 978-1-84826-483-0, 978-1-84826-484-7

Civil Engineering (Vols. 1-2)

Editors: Kiyoshi Horikawa, *Musashi Institute of Technology, Japan*

Qizhong Guo, *Rutgers, The State University of New Jersey,USA*

eISBN: 978-1-905839-73-5, 978-1-905839-74-2
ISBN: 978-1-84826-973-6, 978-1-84826-974-3

Electrical Engineering (Vols. 1-3)

Editor: Kit Po Wong, *Hong Kong Polytechnic University, Hong Kong*

eISBN: 978-1-905839-77-3, 978-1-905839-78-0, 978-1-905839-79-7
ISBN: 978-1-84826-977-4, 978-1-84826-978-1, 978-1-84826-979-8

Transportation Engineering and Planning (Vols. 1-2)

Editor: Tschangho John Kim, *University of Illinois at Urbana-Champaign, USA*

eISBN: 978-1-905839-80-3, 978-1-905839-81-0
ISBN: 978-1-84826-980-4, 978-1-84826-981-1

Telecommunication Systems and Technologies (Vols. 1-2)

Editor: Paolo Bellavista, *DEIS-LIA, Universita degli Studi di Bologna, Italy*

eISBN: 978-1-84826-000-9, 978-1-84826-001-6
ISBN: 978-1-84826-450-2, 978-1-84826-451-9

Structural Engineering and Structural Mechanics

Editor: Xila S. Liu, *Tsinghua University, Beijing, CHINA*

eISBN:
ISBN:

Structural Engineering and Geomechanics

Editor: Sashi K. Kunnath, *University of California, DAVIS,USA*

eISBN:
ISBN:

Crystallography

Editors: John R. Helliwell, *The University of Manchester, Manchester, UK*

Santiago Gracia Granda, *University of Ovieddo, Spain*

eISBN:
ISBN:

Nanosciences and Nanotechnologies

Editors: Valeri Nikolayevich Harkin, *International Commission, Moscow, 115419, Russia*

Chunli Bai, *Institute of Chemistry, Chinese Academy of Sciences, Beijing, 100864, China*

Sae-Chul Kim, *International Union of Pure and Applied Chemistry, South Korea*

eISBN:
ISBN:

Continuum Mechanics

Editors: José Merodio, *Universidad Politécnica de Madrid, Madrid, Spain*

Giuseppe Saccomandi, *Sezione di Ingegneria Industriale, Dipartimento di Ingegneria dell'Innovazione, niversita degli Studi di Lecce, Via per Monteroni, 73100 Lecce, Italy*
eISBN:
ISBN:

Biomechanics

Editors: Manuel Doblare, *Universidad de Zaragoza, Spain*

Jose Merodio, *Universidad Politécnica de Madrid, Madrid, Spain*

José Ma Goicolea Ruigómez, *Ciudad Universitaria, Madrid, Spain*
eISBN:
ISBN:

Building Services Engineering

Editor: Yuguo Li, *The University of Hong Kong, Hong Kong, China.*
eISBN:
ISBN:

Pressure Vessels and Piping Systems

Editors: Yong W.Kwon, *Naval Postgraduate School, Monterey, CA 93943 USA*

Poh-Sang Lam, *Savannah River National Laboratory, Aiken, South Carolina 29808, USA*
eISBN:
ISBN:

Tribology: Friction, Wear, Lubrication

Editor: Roberto Bassani, *Universita di Pisa, Pisa, Italy*
eISBN:
ISBN:

Welding Engineering and Technology

Editors: Slobodan Kralj, *Faculty of Mechanical Engineering and Naval Architecture, Zagreb, Croatia*

Branko Bauer, *Dept. of Welded Structures, Zagreb, Zagreb, 10000, Croatia*
eISBN:
ISBN:

Ships and Offshore Structures

Editors: Jeom Kee Paik, *Pusan National University , Geumjeong-Gu, Busan 609-735, Korea*

Ge(George) Wang, *American Bureau of Shipping Technology,,, USA*

International Editorial Advisory Board (IEAB): Weicheng Cui, *China;* Paul A. Frieze, *UK;* Yoshiaki Kodama, *Japan;* David Molyneux, *Canada;* Anil K. Thayamballi, *USA;* P.A. Wilson, *UK*
eISBN:
ISBN:

Cold Regions Science and Marine Technology

Editor: Hayley Shen, *Clarkson University, Potsdam, USA*
eISBN:
ISBN:

Experimental Mechanics

Editor:José Luiz de França Freire, *Endereço profissional Pontifícia Universidade Católica do Rio de Janeiro, Rio de* Janeiro, *RJ – Brasil.*

eISBN:
ISBN:

CONTROL SYSTEMS, ROBOTICS, AND AUTOMATION

(Vols. 1-22)

Editor: Heinz Unbehauen, *Ruhr-Universität Bochum, Germany*

(The bulleted items in the following represent Topic titles. The total number of chapters is 240. The size of an entry (Chapter) may vary from about 5000 words to about 30000 words.)

Volume 1: System Analysis and Control: Classical Approaches I
- *Control Systems, Robotics and Automation-Introduction and Overview*
- *Elements of Control Systems*
- *Stability Concepts*

eISBN 978-1-84826-140-2
ISBN 978-1-84826-590-5

Volume 2: System Analysis and Control: Classical Approaches II
- *Classical Design Methods for Continuous LTI-Systems*
- *Digital Control Systems*

eISBN 978-1-84826-141-9
ISBN 978-1-84826-591-2

Volume 3: System Analysis and Control: Classical Approaches III
- *Design of State Space Controllers (Pole Placement) for SISO Systems*
- *Basic Nonlinear Control Systems*

eISBN 978-1-84826-142-6
ISBN 978-1-84826-592-9

Volume 4: Modeling and System Identification I
- *Modeling and Simulation of Dynamic Systems*

eISBN 978-1-84826-143-3
ISBN 978-1-84826-593-6

Volume 5: Modeling and System Identification II
- *Frequency Domain System Identification*
- *Identification of Linear Systems in Time Domain*

eISBN 978-1-84826-144-0
ISBN 978-1-84826-594-3

Volume 6: Modeling and System Identification III
- *Identification of Nonlinear Systems*
- *Bound-based Identification*
- *Practical Issues of System Identification*

eISBN 978-1-84826-145-7
ISBN 978-1-84826-595-0

Volume 7: Advanced Control Systems I
- *Control of Linear Multivariable Systems-1*

eISBN 978-1-84826-146-4
ISBN 978-1-84826-596-7

Volume 8: Advanced Control Systems II
- *Control of Linear Multivariable Systems-2*

eISBN 978-1-84826-147-1
ISBN 978-1-84826-597-4

Volume 9: Advanced Control Systems III
- *Robust Control*

eISBN 978-1-84826-148-8
ISBN 978-1-84826-598-1

Volume 10: Advanced Control Systems IV

- *Adaptive Control*
eISBN 978-1-84826-149-5
ISBN 978-1-84826-599-8

Volume 11: Advanced Control Systems V

- *Model-based Predictive Control*
- *Control of Large-Scale Systems*
- *Control of Stochastic Systems*
eISBN 978-1-84826-150-1
ISBN 978-1-84826-600-1

Volume 12: Nonlinear, Distributed, and Time Delay Systems I

- *Control of Nonlinear Systems*
eISBN 978-1-84826-151-8
ISBN 978-1-84826-601-8

Volume 13: Nonlinear, Distributed, and Time Delay Systems II

- *Control of Chaos and Bifurcations*
eISBN 978-1-84826-152-5
ISBN 978-1-84826-602-5

Volume 14: Nonlinear, Distributed, and Time Delay Systems III

- *Distributed Parameter Systems: An Overview*
- *Control of 2-D Systems*
eISBN 978-1-84826-153-2
ISBN 978-1-84826-603-2

Volume 15: Discrete Event and Hybrid Control Systems

- *Discrete Event Systems*
- *Hybrid Control Systems*
eISBN 978-1-84826-154-9
ISBN 978-1-84826-604-9

Volume 16: Fault Diagnosis and Fault-tolerant Control
eISBN 978-1-84826-155-6
ISBN 978-1-84826-605-6

Volume 17: Fuzzy and Intelligent Control Systems

- *Fuzzy Control Systems*
- *Neural Control Systems*
- *Expert Control Systems*
- *Genetic Algorithms in Control Systems Engineering*
eISBN 978-1-84826-156-3
ISBN 978-1-84826-606-3

Volume 18: Industrial Applications of Control Systems I

- *Architectures and Methods for Computer-based Automation*
- *Supervisory Distributed Computer Control Systems*
- *Automation and Control of Thermal Processes*
- *Automation and Control of Electrical Power Generation and Transmission Systems*
eISBN 978-1-84826-157-0
ISBN 978-1-84826-607-0

Volume 19: Industrial Applications of Control Systems II

- *Automation and Control in Process Industries*
- *Automation and Control in Production Processes*
eISBN 978-1-84826-158-7
ISBN 978-1-84826-608-7

Volume 20: Industrial Applications of Control Systems III
- *Automation and Control in Traffic Systems*
eISBN 978-1-84826-159-4
ISBN 978-1-84826-609-4

Volume 21: Elements of Automation and Control
eISBN 978-1-84826-160-0
ISBN 978-1-84826-610-0

Volume 22: Robotics
eISBN 978-1-84826-161-7
ISBN 978-1-84826-611-7

CHEMICAL SCIENCES, ENGINEERING AND TECHNOLOGY RESOURCES

Fundamentals of Chemistry (Vols. 1-2)

Editor: Sergio Carra, *Politecnico di Milano, Italy*
eISBN: 978-1-905839-58-2, 978-1-905839-59-9
ISBN: 978-1-84826-958-3, 978-1-84826-959-0

Environmental and Ecological Chemistry (Vols. 1-3)

Editor: Aleksandar Sabljic, *Rudjer Boskovic Institute, Croatia*
eISBN: 978-1-84826-186-0, 978-1-84826-206-5, 978-1-84826-212-6
ISBN: 978-1-84826-692-6, 978-1-84826-693-3, 978-1-84826-694-0

Chemical Engineering and Chemical Process Technology

Editors: Ryszhard Pohorecki, *Warsaw University of Technology,*
John Bridgwater, *University of Cambridge,*
Rafiqul GANI, *Technical University of Denmark*
eISBN:
ISBN:

Chemical Ecology

Editor: Jorg D. Hardege, *University of Hull, UK*
eISBN: 978-1-84826-179-2
ISBN: 978-1-84826-629-2

Inorganic and Bio-inorganic Chemistry (Vols. 1-2)

Editor: Ivano Bertini, *University of Florence, Florence, Italy*
eISBN: 978-1-84826-214-0, 978-1-84826-215-7
ISBN: 978-1-84826-664-3, 978-1-84826-665-0

Organic and Bio-molecular Chemistry (Vols. 1-2)

Editor: Francesco Nicotra, *University of Milano Bicocca, Italy*
eISBN: 978-1-905839-98-8, 978-1-905839-99-5
ISBN: 978-1-84826-998-9, 978-1-84826-999-6

Radiochemistry and Nuclear Chemistry (Vols. 1-2)

Editor: Sándor Nagy, *Eotvos Lorand University, Hungary*
eISBN: 978-1-84826-126-6, 978-1-84826-127-3
ISBN: 978-1-84826-576-9, 978-1-84826-577-6

Electrochemistry

Editors:Juan M. Feliu-Martinez, *Universidad de Alicante,Departmento de Química Física, Spain*
Victor Climent Payá, *Universidad de Alicante,Departmento de Química Física, Spain*
eISBN:

ISBN:

Catalysis

Editor: Gabriele Centi, *Universita di Messina, Salita Sperone, Messina, Italy.*
eISBN:
ISBN:

Rheology

Editor: Crispulo Gallegos, *Departamento de Ingenieria Quimica, Campus de "El Carmen", Universidad de Huelva, 21071 Huelva, Spain*
eISBN:
ISBN:

Phytochemistry and Pharmacognosy

Editors: John M. Pezzuto, *University of Hawaii at Hilo, Hawaii, USA*
Massuo Jorge Kato, *Instituto de Química USP, Av. Prof. Lineu Prestes, São Paulo, Brasil*
eISBN:
ISBN:

WATER SCIENCES, ENGINEERING AND TECHNOLOGY RESOURCES

The Hydrological Cycle (Vols. 1-4)

Editor: Igor Alekseevich Shiklomanov, *State Hydrological Institute (SHI), Russia*
eISBN: 978-1-84826-024-5, 978-1-84826-025-2, 978-1-84826-026-9, 978-1-84826-193-8
ISBN: 978-1-84826-474-8, 978-1-84826-475-5, 978-1-84826-476-2, 978-1-84826-643-8

Types and Properties of Water (Vols. 1-2)

Editor: Martin Gaikovich Khublaryan, *Water Problems Institute, Russian Academy of Science, Russia*
eISBN: 978-1-90583-922-3, 978-1-90583-923-0
ISBN: 978-1-84826-922-4, 978-1-84826-923-1

Fresh Surface Water (Vols. 1-3)

Editor: James C.I. Dooge, *University College Dublin, Ireland*
eISBN: 978-1-84826-011-5, 978-1-84826-012-2, 978-1-84826-013-9
ISBN: 978-1-84826-461-8, 978-1-84826-462-5, 978-1-84826-463-2

Groundwater (Vols. 1-3)

Editors: Luis Silveira, Eduardo J. Usunoff, *Uruguay Instituto de Hidrologia de Llanuras, Argentina*
eISBN: 978-1-84826-027-6, 978-1-84826-028-3, 978-1-84826-029-0
ISBN: 978-1-84826-477-9, 978-1-84826-478-6, 978-1-84826-479-3

Water Storage, Transport, and Distribution (Vols. 1-2)

Editor: Yutaka Takahasi, *University of Tokyo, Japan*
eISBN: 978-1-84826-176-1
ISBN: 978-1-84826-626-1

Wastewater Recycling, Reuse, and Reclamation (Vols. 1-2)

Editor: Saravanamuthu (Vigi) Vigneswaran, *University of Technology, Sydney, Australia*
eISBN: 978-1-90583-924-7, 978-1-90583-925-4
ISBN: 978-1-84826-924-8, 978-1-84826-925-5

Hydraulic Structures, Equipment and Water Data Acquisition Systems (Vols. 1-4)

Editors: Jan Malan Jordaan, *University of Pretoria, South Africa*
Alexander Bell, *Bell & Associates, South Africa*
eISBN: 978-1-84826-049-8, 978-1-84826-050-4, 978-1-84826-051-1, 978-1-84826-192-1

ISBN: 978-1-84826-499-1, 978-1-84826-500-4, 978-1-84826-501-1, 978-1-84826-642-1

Water Resources Management (Vols. 1-2)

Editors: Hubert H.G. Savenije and Arjen Y. Hoekstra, *International Institute for Infrastructural, Hydraulic and Environmental Engineering (IHE - Delft), The Netherlands*
eISBN: 978-1-84826-177-8, 978-1-84826-224-9
ISBN: 978-1-84826-627-8, 978-1-84826-674-2

Water Quality and Standards (Vols. 1-2)

Editors: Shoji Kubota, *Kyushyu University, Japan*

Yoshiteru Tsuchiya, *Kogakuin University, Tokyo, Japan*
eISBN: 978-1-84826-030-6, 978-1-84826-031-3
ISBN: 978-1-84826-480-9, 978-1-84826-481-6

Environmental and Health Aspects of Water Treatment and Supply

Editors: Shoji Kubota, *Kyushyu University, Japan*

Yasumoto Magara, *Hokkaido University, Japan*
eISBN: 978-1-84826-178-5
ISBN: 978-1-84826-628-5

Water-Related Education, Training and Technology Transfer

Editor: Andre van der Beken, *Vrije Universitiet Brussel (VUB), Belgium*
eISBN: 978-1-84826-015-3
ISBN: 978-1-84826-465-6

Water Interactions with Energy, Environment and Food & Agriculture (Vols. 1-2)

Editor: Maria Concepcion Donoso, *Water Center for the Humid Tropics of Latin America and the Caribbean (CATHALAC), Panama*
eISBN: 978-1-84826-172-3, 978-1-84826-196-9
ISBN: 978-1-84826-622-3, 978-1-84826-646-9

Water and Development (Vols. 1-2)

Editor: Catherine M. Marquette, *Christian Michelsen Institute, Development Studies and Human Rights, Norway*
eISBN: 978-1-84826-173-0, 978-1-84826-197-6
ISBN: 978-1-84826-623-0, 978-1-84826-647-6

Future Challenges of Providing High-Quality Water (Vols. 1-2)

Editors: Jo-Ansie van Wyk, *University of South Africa (UNISA), South Africa*

Richard Meissner, *University of Pretoria, South Africa*

Hannatjie Jacobs, *Independent Research Consultant, South Africa*
eISBN: 978-1-84826-209-6, 978-1-84826-230-0
ISBN: 978-1-84826-659-9, 978-1-84826-680-3

Hydrological Systems Modeling (Vols. 1-2)

Editors: Lev S. Kuchment, *Water Problems Institute, Russian Academy of Sciences, Moscow, Russia*

Vijay P. Singh, *Texas A & M University*
eISBN: 978-1-84826-198-3, 978-1-84826-187-7
ISBN: 978-1-84826-648-3, 978-1-84826-637-7

Wastewater Treatment Technologies (Vols. 1-3)

Editor: Saravanamuthu (Vigi) Vigneswaran, *University of Technology, Sydney, Australia*
eISBN: 978-1-84826-188-4, 978-1-84826-194-5, 978-1-84826-250-8
ISBN: 978-1-84826-638-4, 978-1-84826-644-5, 978-1-84826-700-8

ENERGY SCIENCES, ENGINEERING AND TECHNOLOGY RESOURCES

Coal, Oil Shale, Natural Bitumen, Heavy Oil and Peat (Vols. 1-2)

Editor: Gao Jinsheng, *East China University of Science and Technology(ECUST), China*
eISBN: 978-1-84826-017-7, 978-1-84826-018-4
ISBN: 978-1-84826-467-0, 978-1-84826-468-7

Nuclear Energy Materials and Reactors

Editors: Yassin A. Hassan, *Texas A&M University, USA*

Robin A. Chaplin, *University of New Brunswick, Canada*
eISBN:
ISBN:

Renewable Energy Sources Charged with Energy from the Sun and Originated from Earth-Moon Interaction (Vols. 1-2)

Editor: Evald Emilievich Shpilrain, *IVTAN Institute for High Temperatures, Russian Academy of Sciences, Russia*
eISBN: 978-1-84826-019-1, 978-1-84826-020-7
ISBN: 978-1-84826-469-4, 978-1-84826-470-0

Thermal Power Plants (Vols. 1-3)

Editor: Robin A. Chaplin, *University of New Brunswick, Canada*
eISBN: 978-1-905839-26-1, 978-1-905839-27-8, 978-1-905839-28-5
ISBN: 978-1-84826-926-2, 978-1-84826-927-9, 978-1-84826-928-6

Thermal to Mechanical Energy Conversion(Vols. 1-3)

Editor: Oleg N. Favorsky, *Department of Physical and Technical Problems of Energetics, Russian Academy of Sciences, Russia*
eISBN: 978-1-84826-021-4, 978-1-84826-022-1, 978-1-84826-023-8
ISBN: 978-1-84826-471-7, 978-1-84826-472-4, 978-1-84826-473-1

Energy Carriers and Conversion Systems with Emphasis on Hydrogen (Vols. 1-2)

Editor: Tokio Ohta, *Yokohama National University and Chairman of Frontier Information and Learning Organization (FILO), Japan*
eISBN: 978-1-905839-29-2, 978-1-905839-30-8
ISBN: 978-1-84826-929-3, 978-1-84826-930-9

Energy Storage Systems (Vols. 1-2)

Editor: Yalsin Gogus, *Middle East Technical University Ankara, Turkey*
eISBN: 978-1-84826-162-4, 978-1-84826-163-1
ISBN: 978-1-84826-612-4, 978-1-84826-613-1

Air Conditioning - Energy Consumption and Environmental Quality

Editor: Matheos Santamouris, *University of Athens, Greece*
eISBN: 978-1-84826-169-3
ISBN: 978-1-84826-619-3

Efficient Use and Conservation of Energy (Vols. 1-2)

Editor: Clark W. Gellings, *Technology Initiatives Electric Power Research Institute (EPRI), USA*
eISBN: 978-1-905839-31-5, 978-1-905839-32-2
ISBN: 978-1-84826-931-6, 978-1-84826-932-3

Exergy, Energy System Analysis, and Optimization (Vols. 1-3)

Editor: Christos A. Frangopoulos, *National Technical University of Athens, Greece*
eISBN: 978-1-84826-164-8, 978-1-84826-165-5, 978-1-84826-166-2
ISBN: 978-1-84826-614-8, 978-1-84826-615-5, 978-1-84826-616-2

Energy Policy

Editor: Anthony David Owen, *The University of New South Wales, Australia*
eISBN: 978-1-905839-33-9
ISBN: 978-1-84826-933-0

Solar Energy Conversion and Photoenergy Systems (Vols. 1-2)

Editors: Julian Blanco Gálvez and Sixto Malato Rodríguez, *Plataforma Solar de Almería, Spain*
eISBN: 978-1-84826-285-0, 978-1-84826-286-7
ISBN: 978-1-84826-735-0, 978-1-84826-736-7

Heat and Mass Transfer

Editors: Tao Wen-Quan, *Xian Jiaotong University, Xian, 710049, PR China*
Qianjin J. Yue, *Dalian U.Technology, China*
eISBN:
ISBN:

Petroleum Engineering – Downstream

Editor: Pedro de Alcantara Pessoa Filho, *University of Sao Paulo, Brazil*
eISBN:
ISBN:

Petroleum: Refining, Fuels and Petrochemicals

Editor: James G. Speight, *Western Research Institute, Laramie, WY, USA*
eISBN:
ISBN:

Pipeline Engineering

Editor: Yufeng F. Cheng, *University of Calgary, Calgary, Alberta, Canada*
eISBN:
ISBN:

Petroleum Engineering – Upstream

Editors: Ezio Mesini and Paolo Macini, *University of Bologna, Italy*
eISBN:
ISBN:

Energy and Fuel Sciences

Editor: Semih Eser, *Pennsylvania State University, Pennsylvania, USA*
eISBN:
ISBN:

ENVIRONMENTAL AND ECOLOGICAL SCIENCES, ENGINEERING AND TECHNOLOGY RESOURCES

Hazardous Waste

Editors: Domenico Grasso, Timothy M. Vogel and Barth Smets, *Smith College, USA*
eISBN: 978-1-84826-055-9
ISBN: 978-1-84826-505-9

Natural and Human Induced Hazards and Environmental Waste Management (Vols. 1-4)

338

Editor: Domenico Grasso, *Smith College, USA*
eISBN: 978-1-84826-299-7, 978-1-84826-300-0, 978-1-84826-301-7, 978-1-84826-302-4
ISBN: 978-1-84826-749-7, 978-1-84826-750-3, 978-1-84826-751-0, 978-1-84826-752-7

Coastal Zones and Estuaries

Editors: Federico Ignacio Isla, and Oscar Iribarne, *Universidad Nacional de Mar del Plata, Argentina*
eISBN: 978-1-84826-016-0
ISBN: 978-1-84826-466-3

Marine Ecology

Editors: Carlos M. Duarte, *CSIC-Univ. Illes Balears, Spain*

Antonio Lot, *Universidad Nacional Autonoma de Mexico (UNAM), México*
eISBN: 978-1-84826-014-6
ISBN: 978-1-84826-464-9

Point Sources of Pollution: Local Effects and Their Control (Vols. 1-2)

Editor: Qian Yi, *Tsinghua University, China*
eISBN: 978-1-84826-167-9, 978-1-84826-168-6
ISBN: 978-1-84826-617-9, 978-1-84826-618-6

Environmental Toxicology and Human Health (Vols. 1-2)

Editor: Tetsuo Satoh, *Biomedical Research Institute and Chiba University, Japan*
eISBN: 978-1-84826-251-5, 978-1-84826-252-2
ISBN: 978-1-84826-701-5, 978-1-84826-702-2

Waste Management and Minimization

Editors: Stephen R. Smith, Chris Cheeseman, and Nick Blakey, *Imperial College, UK.*
eISBN: 978-1-84826-119-8
ISBN: 978-1-84826-569-1

Pollution Control Technologies (Vols. 1-3)

Editor: Georgi Stefanov Cholakov, *University of Chemical Technology and Metallurgy, Bulgaria*

Bhaskar Nath, *European Centre for Pollution Research, UK*
eISBN: 978-1-84826-116-7, 978-1-84826-117-4, 978-1-84826-118-1
ISBN: 978-1-84826-566-0, 978-1-84826-567-7, 978-1-84826-568-4

Environmental Systems (Vols. 1-2)

Editor: Achim Sydow, *GMD FIRST, Berlin, Germany*
eISBN: 978-1-84826-210-2, 978-1-84826-211-9
ISBN: 978-1-84826-660-5, 978-1-84826-661-2

Environmental Regulations and Standard Setting

Editor: Bhaskar Nath, *European Centre for Pollution Research, UK*
eISBN: 978-1-84826-103-7
ISBN: 978-1-84826-553-0

Interactions: Energy/Environment

Editor: Jose' Goldemberg, *The University of Sao Paulo, Brazil*
eISBN: 978-1-84826-090-0
ISBN: 978-1-84826-540-0

Biodiversity: Structure and Function (Vols. 1-2)

Editors: Wilhelm Barthlott, *Rheinische Friedrich-Wilhelms-Universitat Bonn, Germany*

K. Eduard Linsenmair, *Universität Würzburg, Germany*

Stefan Porembski, *Universität Rostock, Germany*

eISBN: 978-1-905839-34-6, 978-1-905839-35-3

ISBN: 978-1-84826-934-7, 978-1-84826-935-4

Environmental Monitoring (Vols. 1-2)

Editors: Hilary I. Inyang and John L. Daniels, *The University of North Carolina at Charlotte, USA*

eISBN: 978-1-905839-75-9, 978-1-905839-76-6

ISBN: 978-1-84826-975-0, 978-1-84826-976-7

Engineering Geology, Environmental Geology & Mineral Economics

Editors: Syed E. Hasan, *University of Missouri, USA*

Syed M. Hasan, *Canada*

eISBN:

ISBN:

Ecology (Vols. 1-2)

Editors: Antonio Bodini, *University of Parma, Italy*

Stefan Klotz, *Centre for Environmental Research Leipzig-Halle Ltd., Germany*

eISBN: 978-1-84826-290-4, 978-1-84826-291-1

ISBN: 978-1-84826-740-4, 978-1-84826-741-1

Ethnobiology

Editor: John Richard Stepp, *University of Florida, USA*

eISBN:

ISBN:

Economic Botany

Editor: Brad Bennett, *Florida International University, Miami, Florida, USA*

eISBN:

ISBN:

Air, Water, Sediment and Soil Pollution Modeling

Editor: Bhaskar Nath, *European Center for Pollution Research, UK*

eISBN:

ISBN:

Phytotechnologies solutions for sustainable land management

Editor: Tomas Vanek, *Institute of Experimental Botany Academy of Sciences of the Czech Republic and Research Institute of Crop Sciences, Prague Czech Republic.*

eISBN:

ISBN:

World Environmental History

Editors: Mauro Agnoletti, *University of Florence, Florence Italy; Universitá degli Studi di Firenze, Italy*

Elizabeth Johann, *International Union of Forest Research Organization, Vienna, 1130, Austria*

Simone Neri Serneri, *University of Siena, 10-53100, Italy*

eISBN:

ISBN:

FOOD AND AGRICULTURAL SCIENCES, ENGINEERING AND

TECHNOLOGY RESOURCES

Soils, Plant Growth and Corp Production

Editor: Willy H. Verheye, *University of Gent, Belgium*

eISBN:

ISBN:

Interactions: Food and Agriculture/ Environment

Editor: G.Lysenko, *Division of Economics and Landscape Relationships, RAS, Moscow, Russia*

eISBN:

ISBN:

The Role of Food, Agriculture, Forestry and Fisheries in Human Nutrition (Vols. 1-4)

Editor: Victor R. Squires, *Dry Land Management Consultant, Australia*

eISBN: 978-1-84826-134-1, 978-1-84826-135-8, 978-1-84826-136-5, 978-1-84826-195-2

ISBN: 978-1-84826-584-4, 978-1-84826-585-1, 978-1-84826-586-8, 978-1-84826-645-2

Cultivated Plants, Primarily as Food Sources (Vols. 1-2)

Editor: Gyorgy Fuleky, *University of Agricultural Sciences, Hungary*

eISBN: 978-1-84826-100-6, 978-1-84826-101-3

ISBN: 978-1-84826-550-9, 978-1-84826-551-6

Forests and Forest Plants (Vols. 1-3)

Editors: John N. Owens, *Centre for Forest Biology, University of Victoria, Canada*

H. Gyde Lund, *Forest Information Services, USA*

eISBN: 978-1-905839-38-4, 978-1-905839-39-1, 978-1-905839-40-7

ISBN: 978-1-84826-938-5, 978-1-84826-939-2, 978-1-84826-940-8

Fisheries and Aquaculture (Vols. 1-5)

Editor: Patrick Safran, *Centre de Coopération Internationale en Recherche Agronomique pour le Développement (CIRAD), Paris, France, France*

eISBN: 978-1-84826-108-2, 978-1-84826 109-9, 978-1-84826-110-5, 978-1-84826-111-2, 978-1-84826-112-9

ISBN: 978-1-84826-558-5, 978-1-84826-559-2, 978-1-84826-560-8, 978-1-84826-561-5, 978-1-84826-562-2

Food Quality and Standards (Vols. 1-3)

Editor: Radomir Lasztity, *Budapest University of Technology and Economics, Hungary*

eISBN: 978-1-905839-41-4, 978-1-905839-42-1, 978-1-905839-43-8

ISBN: 978-1-84826-941-5, 978-1-84826-942-2, 978-1-84826-943-9

Agricultural Land Improvement: Amelioration and Reclamation (Vols. 1-2)

Editor: Boris Stepanovich Maslov, *Russian Academy of Agriculture Sciences, Russia*

eISBN: 978-1-84826-098-6, 978-1-84826-099-3

ISBN: 978-1-84826-548-6, 978-1-84826-549-3

Food Engineering (Vols. 1-4)

Editor: Gustavo V. Barbosa-Cánovas, *Washington State University, USA*

eISBN: 978-1-905839-44-5, 978-1-905839-45-2, 978-1-905839-46-9, 978-1-905839-47-6

ISBN: 978-1-84826-944-6, 978-1-84826-945-3, 978-1-84826-946-0, 978-1-84826-947-7

Agricultural Mechanization & Automation (Vols. 1-2)

Editors: Paul McNulty and Patrick M. Grace, *University College Dublin, Ireland*

eISBN: 978-1-84826-096-2, 978-1-84826-097-9

ISBN: 978-1-84826-546-2, 978-1-84826-547-9

Management of Agricultural, Forestry and Fisheries Enterprises (Vols. 1-2)

Editor: Robert J. Hudson, *University of Alberta, Canada*
eISBN: 978-1-84826-199-0, 978-1-84826-200-3
ISBN: 978-1-84826-649-0, 978-1-84826-650-6

Animal and Plant Productivity
Editor: Robert J. Hudson,*University of Alberta, Canada*
eISBN:
ISBN:

Systems Analysis and Modeling in Food and Agriculture
Editors: K. C. Ting, *The Ohio State University, USA*
David H. Fleisher, *United States Department of Agriculture, Alternate Crops and Systems Laboratory, USA*
Luis F. Rodriguez, *National Research Council, NASA Lyndon B. Johnson Space Center, USA*
eISBN: 978-1-84826-133-4
ISBN: 978-1-84826-583-7

Public Policy in Food and Agriculture
Editor: Azzeddine Azzam, *University of Nebraska, Lincoln, USA*
eISBN: 978-1-84826-095-5
ISBN: 978-1-84826-545-5

Impacts of Agriculture on Human Health and Nutrition (Vols. 1-2)
Editors: Ismail Cakmak, *Sabanci University, Turkey*
Ross M. Welch, *Cornell University, USA*
eISBN: 978-1-84826-093-1, 978-1-84826-094-8
ISBN: 978-1-84826-543-1, 978-1-84826-544-8

Interdisciplinary and Sustainability Issues in Food and Agriculture
Editor: Olaf Christen,*Martin-Luther-University, Germany*
eISBN:
ISBN:

Agricultural Sciences (Vols. 1-2)
Editor: Rattan Lal, *The Ohio State University, USA*
eISBN: 978-1-84826-091-7, 978-1-84826-092-4
ISBN: 978-1-84826-541-7, 978-1-84826-542-4

Range and Animal Sciences and Resources Management
Editor: Victor R. Squires, *Dry Land Management Consultant, Australia*
eISBN:
ISBN:

Shaaka Paakam- Vegetarian Culinary Culture of Telugu (Andhra) Draavida Community of South India
Author: Meenakshi Ganti, *PO Box 2623, Abu Dhabi, UAE*
eISBN: 978-1-84826-253-9
ISBN: 978-1-84826-703-9

HUMAN RESOURCES POLICY, DEVELOPMENT AND MANAGEMENT

Human Resources and Their Development (Vols. 1-2)
Editor: Michael J. Marquardt, *The George Washington University, USA*
eISBN: 978-1-84826-056-6, 978-1-84826-057-3

ISBN: 978-1-84826-506-6, 978-1-84826-507-3

Social and Cultural Development of Human Resources

Editor: Tomoko Hamada, *College of William and Mary, USA*
eISBN: 978-1-84826-087-0
ISBN: 978-1-84826-537-0

Quality of Human Resources: Education (Vols. 1-3)

Editors: Natalia Pavlovna Tarasova, *D. Mendeleyev University of Chemical Technology of Russia, Russia*
eISBN: 978-1-905839-07-0, 978-1-905839-08-7, 978-1-905839-09-4
ISBN: 978-1-84826-907-1, 978-1-84826-908-8, 978-1-84826-909-5

Population and Development: Challenges and Opportunities

Editor: Anatoly G. Vishnevsky, *Institute for Economic Forecasting, Russian Academy of Sciences, Russia*
eISBN: 978-1-84826-086-3
ISBN: 978-1-84826-536-3

Quality of Human Resources: Gender and Indigenous Peoples

Editor: Eleonora Barbieri-Masini, *Faculty of Social Sciences, Gregorian University, Italy*
eISBN: 978-1-905839-10-0
ISBN: 978-1-84826-753-4

Environmental Education and Awareness (Vols. 1-2)

Editor: Bhaskar Nath, *European Centre for Pollution Research, UK*
eISBN: 978-1-84826-114-3, 978-1-84826-115-0
ISBN: 978-1-84826-564-6, 978-1-84826-565-3

Sustainable Human Development in the Twenty-First Century (Vols. 1-2)

Editor: Ismail Sirageldin, *The Johns Hopkins University, USA*
eISBN: 978-1-905839-84-1, 978-1-905839-85-8
ISBN: 978-1-84826-984-2, 978-1-84826-985-9

Education for Sustainability

Editors: Robert V. Farrell, *Florida International University (FIU), USA*
eISBN: 978-1-84826-124-2
ISBN: 978-1-84826-574-5

NATURAL RESOURCES POLICY AND MANAGEMENT

Earth System: History and Natural Variability (Vols. 1-4)

Editor: Vaclav Cilek, *Geological Institute, Czech Academy of Sciences, Czech Republic,*
eISBN: 978-1-84826-104-4, 978-1-84826-105-1, 978-1-84826-106-8, 978-1-84826-107-5
ISBN: 978-1-84826-554-7, 978-1-84826-555-4, 978-1-84826-556-1, 978-1-84826-557-8

Climate Change, Human Systems and Policy (Vols. 1-3)

Editor: Antoaneta Yotova, *National Institute of Meteorology and Hydrology, Bulgaria*
eISBN: 978-1-905839-02-5, 978-1-905839-03-2, 978-1-905839-04-9
ISBN: 978-1-84826-902-6, 978-1-84826-903-3, 978-1-84826-904-0

Oceans and Aquatic Ecosystems (Vols. 1-2)

Editor: Eric Wolanski, *FTSE, Australian Institute of Marine Science, Australia*
eISBN: 978-1-905839-05-6, 978-1-905839-06-3
ISBN: 978-1-84826-905-7, 978-1-84826-906-4

Natural and Human Induced Hazards (Vols. 1-2)

Editor: Chen Yong, *China Seismological Bureau, China*
eISBN: 978-1-84826-231-7, 978-1-84826-232-4
ISBN: 978-1-84826-681-0, 978-1-84826-682-7

Biodiversity Conservation and Habitat Management (Vols. 1-2)

Editors: Francesca Gherardi, Claudia Corti, and Manuela Gualtieri, *Universita di Firenze, Italy*
eISBN: 978-1-905839-20-9, 978-1-905839-21-6
ISBN: 978-1-84826-920-0, 978-1-84826-921-7

Wildlife Conservation and Management in Africa
International Commission
President: Emmanuel K. Boon, *Free University of Brussels, Laarbeeklaan 103, Brussels, Belgium*
Vice Presidents:
Stephen Holness, *South African National Parks, South Africa*
Kai Schmidt-Soltau, *Cameroon*
Julius Kipng'etich, *Director of the Kenya Wildlife Service, Kenya*
Yaa Ntiamoa-Baidu, *Switzerland*
eISBN:
ISBN:

DEVELOPMENT AND ECONOMIC SCIENCES

Socioeconomic Development
Editor: Salustiano del Campo, *Universidad Complutense, Spain*
eISBN: 978-1-84826-085-6
ISBN: 978-1-84826-535-6

Welfare Economics & Sustainable Development (Vols. 1-2)

Editors: Yew-Kwang Ng and Ian Wills, *Monash University, Australia*
eISBN: 978-1-84826-009-2, 978-1-84826-010-8
ISBN: 978-1-84826-459-5, 978-1-84826-460-1

International Economics, Finance and Trade (Vols. 1-2)

Editor: Pasquale Michael Sgro, *Deakin University, Australia*
eISBN: 978-1-84826-184-6, 978-1-84826-185-3
ISBN: 978-1-84826-634-6, 978-1-84826-635-3

Global Transformations and World Futures (Vols. 1-2)

Editor: Sohail Tahir Inayatullah, *University of the Sunshine Coast, Maroochydore; Tamkang University, Taiwan; and, Queensland University of Technology, Australia*
eISBN: 978-1-84826-216-4, 978-1-84826-217-1
ISBN: 978-1-84826-666-7, 978-1-84826-667-4

Introduction to Sustainable Development

Editors: David V. J. Bell, *York University, Canada*
Yuk-kuen Annie Cheung, *York Centre for Applied Sustainability (YCAS); Research Associate, The University of Toronto, York University - Joint Centre for Asia Pacific Studies, Canada*
eISBN: 978-1-84826-222-5
ISBN: 978-1-84826-672-8

Principles of Sustainable Development (Vols. 1-3)

Editor: Giancarlo Barbiroli, *University of Bologna, Italy*
eISBN: 978-1-84826-079-5, 978-1-84826-080-1, 978-1-84826-081-8

ISBN: 978-1-84826-529-5, 978-1-84826-530-1, 978-1-84826-531-8

Dimensions of Sustainable Development (Vols. 1-2)

Editors: Kamaljit S. Bawa and Reinmar Seidler, *University of Massachusetts, Boston, USA*
eISBN: 978-1-84826-207-2, 978-1-84826-208-9
ISBN: 978-1-84826-657-5, 978-1-84826-658-2

Environment and Development (Vols. 1-2)

Editors: Teng Teng and Ding Yifan, *Chinese Academy of Social Sciences, China*
eISBN: 978-1-84826-270-6, 978-1-84826-271-3
ISBN: 978-1-84826-720-6, 978-1-84826-721-3

The Evolving Economics of War and Peace

Editors: James K. Galbraith, *Lyndon B. Johnson School of Public Affairs, The University of Texas at Austin, Jerome Levy Economics Institute and Chair of Economics Allied for Arms Reduction, USA*

Jurgen Brauer, *Augusta State University, USA*

Lucy L. Webster, *Economists Allied for Arms Reduction (ECAAR), USA*
eISBN: 978-1-84826-048-1
ISBN: 978-1-84826-498-4

Economics Interactions with Other Disciplines (Vols. 1-2)

Editor: John M. Gowdy, *Rensselaer Polytechnic Institute, USA*
eISBN: 978-1-84826-037-5, 978-1-84826-038-2
ISBN: 978-1-84826-487-8, 978-1-84826-488-5

International Development Law

Editor: A. F. Munir Maniruzzaman, *University of Kent at Canterbury, UK*
eISBN:
ISBN:

INSTITUTIONAL AND INFRASTRUCTURAL RESOURCES

Human Settlement Development (Vols. 1-4)

Editor: Saskia Sassen, *The University of Chicago, USA*
eISBN: 978-1-84826-044-3, 978-1-84826-045-0, 978-1-84826-046-7, 978-1-84826-047-4
ISBN: 978-1-84826-494-6, 978-1-84826-495-3, 978-1-84826-496-0, 978-1-84826-497-7

Public Administration and Public Policy (Vols. 1-2)

Editor: Krishna K. Tummala, *Kansas State University, USA*
eISBN: 978-1-84826-064-1, 978-1-84826-065-8
ISBN: 978-1-84826-514-1, 978-1-84826-515-8

International Relations (Vols. 1-2)

Editors: Jarrod Wiener, *University of Kent at Canterbury, UK*
Robert A. Schrire, *University of Cape Town, South Africa*
eISBN: 978-1-84826-062-7, 978-1-84826-063-4
ISBN: 978-1-84826-512-7, 978-1-84826-513-4

International Law and Institutions

Editors: Aaron Schwabach, *Thomas Jefferson School of Law, USA*
Arthur John Cockfield, *Queen's University, Canada*
eISBN: 978-1-84826-078-8
ISBN: 978-1-84826-528-8

Institutional Issues Involving Ethics and Justice (Vols. 1-3)

Editor: Robert Charles Elliot, *Sunshine Coast University, Australia*
eISBN: 978-1-905839-14-8, 978-1-905839-15-5, 978-1-905839-16-2
ISBN: 978-1-84826-914-9, 978-1-84826-915-6, 978-1-84826-916-3

International Security, Peace, Development, and Environment (Vols. 1-2)

Editor: Ursula Oswald Spring, *National Autonomous University of Mexico-UNAM/CRIM,*
eISBN: 978-1-84826-082-5, 978-1-84826-083-2
ISBN: 978-1-84826-532-5, 978-1-84826-533-2

Conflict Resolution (Vols. 1-2)

Editor: Keith William Hipel, *University of Waterloo, Canada*
eISBN: 978-1-84826-120-4, 978-1-84826-121-1
ISBN: 978-1-84826-570-7, 978-1-84826-571-4

Democratic Global Governance

Editor: Irene Lyons Murphy, *Consultant, USA*
eISBN: 978-1-84826-181-5
ISBN: 978-1-84826-631-5

National, Regional Institutions & Infrastructures

Editor: Neil Edward Harrison, *University of Wyoming, USA*
eISBN:
ISBN:

Conventions, Treaties and other Responses to Global Issues (Vols. 1-2)

Editor: Gabriela Maria Kutting, *University of Aberdeen, UK*
eISBN: 978-1-84826-233-1, 978-1-84826-234-8
ISBN: 978-1-84826-683-4, 978-1-84826-684-1

TECHNOLOGY, INFORMATION, AND SYSTEMS MANAGEMENT RESOURCES

Information Technology and Communications Resources for Sustainable Development

Editor: Ashok Jhunjhunwala, *Indian Institute of Technology, Madras, India*
eISBN:
ISBN:

Systems Analysis and Modeling of Integrated World Systems (Vols. 1-2)

Editors: Veniamin N. Livchits, *Institute for Systems Analysis, Russian Academy of Sciences, Russia*
Vladislav V. Tokarev, *Moscow State University, Russia*
eISBN: 978-1-84826-088-7, 978-1-84826-089-4
ISBN: 978-1-84826-538-7, 978-1-84826-539-4

Systems Engineering and Management for Sustainable Development (Vols. 1-2)

Editor: Andrew P. Sage, *George Mason University, USA*
eISBN: 978-1-905839-00-1, 978-1-905839-01-8
ISBN: 978-1-84826-900-2, 978-1-84826-901-9

Knowledge Management, Organizational Intelligence and Learning, and Complexity (Vols. 1-3)

Editor: L. Douglas Kiel, *The University of Texas at Dallas, USA*
eISBN: 978-1-905839-11-7, 978-1-905839-12-4, 978-1-905839-13-1
ISBN: 978-1-84826-911-8, 978-1-84826-912-5, 978-1-84826-913-2

Science and Technology Policy (Vols. 1-2)

Editor: Rigas Arvanitis, *IRD - France & Zhongshan (Sun Yatsen) University, Guangzhou,*
eISBN: 978-1-84826-058-0, 978-1-84826-059-7
ISBN: 978-1-84826-508-0, 978-1-84826-509-7

Globalization of Technology

Editor: Prasada Reddy, *Lund University, Sweden*
eISBN: 978-1-84826-084-9
ISBN: 978-1-84826-534-9

Sustainable Built Environment (Vols. 1-2)

Editors: Fariborz Haghighat *Concordia University, Canada*
Jong-Jin Kim, *The University of Michigan, USA*
eISBN: 978-1-84826-060-3, 978-1-84826-061-0
ISBN: 978-1-84826-510-3, 978-1-84826-511-0

Integrated Global Models of Sustainable Development (Vols. 1-3)

Editor: Akira Onishi, *Centre for Global Modeling, Japan*
eISBN: 978-1-905839-17-9, 978-1-905839-18-6, 978-1-905839-19-3
ISBN: 978-1-84826-917-0, 978-1-84826-918-7, 978-1-84826-919-4

Artificial Intelligence

Editor: Joost Nico Kok, *Leiden University, The Netherlands.*
eISBN: 978-1-84826-125-9
ISBN: 978-1-84826-575-2

Computer Science and Engineering

Editors: Zainalabedin Navabi, *University of Tehran, Iran*
David R. Kaeli, *Northeastern University,*
eISBN: 978-1-84826-227-0
ISBN: 978-1-84826-677-3

System Dynamics (Vols. 1-2)

Editor: Yaman Barlas, *Bogazici University, Turkey*
eISBN: 978-1-84826-137-2, 978-1-84826-138-9
ISBN: 978-1-84826-587-5, 978-1-84826-588-2

LAND USE, LAND COVER AND SOIL SCIENCES

(Vols. 1-7)

Editor: Willy H. Verheye, *University of Gent, Belgium*

(Each Volume in the following covers a set of chapters, the total number of which is 78. The size of a chapter may vary from about 5000 words to about 30000 words.)

Volume 1: Land Cover, Land Use and the Global Change
eISBN: 978-1-84826-235-5
ISBN: 978-1-84826-685-8
Volume 2: Land Evaluation
eISBN: 978-1-84826-236-2
ISBN: 978-1-84826-686-5
Volume 3: Land Use Planning
eISBN: 978-1-84826-237-9
ISBN: 978-1-84826-687-2
Volume 4: Land Use Management and Case Studies
eISBN: 978-1-84826-238-6

ISBN: 978-1-84826-688-9
Volume 5: Dry Lands and Desertification
eISBN: 978-1-84826-239-3
ISBN: 978-1-84826-689-6
Volume 6: Soils and Soil Sciences - I
eISBN: 978-1-84826-240-9
ISBN: 978-1-84826-690-2
Volume 7: Soils and Soil Sciences - II
eISBN: 978-1-84826-241-6
ISBN: 978-1-84826-691-9

AREA STUDIES (REGIONAL SUSTAINABLE DEVELOPMENT REVIEWS)

Area Studies (Regional Sustainable Development Review): Africa (Vols. 1-2)
Editor: Emmanuel Kwesi Boon, *Free University Brussels, Belgium*
eISBN: 978-1-84826-066-5, 978-1-84826-067-2
ISBN: 978-1-84826-516-5, 978-1-84826-517-2

Area Studies (Regional Sustainable Development Review): Brazil
Editor: Luis Enrique Sanchez, *University of Sao Paulo, Brazil*
eISBN: 978-1-84826-171-6
ISBN: 978-1-84826-621-6

Area Studies (Regional Sustainable Development Review): USA (Vols. 1-2)
Editors: Lawrence C. Nkemdirim, *University of Calgary, Canada*
eISBN: 978-1-84826-069-6, 978-1-84826-070-2
ISBN: 978-1-84826-519-6, 978-1-84826-520-2

Area Studies (Regional Sustainable Development Review): China (Vols. 1-3)
Editor: Sun Honglie, *Chinese Academy of Sciences, China*
eISBN: 978-1-84826-071-9, 978-1-84826-072-6, 978-1-84826-073-3
ISBN: 978-1-84826-521-9, 978-1-84826-522-6, 978-1-84826-523-3

Area Studies (Regional Sustainable Development Review): Europe
Editors: Alexander Mather and John Bryden, *University of Aberdeen, UK*
eISBN: 978-1-84826-068-9
ISBN: 978-1-84826-518-9

Area Studies (Regional Sustainable Development Review): Japan
Editor: Yukio Himiyama, *Hokkaido University of Education, Japan*
eISBN: 978-1-84826-170-9
ISBN: 978-1-84826-620-9

Area Studies (Regional Sustainable Development Review): Russia (Vols. 1-2)
Editor: Nicolay Pavlovich Laverov, *RAS, Russia*
eISBN: 978-1-84826-074-0, 978-1-84826-075-7
ISBN: 978-1-84826-524-0, 978-1-84826-525-7

INTERNATIONAL EDITORIAL COUNCIL (IEC)

The EOLSS International Editorial Council was established thus: Board of General Advisors (BGA), Honorary Editorial Advisory Board (HEAB), and Honorary Academic Editorial Board (HAEB) as a large editorial advisory body with several hundreds of expert members chosen to advice on the development of the Encyclopedia. The Configuration Control Board is a high-level body with membership chosen from the International Editorial Council (IEC). Several individuals advised and contributed to the development of the EOLSS Body of Knowledge through meetings, discussions and communications. The list is given below with apologies for any inadvertent omissions

The Board of General Advisors (BGA)
Laureates: Nobel Prize, Japan Prize, Kalinga Prize, and World Food Prize

Beachell, H. M.	*USA*	Knipling, E.F.	*USA*	Ramsey, N.F.	*USA*
Chandler, R. Jr	*USA*	Kurien, V	*India*	Richter, B.	*USA*
Charpak, G.	*Switzerland*	Lederberg, J.	*USA*	Scrimshaw, N	*USA*
Crutzen, P.J.	*Germany*	Lederman, L.M.	*USA*	Shull, C.G.	*USA*
Ernst, R.R.	*Switzerland*	Lee, Y.T	*China*	Swaminathan, M.S.	***India***
Fukui, K.	*Japan*	Lehn, J.-M	*France*	Yang Chen Ning	*USA*
Herren, H.	*Kenya*	Lions, J.	*France*	Zadeh, L.A	*USA*
Kapitza, S.	*Russia*	Niederhauser, J	*USA*		
Klein, L.R.	***USA***	Prigogine, I.	***Belgium***		

Other IEC members:

Aguirre, L.M.	*Chile*	Bates, D.G.	*USA*
Aharoni, A.	*Israel*	Bauzer Medeiros, C.M.	*Brazil*
Aidarov. I.P.	*Russia*	Bawa, K.S.	*USA*
Albou, L.	*France*	Baydack, R.K.	*Canada*
Al Gobaisi, D.	*UAE*	Bela Kanyar	*Hungary*
Alonso Ramirez	*USA*	Bell, A.	*South Africa*
Aluwihare, A.P.R.	*Sri Lanka*	Bell, D.V.J.	*Canada*
Ana Angélica Almeida Barbosa	*Brazil*	Bellavista, P.	*Italy*
		Bergles, A.E.	*USA*
Anas, S. M.	*India*	Bhattacharya, S.C.	*Thailand*
Andre Pires da Silva	*USA*	Blakey, N.	*UK*
Andrea de Gaetano	*Italy*	Blancarte Pimentel, R.	*México*
Angelini, M	*Italy*	Block de Behar, L.	*Uruguay*
Araki, H.	*Japan*	Blok, K.	*The Netherlands*
Arata, A. A.	*USA*	Bobrovnitskii, Y.I.	*Russia*
Arima, A.	*Japan*	Bodini, A.	*Italy*
Arlt, H.	*Austria*	Boon, E.K.	*Belgium*
Arocena, J.	*Canada*	Bozena Czerny	*Poland*
Arturo Bonet	*México.*	Brad Bennett	*USA*
Arvanitis, R.	*France*	Brauer, G.W	*Canada*
Ashok Jhunjhunwala	*India*	Brauer, J.	*USA*
Assaad, K. A..	*Lebanon*	Bridgwater, J.	*UK*
Atalay, M.	*Finland*	Burden, C.J.	*Australia*
Atherton, D.P.	*UK*	Cakmak, I.	*Turkey*
Atkinson, P.	*UK*	Caporaso, J.A.	*USA*
Attiga, A.A.	*Jordan*	Carabias, J.	*Mexico*
Aubin, J.-P.	*France*	Carpenter, A.	*New Zealand*
Ayala Luis	*USA*	Carra, S.	*Italy*
Ayres, R.U.	*France*	Carroll, R.G.	*USA*
Azzam, A.	*USA*	Carta, S.	*Italy*
Barbieri-Masini, E.	*Italy*	Carvalhal, T.F.	*Brazil*
Barbiroli, G.	*Italy*	Catanzaro, E.	*USA*
Barbosa-Canovas, G.	*USA*	Ceballos, G.	*Mexico*
Barlas, Y.	*Turkey*	Chaplin, R.	*Canada*
Barritt, E.R.	*USA*	Chapman, D.	*Ireland*
Barron, E.J.	*USA*	Cheeseman, C.	*UK*
Barthlott, W.	*Germany*	Chen Han Fu	*China*

Horikawa, K.	*Japan*	Kubota, S.	*Japan*
Htun Nay	*Myanmar*	Kumar, S.	*Thailand*
Hu Ying	*China*	Kupitz, J.	*Austria*
Huang, C.H.	*Taiwan*	Kurzhanski, A.B.	*Russia*
Hudson, R.J.	*Canada*	Kutting, G.M.	*UK*
Imura, H.	*Japan*	Lagowski, J.J.	*USA*
Inayatullah, S.T.	*Australia*	Lal, R.	*USA*
Inyang, H.I.	*USA*	Lara, F.	*Mexico*
Iribarne, O.	*Argentina*	Last, J.	*Canada*
Ishida, M.	*Japan*	Lastovicka, J.	*Czech Republic*
Ishii, R.	*Japan*	Laszlo Wojnarovits	*Hungary*
Isla, F. I.	*Argentina*	Lasztity, R.	*Hungary*
Ivano Bertini	*Italy*	Lattanzio, John	*Australia*
Izrael, Y.A.	*Russia*	Laverov, N.P.	*Russia*
Jackson, M.	*UK*	Lee Newman	*USA*
Jacobson, H.K.	*USA*	Lee, Han Bee	*Republic of*
Jain, A.	*India*		*Korea*
Jakeman, A.J.	*Australia*	Lee, S.	*USA*
James G. Speight	*USA*	Leemans, R.	*Netherlands*
Jeom Kee Paik	*Korea*	Lefevre, J.M.	*France*
Jeremy Gray	*UK*	Lefevre, T.	*Thailand*
Jeyaratnam, J.	*Singapore*	Leff, E.	*Mexico*
Jhunjhunwala, A.	*India*	Lelija Socanac	*Croatia*
Jimeno, M.	*Colombia*	Lester R. Morss	*USA*
Jimi Naoki Nakajima	*Brazil*	Levashov, V.K.	*Russia*
Jinsheng, G.	*China*	Levine, A.D.	*USA*
Jintai Ding	*USA*	Lichnerovicz, A.	*France*
John Lattanzio	*Australia*	Liebowitz, H.	*USA*
John R. Helliwell	*UK*	Lin, Qun	*China*
John M. Pezzuto	*USA*	Lin, Zhi-Qing	*USA*
John Richard Stepp	*USA*	Lindsay Dent	*Australia*
Jordaan, J.M.	*South Africa*	Linus U. Opara	*Sultanate of*
José Luiz de França Freire	*Brazil.*		*Oman*
Jose M. Melero-Vara	*Spain*	Liss, P.	*UK*
José Ma Goicolea Ruigómez	*Spain*	Liu, Bao	*China*
Jose Merodio	*Spain*	Livchits, V.N.	*Russia*
José Masdeu	*Spain*	Llorens, P.	*Spain*
Joseph Magill	*Germany*	Lomnitz, C.	*Mexico*
Juan M. Feliu-Martinez	*Spain*	Lorenzano, P.	*Argentina*
Julian Blanco Gálvez	*Spain*	Loyo, C.	*Mexico*
Julio Alberto Costello	*Argentina*	Lozano, F.	*Mexico*
Julius Kipng'etich	*Kenya*	Lozano, M.	*Mexico*
Jureckova, J.	*Czech Republic*	Lu, G.Q.M.	*Australia*
Kaeli, D.R.	*USA*	Lund, H.G.	*USA*
Kai Schmidt-Soltau	*Cameroon*	Luthra, R.	*USA*
Kakihana, H.	*Japan*	Lysenko, E.G.	*Russia*
Kálmán Varga	*Hungary*	Mack, A.	*Australia*
Kanwar, R.S.	*USA*	Mackey, B.G.	*Australia*
Kao, D.T.	*USA*	Magara, Y.	*Japan*
Kapitza, S.	*Russia*	Mageed, Y.A.	*Sudan*
Karoly Suvegh	*Hungary*	Mahfouz, S.	*Egypt*
Kathleen M. Thiessen	*USA*	Majewski, H.	*Australia*
Kawabe, T.	*Japan*	Majumdar, M.	*USA*
Kayis, A.B.	*Australia*	Manciaux, M.R.G.	*France*
Kerekes, J.	*Hungary*	Maniruzzaman, A.F.M.	*UK*
Khosla, A.	*India*	Mansourian, B.P.	*Switzerland*
Khublaryan, M.G.	*Russia*	Manuel Doblare	*Spain*
Kiel, L.D.	*USA*	Mauro Agnoletti	*Italy*
Kim, J.J.	*USA*	Marcus Vinicius	*Brazil*
Kim, T.J.	*USA*	Sampaio	
Kirby, R.	*South Africa*	Maria Wallenius	*Germany*
Kit Keith L. Yam	*USA*	Markandya, A.	*UK*
Klaus Mayer	*Germany*	Marquardt, M.J.	*USA*
Kleber Del Claro	*Brazil*	Marquette, C.M.	*Norway*
Klir, G.	*USA*	Maslov, B.S.	*Russia*
Klotz, S.	*Germany*	Mather, A.	*UK*
Kok, J.N.	*The Netherlands*	Massuo Jorge Kato	*USA*
Kolmer, L.	*USA*	Mauricio Quesada	*Mexico*
Kondo, J.	*Japan*	McCarthy, O.J.	*New Zealand*
Kondratyev, K.Ya.	*Russia*	McDowell, D.	*Switzerland*
Kotlyakov, V.M.	*Russia*	McMichael, A.J.	*UK*
Krashnoschokov, N.N.	*Russia*	McMullan, J.T.	*UK*
Krawczyk, J.B.	*New Zealand*	McMurtry, J.	*Canada*

McNulty, P.	*Ireland*	Petrov, R.V.	*Russia*
Medina, E.	*Venezuela*	Pinar Bilgin	*USA*
Meissner, R.	*South Africa*	Pironneau, O.	*France*
Meller, F.	*USA*	Plate, N.A.	*Russia*
Menaut, J.C.	*France*	Poh-Sang Lam	*USA*
Mendes, C.	*Brazil*	Porembski, S.	*Germany*
Mesarovic, M.D.	*USA*	Prabhu Nath	*UK*
Mildonian, P.	*Italy*	P. Singh	
Minelli, A.	*Italy*	Prinn, R.G.	*USA*
Mitchell, C.M.	*USA*	Puel, J.P.	*France*
Mitsumori, S.	*Japan*	Qi, L.	*China*
Mohamed I.El	*Egypt*	Qian, Y.	*China*
Gohary		Qureshy, M.N.	*USA*
Mohamed Henini	*UK*	Rajeswar J.	*India*
Molly R. Morris	*USA*	Ralph V. Summy	*Australia*
Molzahn, M.	*Germany*	Ramallo, L.I.	*Spain*
Momoh, A.	*Lagos*	Rao, C.N.R.	*India*
Monica M.C. Muelbert	*Brazil*	Rao, C.R.	*USA*
Mônica Palácios Rios	*México,*	Rao, G.M.	*India*
Moore, B. III	*USA*	Rao, G.P.	*India*
Moran-Lopez, J.L.	*Mexico*	Rao, S.S.	*India*
Muhammed, A.	*Pakistan*	Rapport, D.J.	*Canada.*
Muhvi?-Dimanovski, V	*Croatia*	Rasool, I.	*France*
Muller, P.	*Germany*	Raumolin, J.	*Finland*
Munasinghe, M.	*Sri Lanka*	Rawlings, R.D.	*UK*
Murphy, I.L.	*USA*	Reddy, P.	*Sweden*
Naito, M.	*Japan*	Regina Helena	*Brazil*
Napalkov, N.P.	*Russia*	Ferraz Macedo	
Nasson, W.R.	*South Africa*	Rein Munter	*Estonia*
Nath, B.	*UK*	Reza, F.	*Canada*
Natori, K.	*Japan*	Robert A. Denemark	*USA*
Navabi, Z.	*Iran*	Robert J. Marquis	*USA*
Nefedov, O.M.	*Russia*	Roberto Bassani	*Italy*
Nelson Ramírez	*Venezuela*	Roberto Blancarte	*México*
Ng, Y.K.	*Australia*	Pimentel	
Nicotra, F.	*Italy*	Rodriguez, L.F.	*USA*
Nihoul, J.C.J.	*Belgium*	Rogério Parentoni	*Brazil*
Nishizawa, J.	*Japan*	Martins	
Nishizuka, Y.	*Japan*	Rolf Stabell	*Norway*
Nkemdirim, L.C.	*Canada*	Rudan, P.	*Croatia*
Nkwi, P.N.	*Cameroon*	Rydzynski, K.J.	*Poland*
Norman M. Edelstein	*USA*	Sabljic, A.	*Croatia*
Novitsky, E.G.	*Russia*	Safran, P.	*Philippines*
Obasi, G.O.P.	*Switzerland*	Sajeev,, V.	*India*
Oddbjørn Engvold	*Norway*	Sala, M.	*Spain*
Odhiambo, T.	*Kenya*	Sala, O.	*Argentina*
Ohta, T.	*Japan*	Salim, E.	*Indonesia*
Ojo, S.O.	*Nigeria*	Sanchez, L.E.	*Brazil*
Okada, N.	*Japan*	Sándor Nagy	*Hungary*
Oki, T.	*Japan*	Santamouris, M.	*Greece*
Olden, K.	*USA*	Santiago Gracia	*Spain*
Olembo, R.	*Kenya*	Granda	
Onishi, A.	*Japan*	Sashi K. Kunnath	*DAVIS ,USA*
Osipov, Yu.S.	*Russia*	Sassen, S.	*USA*
Oswald Spring, U.	*México*	Satoh, T.	*Japan*
Oswaldo Marcal Júnior	*Brazil*	Savenije, H.H.G.	*The Netherlands*
Oteiza, E.	*Argentina*	Sayers, B. McA.	*UK*
Owen, A.D.	*Australia*	Schrire, R.A.	*South Africa*
Owens, J.N.	*Canada*	Schwabach, A.	*USA*
Oyebande, L.	*Nigeria*	Seidler, R.	*USA*
Pablo Lorenzano	*Argentina*	Sekiguchi, M.	*Japan*
Pachauri, R.K.	*India*	Semih Eser	*USA*
Padoch, C.	*USA*	Sen, G.	*India*
Paolo Macini	*Italy*	Serageldine, I.	*USA*
Parra-Luna, F.	*Spain*	Sgro, P.M.	*Australia*
Paterno, L.	*Italy*	Shaib, B.	*Nigeria*
Patricia A. Baisden	*USA*	Shaidurov, V.V.	*Russia*
Paulette L.Ford.	*USA*	Shidong, Z.	*China*
Paulo S. Oliveira	*Brazil*	Shiklomanov, I.A.	*Russia*
Payne, W.A.	*USA*	Shpilrain, E.E.	*Russia*
Pedro de Alcantara	*Brazil*	Shuttleworth, W.J.	*USA*
Pessoa Filho		Silveira, L.	*Uruguay*
Peter van den Haute	*Belgium*		

Silvio Carlos Rodrigues	*Brazil*
Sinding-Larsen, R.	*Norway*
Singh, H.	*New Zealand*
Singh, M.G.	
Sirageldin, I.	*USA*
Siva Sivapalan, M.	*Australia*
Sivov, N.A.	*Russia*
Sixto Malato Rodríguez	*Spain*
Skogseid, H.	*Norway*
Slobodan Kralj	*Croatia*
Smeers, Y.	*Belgium*
Smets, B.	*USA*
Smith, S.	*UK*
Socanac, L.	*Croatia*
Sollars, C.J.	*UK*
Sonntag, H.	*Venezuela*
Soria Escoms, B.	*Spain*
Southwood, R.	*UK*
Squires, V.R.	*Australia*
Stephen Holness	*South Africa*
Stepp, J.R.	*USA*
Stuewe, K.	*Austria*
Sujoldzic, A.	*Croatia*
Sullivan, L.	*Australia*
Sun Honglie	*China*
Susheela, A.K.	*India*
Sydow, A.	*Germany*
Szczerban, J.	*Poland*
Takahasi, Y.	*Japan*
Takeda, G.	*Japan*
Takle, G.S.	*USA*
Tao, W-Q	*China*
Tarasova, N.P.	*Russia*
Tarlock, A.D.	*USA*
Teller, E.	*USA*
Teng, T.	*China*
Ting, K.C.	*USA*
Titli, A.	*France*
Toschenko, J.T.	*Russia*
Tsipenyuk, Y.M.	*Russia*
Ul Haq, M	*Pakistan*
Ulrich Lüttge	*Denmark*
Unbehauen, H.D.	*Germany*
Vagn Lundsgaard Hansen	*Denmark*
Valero-Capilla, A.	*Spain*
van der Beken, A.	*Belgium*
van Wyk, J.A.	*South Africa*
Vanek Tomas	*Czech Republic.*
Vazquez, C.	*Mexico*
Verhasselt, Y.L.G.	*Belgium*
Victor Climent Payá	*Spain*
Victor Rico-Gray	*México*
Victor R. Squires	*Australia*
Viertl, R.	*Austria*
Vigneswaran, S.	*Australia*
Vishnevsky, A.G.	*Russia*
Volgy, T.J.	*USA*
Wang, Mingxing	*China*
Wang, W.C.	*USA*
Wang, Xingyu	*China*
Webster, L.	*USA*
Weder, R.	*Mexico*
Weffort, F.	*Brazil*
Wei, Zhou	*Hungary*
Wei-Bin Zhang	*Japan*
Weiss, T.G.	*USA*
Weizsacker, E.U. von	*Germany*
Welch, R.M.	*USA*
Wiener, J.	*UK*
Wills, I.	*Australia*
Wilson, S.R.	*Australia*
Willy Verheye	*Belgium*
Wohnlich, S.	*Germany*
Wojtczak, A.	*USA*
Wojtczak, A.M.	*USA*
Wolanski, E.	*Australia*
Woldai, A.	*Eritrea*
Wolfgang Siebel	*Germany*
Wong, K.P.	*Australia*
Xila S. Liu	*China*
Yaa Ntiamoa-Baidu	*Switzerland*
Yagodin, G.A.	*Russia*
Yam, K.	*USA*
Yam, K.K.L.	*USA*
Yan, S.	*China*
Yao, Y.L.	*USA*
Yatim, B.	*Malaysia*
Young W.Kwon	*USA*
Yong, C.	*China*
Yotova, A.	*Bulgaria*
Young Binglin	*China*
Yuan Weikang	*China*
Yufeng F. Cheng	*Canada*
Yuguo Li	*China.*
Yuichiro Nagame	*Japan*
Yuqing Wang	*USA*
Zaitsev, V.A.	*Russia*
Zeng Yi	*China*
Zhang Xinshi	*China*
Zoltan Homonnay	*Hungary*
Zoltan Nemeth	*Hungary*
Zsolt Fulop	*Hungary*

UNESCO-EOLSS JOINT COMMITTEE

A UNESCO-EOLSS Joint Committee was established with the objectives of (a) seeking, selecting, inviting and appointing Honorary Theme Editors (HTEs)/International Commissions for the Themes, (b) providing assistance to them, (c) obtaining appropriate contributions for the different levels of the Encyclopedia, and (d) monitoring the text development. The membership of the Committee is as follows:

Badran, A.	*President, Arab Academy of Sciences, (Co-Chairman)*
Al Gobaisi, D.	*EOLSS Editor-in-Chief (Co-Chairman)*
El Tayeb, M.	*Director, Division for Science Policy & Sustainable Development, UNESCO,(Secretary)*
Tolba, M.K.	*President, International Center for Environment and Development, Former Executive Director of UNEP*
Sage, A.P.	*George Mason University, USA (Chairman, EOLSS Configuration Control Board)*
Marchuk, G.I.	*Russian Academy of Sciences*
Szollosi-Nagy, A.	*Deputy ADG/SC and Director, Division of Water Sciences , UNESCO*
Chesters, G. (Deceased)	*University of Wisconsin, USA*
Johns, A.T.	*University of Bath, UK*
Lundberg, H.D.	*Sweden*
Younes, T.	*International Union of Biological Sciences, France*
Dempsey, J.	*UK*
Rao, G. P.	*India*
Joundi, S.	*France*
Makkawi, B.	*Sudan*
Woldai, A.	*Eritrea*
Agoshkov, V.I.	*Russia*
Hornby, R. J.	*UK*
Wall, G.	*Sweden*
Watt, H.M.	*USA*
Kotchetkov, V.	*Consultant, UNESCO*
Al-Radif, A.	*Canada*
Sasson, A.	*Morocco*
Bruk, S.	*Consultant UNESCO*

Secretariat

Huynh, H.	*Assistant Project Coordinator, SC/PSD, UNESCO*
Barbash A.	*Systems Manager, SC/PSD, UNESCO*

Further details may be obtained from

www.eolss.net

helpdesk@eolssonline.net

DISCLAIMER

The information, ideas, and opinions presented in these publication are those of the authors and do not represent those of UNESCO and/or Eolss Publishers. UNESCO and/or Eolss Publishers make every effort to ensure, but do not guarantee, the accuracy of the information in these publications and shall not be liable whatsoever for any damages incurred as a result of its use. Hyperlinks to other web sites imply neither responsibility for, nor approval of, the information contained in those other web sites on the part of UNESCO and Eolss Publishers.